Optoelectronics: A Practical Approach

Optoelectronics: A Practical Approach

Kate Timberlake

MURPHY & MOORE
www.murphy-moorepublishing.com

Optoelectronics: A Practical Approach
Kate Timberlake
ISBN: 978-1-63987-413-2 (Hardback)

MURPHY & MOORE
Published by Murphy & Moore Publishing,
1 Rockefeller Plaza,
New York City, NY 10020, USA

Cataloging-in-Publication Data

Optoelectronics : a practical approach / Kate Timberlake.
p. cm.
Includes bibliographical references and index.
ISBN 978-1-63987-413-2
1. Optoelectronics. 2. Photonics. 3. Electronics. I. Timberlake, Kate.
TA1750 .O68 2022
621.381 045--dc23

For more information regarding Murphy & Moore Publishing and its products, please visit the publisher's website www.murphy-moorepublishing.com

Table of Contents

Preface

Optoelectronics is the branch of science which deals with the study and application of various electronic systems and devices which are capable of sourcing, detecting and controlling light. This encompasses visible as well as invisible form of light such as infrared rays, ultraviolet rays and gamma rays. Optoelectronic devices are transducers which possess either optical-to-electrical or electrical-to-optical nature. Such devices are based on the quantum mechanical effects of light on different electronic materials. Some of the important aspects of the discipline of optoelectronics are photovoltaic effect, photoconductivity, stimulated emission, radiative recombination and photoemissivity. Optical fibers are one of the most prominent applications of optoelectronics. This book presents the complex subject of optoelectronics in the most comprehensible and easy to understand language. The topics included herein on optoelectronics are of utmost significance and bound to provide incredible insights to readers. This book will serve as a valuable source of reference for those interested in this field.

A foreword of all Chapters of the book is provided below:

Chapter 1 - The branch of photonics which deals with the study of electronic devices and systems that use light as well as their applications is known as optoelectronics. Some of the topics studied under optoelectronics are electromagnetic spectrum, optoelectronic devices, fiber optics, etc. This is an introductory chapter which will briefly introduce all these significant aspects of optoelectronics; **Chapter 2 -** Optical fibre refers to a flexible and transparent fiber made by silica or plastic that is used to transmit light between two ends of the fibre. A few of the important aspects of optical fibres are propagation of light through optical fibre, characteristics of optical fibres and comparison of optical fibers. This chapter has been carefully written to provide an easy understanding of these varied facets of optical fibres; **Chapter 3 -** Fiber optic link refers to a part of the optical fiber communication systems that provides a point-to-point data connection. There are some significant aspects of fiber optic link design such as design criteria, design of point-to-point optical link, factors for evaluation of fiber optic link, calculations for fiber optic link, etc. These topics are crucial for a complete understanding of the subject; **Chapter 4 -** An electronic component that is used to transfer electric signals between two circuits by using light is termed as an optical coupler. There are various types of optical couplers such as directional couplers, star couplers and T couplers. The topics elaborated in this chapter will help in gaining a better perspective about these types of optical couplers; **Chapter 5 -** Laser refers to a device that generates an intense beam of light through the process of optical amplification by emitting photons from excited atoms. Some of the different types of lasers are solid laser, gas laser and liquid laser. All the diverse principles related to lasers have been carefully analyzed in this chapter.

I would like to thank the entire editorial team who made sincere efforts for this book and my family who supported me in my efforts of working on this book. I take this opportunity to thank all those who have been a guiding force throughout my life.

Kate Timberlake

1

Introduction to Optoelectronics

The branch of photonics which deals with the study of electronic devices and systems that use light as well as their applications is known as optoelectronics. Some of the topics studied under optoelectronics are electromagnetic spectrum, optoelectronic devices, fiber optics, etc. This is an introductory chapter which will briefly introduce all these significant aspects of optoelectronics.

Optoelectronics is branch of science which deals with optics and electronics. The term optoelectronics is a specific discipline of electronics that focuses on light-emitting or light-detecting devices. Light-emitting devices use voltage and current to produce electromagnetic radiation (i.e., light). Such light-emitting devices are commonly used for purposes of illumination or as indicator lights.

In contrast, light-detecting devices, such as phototransistors, are designed to convert received electromagnetic energy into electric current or voltage. Light-detecting devices can be used for light sensing and communication. Examples of these include darkness-activated switches and remote controls. In general terms, light-detecting devices work by using photons to liberate bound electrons within semiconductor materials. Figures below shows a variety of optoelectronic devices.

Incandescent lamp

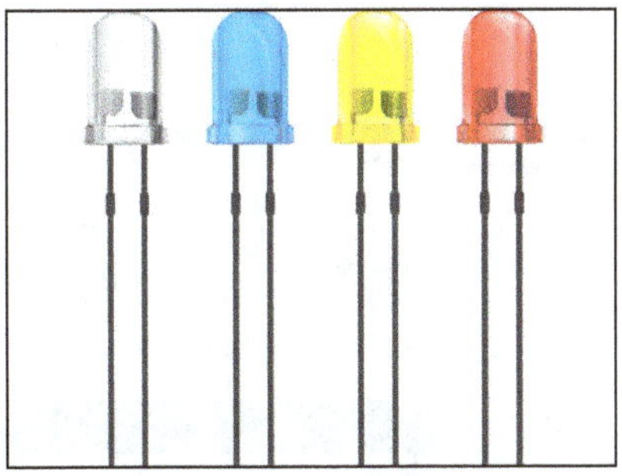

LEDs

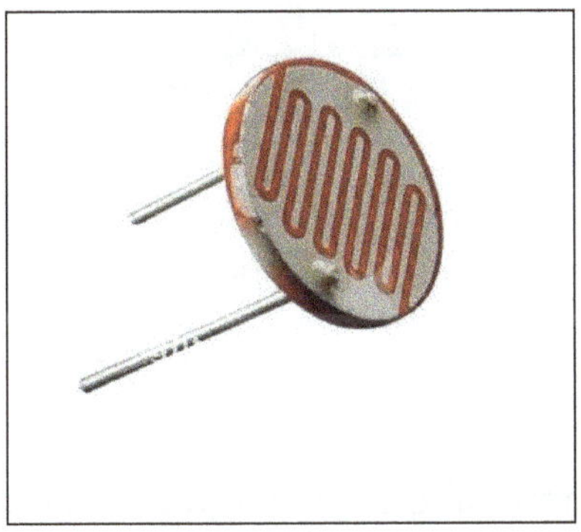

Photoresistor

Solar cell

Photons

Photons are the fundamental units of electromagnetic radiation (EMR). Photons have a frequency of propagation, and we classify EMR based on this frequency—microwave EMR, infrared EMR, optical EMR, and so forth. The human eye is sensitive to optical EMR, which is further categorized into colors. Color is not an inherent property of photons; rather, photons have frequency, and human beings interpret these different frequencies as different colors.

Some Physics behind Photons

The relationship between a photon's frequency and its wavelength (λ) is given by:

$$\lambda = v/f \text{ (in units of meters)}$$

where,

- v = velocity, or speed, of the photon (units of m/s),
- f = frequency (in units of Hz).

In free space, v is the speed of light ($c = 3.0 \times 10^8$ m/s). But in other media, such as glass, v becomes slower than the speed of light.

A photon with a longer wavelength (i.e., a lower frequency) has less energy than a photon with a shorter wavelength (i.e., a higher frequency).

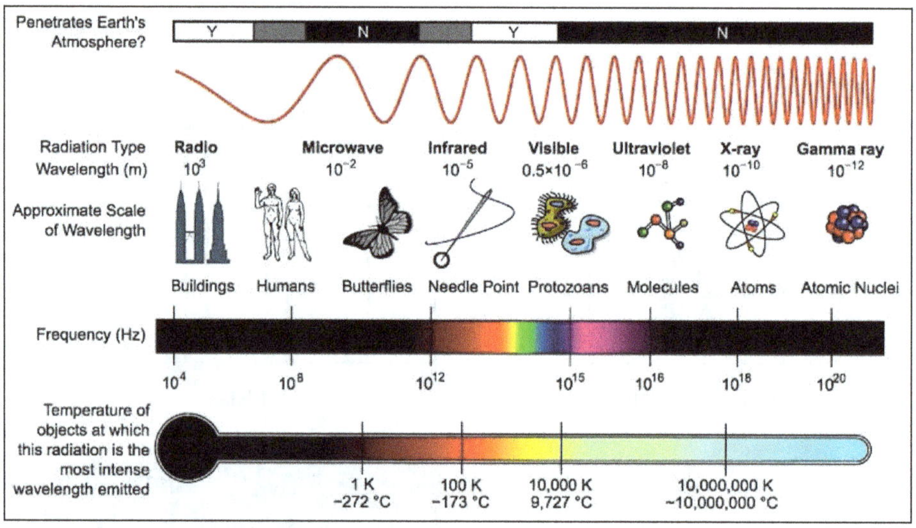

The Electromagnetic (EM) spectrum.

Electromagnetic Spectrum

Electromagnetic spectrum is the entire distribution of electromagnetic radiation according to frequency or wavelength. Although all electromagnetic waves travel at the speed of light in a vacuum, they do so at a wide range of frequencies, wavelengths, and photon energies. The electromagnetic spectrum comprises the span of all electromagnetic radiation and consists of many subranges, commonly referred to as portions, such as visible light or ultraviolet radiation. The various portions bear different names based on differences in behaviour in the emission, transmission, and absorption of the corresponding waves and also based on their different practical applications. There are no precise accepted boundaries between any of these contiguous portions, so the ranges tend to overlap. The entire electromagnetic spectrum, from the lowest to the highest frequency (longest to shortest wavelength), includes all radio waves (e.g., commercial radio and television, microwaves, radar), infrared radiation, visible light, ultraviolet radiation, X-rays, and gamma rays. Nearly all frequencies and wavelengths of electromagnetic radiation can be used for spectroscopy.

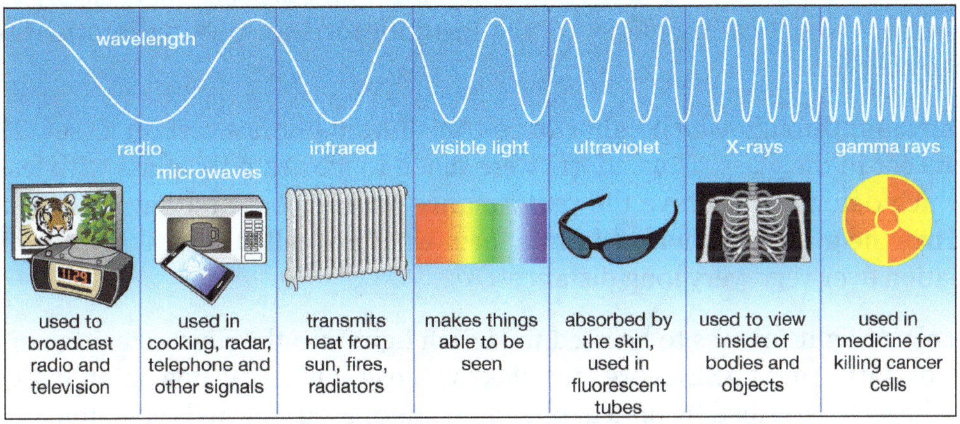

Types of electromagnetic radiation: Radio waves, infrared rays, visible light, ultraviolet rays, X-rays, and gamma rays are all types of electromagnetic radiation. Radio waves have the longest wavelength, and gamma rays have the shortest wavelength.

Let us consider the electromagnetic spectrum as shown in Figure. It has wavelengths ranging from 10 meter, 1 meter, 10 centimeter, 1 centimeter, 1 millimeter, 0.1 millimeter, 10 micrometer, 1 micrometer and 0.1 micrometer. This corresponds to the frequencies for 30 megahertz, 300 megahertz, 3 gigahertz going all the way up to 300 terahertz. When the communication started the communication started in low frequency bands; that means, in the range of few kilohertz then the frequency was increased to few megahertz. Then the frequency was increased to few 100 megahertz then the frequency was increased to few gigahertz and so on. Over last hundred years or so, the frequency of operation for communication has consistently increased towards higher end.

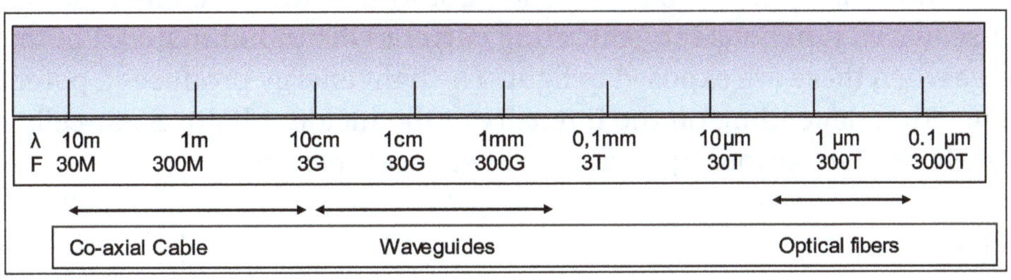

Signal to noise ratio and bandwidth of operation are two important factors which causes this increase in frequency of operation. In other words for a given input signal, if the medium has very low loss then at the receiving end the signal amplitude will be reasonably large and the signal to noise ratio will be large. So we can say that if you have a high signal to noise ratio, we must have a medium of transmission which has as low loss as possible. The bandwidth requirement is proportional to number of users if all of them are using the voice information. Therefore, to have a reliable communication where large number of people can send information from one point to another, we must have a medium which has as low loss as possible and as large bandwidth as possible. Therefore one has to design electrical circuit such that the quality factor of an electrical circuit is practically independent of frequency.

The quality factor is proportional to center frequency or for a given quality factor the bandwidth is proportional to the center frequency. It was observed that in optical window one can get more bandwidth which is 10000 times more than what microwave and millimeter wave circuits could provide. Hence optical window was used for communication purpose. So, basically to use optical communication or the optical frequency as the medium for transporting information from one point to one to another.

Glass is the medium through which light can be transmitted but has certain losses. This is due to impurities present in the glass. Thus efforts were made to fabricate the glass with less number of impurities which definitely improve the performance i.e. less loss and have more bandwidth at faster rate. Thus the glass was molded in to a form what are called optical fibers which now can carry information over very, very long distances.

One has to look for light sources to be used to launch light into the fiber so as transmit light over larger distances. LEDs and LASERs are identified as those sources. Propagation light through optical fiber can be well described by optical phenomenon of total internal reflection.

Optoelectronic Devices

Optoelectronic devices find applications in industry, telecommunication, military, aerospace etc. Optoelectronic devices produce electric energy when exposed to light. They use light in visible or infrared regions of electromagnetic spectrum. Solid state devices like light emitters, IR sensors, LASERS, LEDS are used in optoelectronic applications.

Photoconductive devices such as photo resistors are widely used in counting systems, twilight switches, house security systems, etc. These detect variations in the light intensities and activate or deactivate electronic circuits. Photodiodes and phototransistors also fall in this category. These utilise the reverse biased junctions for generating current when illuminated. Photovoltaic devices produce a voltage when these are exposed to light. The light energy produces a potential difference across the p-n junction depending on the intensity of the incident light. Solar cells and photovoltaic cells are widely used in various applications to generate electricity.

Light Sources

Optical sources are generally light sources which emit light may be in visible region or infrared region. Electric bulbs, incandescent lamps, tube lights are well known sources of light. If one uses one of these sources then some property of light should change with information. This helps in carrying the information up to large distances. Intensity, phase, polarization etc. are the parameters associated with light source. For optical communication, one should choose the light source whose intensity varies exactly at the rate at which information is changing. Electric bulbs, tube lights do not have such ability. Hence white light is the only option which has wider spectral width but operates at smaller frequencies.

Co-incidentally, LASER sources were invented during the same period which has smaller spectral bandwidth but operates at optical frequencies. LASERs happened to have sufficiently narrow spectral widths and high beam directivity, adequate to be used as optical signal sources. So we then had a compatibility of optical media like and optical sources. Initial LASER Diodes emitted lights of wavelengths of about 800 nm. And so, initial optical communications started with 800 nm wavelength due to which it was called the "First Optical Window" of optical communication. The above discussion, hence, gives a very brief introduction to a very interesting and fascinating communication technology called the Optical Communication. Another light source which is useful in optical communication is Light Emitting Diode (LED). It has wider spectral width but operates at optical frequencies.

Lamps

Lamps, such as incandescent light bulbs, are devices that convert electric current into visible light energy. Incandescent lamps have a filament made from tungsten wire. As current flows through this filament, the current collides with the filament's atoms causing the filament to generate heat resulting in photons being emitted. This particular process produces photons with a variety of wavelengths, resulting in emitted light that appears whitish in color.

Halogen lamps are similar to incandescent lamps. In fact, halogen bulbs are considered to be an advanced form of the incandescent bulb. Halogen lamps are commonly known for both their brilliant light and their very hot-to-the-touch bulbs. A halogen lamp uses a filament that resides inside a gas-pressurized bulb. The pressurized gas consists of an inert gas and a small amount of a halogen element such as bromine or iodine. Also, the glass of a halogen bulb is stronger than the glass in a regular incandescent bulb.

Fluorescent lamps are quite different. They consist of a mercury vapor–filled glass tube whose inner wall is coated with a material that fluoresces. When electrons, which are emitted from the fluorescent bulbs cathode electrode, collide with the mercury atoms, UV (ultraviolet) radiation is emitted. This UV radiation is absorbed by the lamp's fluorescent coating, which in turn releases visible light.

LEDs

Light-emitting diodes (LEDs) are two-lead semiconductor devices that are similar to normal diodes except that they emit light that can be visible, infrared, or ultraviolet. When an LED's anode lead becomes more positive in voltage than its cathode lead (typically by a voltage ranging from of 0.6 to 2.2V) current flows through the LED device resulting in emitted light. In historical terms, red was the first LED color; it was developed in 1962 and then mass produced in 1968. Next came yellow, green, and infrared LEDs. It wasn't until 1989 that the blue LED became commercially available. Nowadays, virtually any LED color, including white, is commercially available.

Laser Diodes

A laser diode is a semiconductor laser device that is very similar, in both form and operation, to a light-emitting diode (LED).

Photoresistor

Photoresistor are nothing more than light-controlled variable resistors, also known as light dependent resistors (LDRs). Typically, when a photoresistor is placed in the dark its resistance is very high (in the megaohms). In contrast, when a photoresistor is illuminated, its resistance drops dramatically—depending on the intensity of the light shining on it, the resistance may be as low as hundreds of ohms. Photoresistors are used in light-sensitive switching devices.

Photodiodes

Photodiodes are two-lead semiconductor devices that convert light energy (i.e., photons) directly into electric current. A photodiode is constructed using a very thin n-type semiconductor together with a thicker p-type semiconductor. The n side has an abundance of electrons and is considered the cathode while the p side has an abundance of holes and is considered the anode. When a photon (i.e., light) of sufficient energy strikes the diode, it creates an electron-hole pair. The holes move toward the anode while the electrons move toward the cathode, thus creating a light-induced current (i.e., a photocurrent).

Solar Cells

Solar cells are simply photodiodes with exceptionally large surface areas. These larger areas allow the solar cells to be more sensitive to incoming light as well as more powerful—in terms of both voltage and current—than photodiodes. Solar cells are commonly used in solar panels, but they are also often used as light-sensitive elements in detectors of visible light. Examples include light meters and light-sensitive relays.

Phototransistors

As you may have guessed, phototransistors are light-sensitive transistors. There are two common types: The first resembles the BJT (bipolar junction transistor) and the second is similar to a FET (field-effect transistor). The BJT-type phototransistor has its base replaced by a light-sensitive area; when this surface is kept dark the device remains off. The FET-type phototransistor, sometimes referred to as a photo-FET, uses light to generate a gate voltage that controls a drain-source current. FET-type phototransistors are more sensitive to variations in light compared to BJT-type phototransistors.

Optoisolators

Optoisolators (also known as optocouplers) are electrical devices that interconnect two circuits by means of an optical interface. For instance, a typical optoisolator is composed of an LED and a phototransistor, both of which are encased in a light-tight enclosure. The LED portion of an optoisolator is connected to a driver circuit, and the phototransistor is the output device. Accordingly, when the LED is energized it emits photons that are detected by the phototransistor. A typical application of an optoisolator is to provide electrical isolation between two separate circuits.

Fiber Optics

Fiber optics, which is the science of light transmission through very fine glass or plastic fibers, continues to be used in more and more applications due to its inherent advantages over copper conductors.

Advantages of Fiber Optics

Fiber optics has many advantages over copper wire including:

- Increased bandwidth: The high signal bandwidth of optical fibers provides significantly greater information carrying capacity. Typical bandwidths for multimode (MM) fibers are between 200 and 600MHz-km and >10GHz-km for single mode (SM) fibers. Typical values for electrical conductors are 10 to 25MHz-km.
- Electromagnetic/Radio Frequency Interference Immunity: Optical fibers are immune to electromagnetic interference and emit no radiation.
- Decreased cost, size and weight: Compared to copper conductors of equivalent signal carrying capacity, fiber optic cables are easier to install, require less duct space, weigh 10 to 15 times less and cost less than copper.
- Lower loss: Optical fiber has lower attenuation (loss of signal intensity) than copper conductors, allowing longer cable runs and fewer repeaters.
- No sparks or shorts: Fiber optics does not emit sparks or cause short circuits, which is important in explosive gas or flammable environments.
- Security: Since fiber optic systems do not emit RF signals, they are difficult to tap into without being detected.
- Grounding: Fiber optic cables do not have any metal conductors; consequently, they do not pose the shock hazards inherent in copper cables.
- Electrical isolation: Fiber optics allows transmission between two points without regard to the electrical potential between them.

Table: Advantages of Fiber Optics over Copper.

	Coaxial Cable	Fiber Optic Cable (MM)	Fiber Optic Cable (SM)
Representative distance bandwidth products	100 MHz km	500 MHz km	100,000+MHz km
Attenuation/km @ 1 GHz	>45 dB	1 dB	0.2 dB
Cable cost ($/m)	$$$$$$$$$	$	$
Cable diameter (in.)	1	1/8	1/8
Data security	Low	Excellent	Excellent
EMI immunity	OK	Excellent	Excellent

Fiber Optic Link Components

In order to comprehend how fiber optic applications work, it is important to understand the components of a fiber optic link. Simplistically, there are four main components in a fiber optic link are:

- Optical Transmitter,
- Optical Fiber/Cable,
- Connectors,
- Optical Receiver.

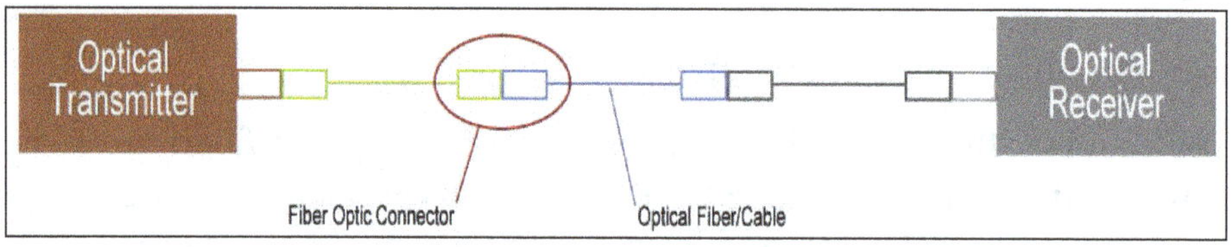

Simple Fiber Optic Link.

Transmitter

The transmitter converts the electrical signals to optical. A transmitter contains a light source such as a Light Emitting Diode (LED) or a Laser (Light Amplification by Stimulated Emission of Radiation) diode, or a Vertical Cavity Surface Emitting Laser (VCSEL).

- LED: Is used in multimode applications and has the largest spectral width that carries the least amount of bandwidth.
- VCSEL: Is also used in multimode applications with a narrower spectral width that can carry more bandwidth than the LED.
- LASER: Has the smallest spectral width, carries the most bandwidth, and is used in single-mode applications.

These sources produce light at certain wavelengths depending upon the materials from which they are made. Most fiber optic sources use wavelengths in the infrared band, specifically 850nm (1nm = 10^{-9}m), 1300nm and 1550nm. For reference, visible light operates in the 400-700nm range.

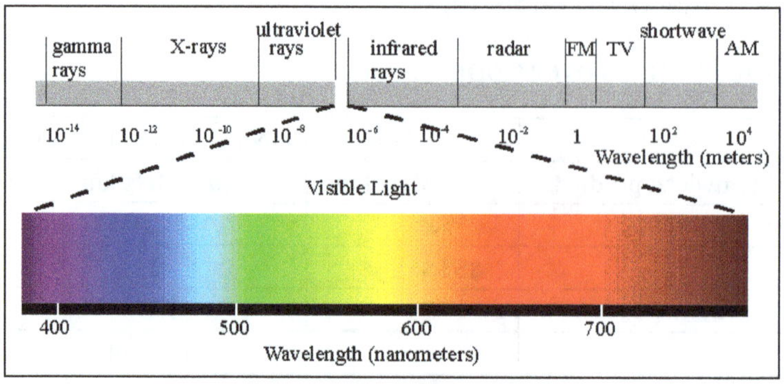

Electromagnetic Spectrum

Optical Fiber/Cable

Here, we discuss the structure and properties of an optical fiber, how it guides light, and how it is cabled for protection. An optical fiber is made of 4 concentric layers:

- Core: This central section made of silica or doped silica, is the light transmitting region of the fiber.
- Cladding: This is the first layer around the core. It is also made of silica, but not the same composition as the core. This creates an optical waveguide which confines the light in the core by total internal reflection at the core-cladding interface.
- Coating: The coating is the first non-optical layer around the cladding. The coating typically consists of one or more layers of polymer that protect the silica structure against physical or environmental damage. The coating is stripped off when the fiber is connectorized or fusion spliced.
- Buffer: The buffer is an important feature of the fiber. It is 900 microns and helps protect the fiber from breaking during installation and termination and is located outside of the coating.

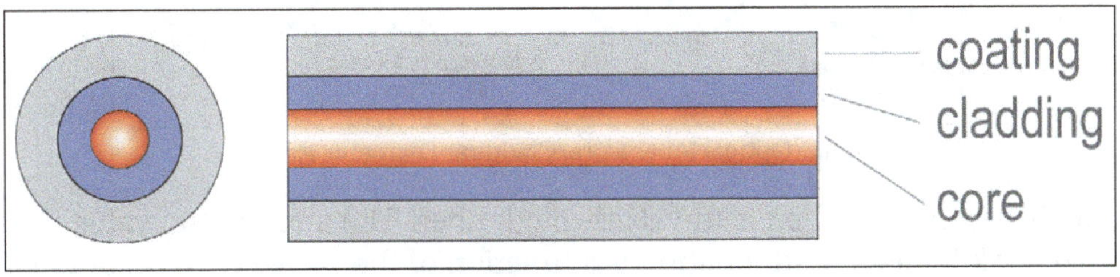

Optical Fiber Construction.

The light is "guided" down the core of the fiber by the optical "cladding" which has a lower refractive index (the ratio of the velocity of light in a vacuum to its velocity in a specified medium) that traps light in the core through "total internal reflection."

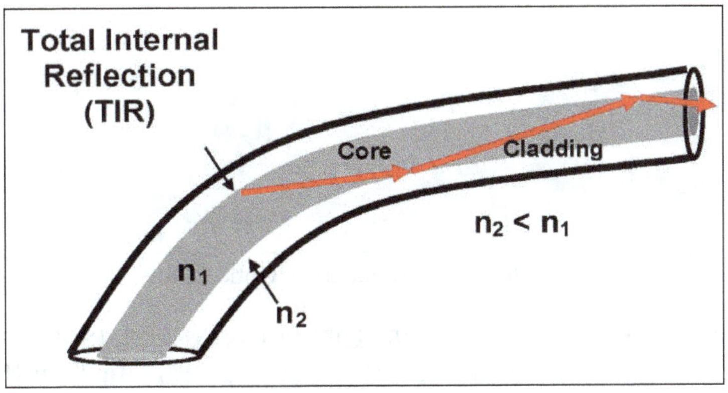

Total Internal Reflection.

In fiber optic communications, single mode and multimode fiber constructions are used depending on the application. In multimode fiber, light travels through the fiber following different light paths called "modes." In single mode fiber, only one mode is propagated "straight" through the fiber.

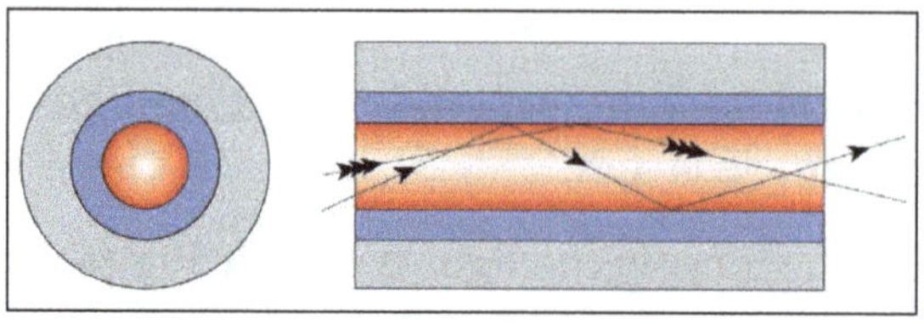

Multimode Fiber Light Propagation.

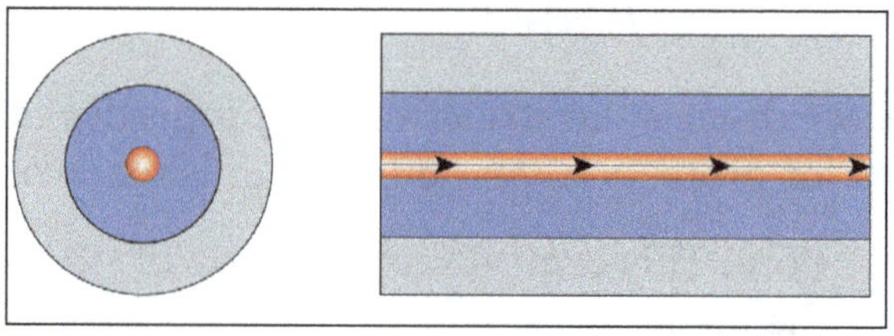

Single Mode Fiber Light Propagation.

Typical multimode fibers have a core diameter/cladding diameter ratio of 50 microns/125 microns (10^{-6} meters) and 62.5/125 (although 100/140 and other sizes are sometimes used depending on the application). Single mode fibers have a core/cladding ratio of 9/125 at wavelengths of 1300nm and 1550nm.

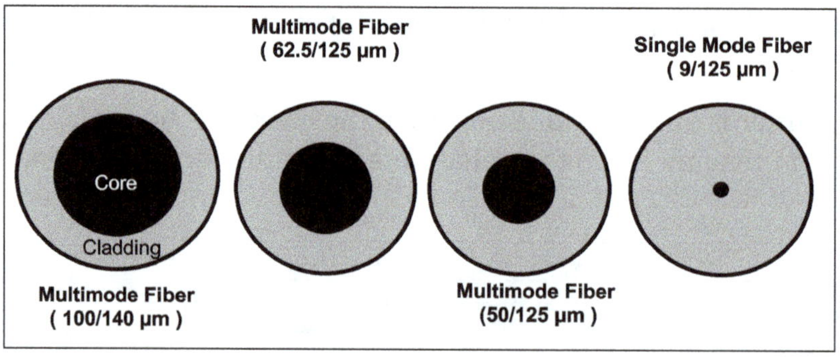

Popular Optical Fiber Core/Cladding Diameter Ratios.

Light is gradually attenuated when it travels through fiber. The attenuation value is expressed in dB/km (decibel per kilometer). Attenuation is a function of the wavelength (λ) of the light. Figure shows the attenuation as a function of the wavelength.

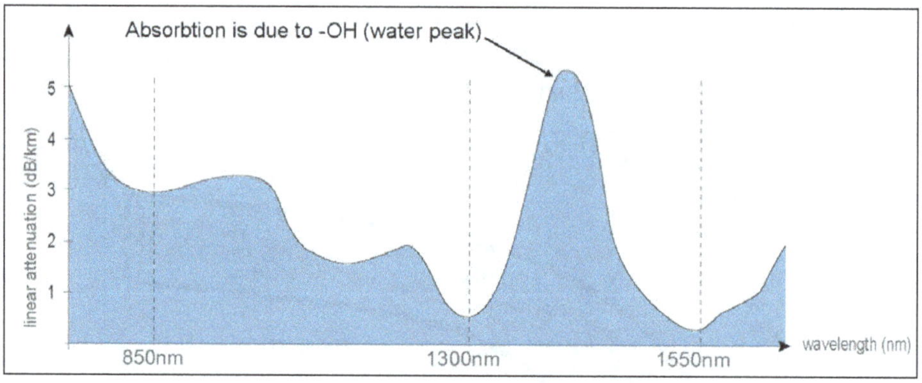

Attenuation vs. Wavelength of Optical Fiber.

The typical operating wavelengths are 850nm (nanometers) and 1300nm in multimode, and 1300nm or 1550nm in single mode. Note that there are natural "dips" in the attenuation graph at these wavelengths. For example, at an 850nm operating wavelength, there is 3dB attenuation after 1km propagation (according to the graph). 3dB of attenuation means that 50% of light has been lost.

Bandwidth is a measure of the data-carrying capacity of an optical fiber. It is expressed as the product of frequency and distance. For example, a fiber with a bandwidth of 500MHz-km

(Megahertz kilometer) can transmit data at a rate of 500MHz along one kilometer of fiber. The bandwidth of single mode fibers is much higher than in multimode fibers. The main reason for the lower bandwidth in multimode fibers is modal dispersion.

In multimode fibers, information (ABC) is propagated in fiber according to N modes or paths, as if it were "duplicated" N times (for example, in the diagram, the mode 3 path is longer than the mode 2 path, which are both longer than the mode 1 path). If information is too close, there is a risk of overlapping ("smearing") the information, and then it will not be recoverable at the end of the fiber. It is necessary to space the data sufficiently to avoid overlap, i.e., to limit the bandwidth.

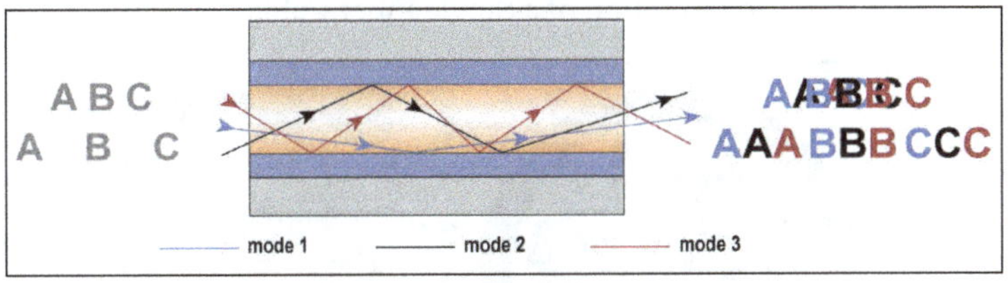

Modal Dispersion in Multimode Fibers.

Modal dispersion can be alleviated to a large extent by grading the index of refraction from the middle of the core to the cladding (graded index fiber), thereby equalizing the paths. In a step index fiber, the index of refraction changes abruptly from the core to the cladding. To help reduce modal dispersion, fiber manufacturers created graded-index fiber. Graded-index fiber has an index of refraction which gradually increases as it progresses to the center of the core. Light travels slower as the index of refraction increases. Thus, a light path propagating directly down the center of the fiber has the shortest path but will arrive at the receiver at the same time as light that took a longer path due to the graded-index of the fiber.

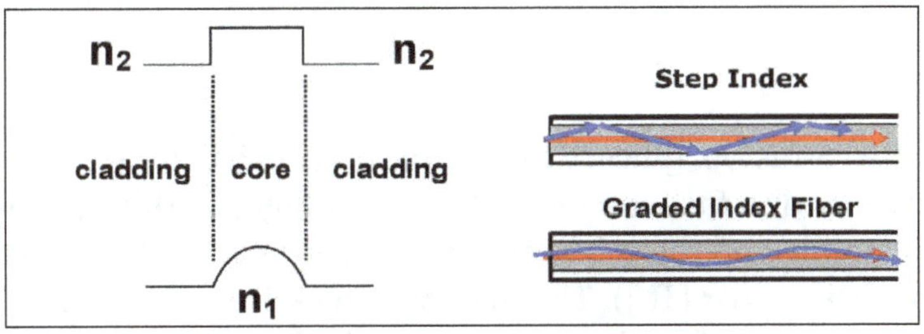

Graded Index in Multimode Fibers.

Of course, modal dispersion is not an issue in single mode fiber because only a single mode is propagated.

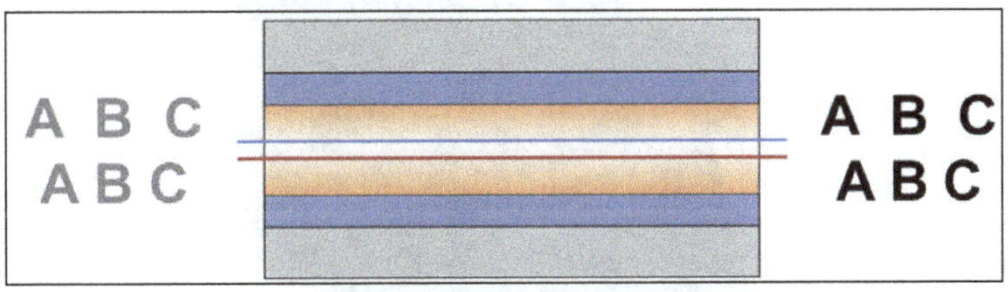

Single Mode Propagation.

Unfortunately, the optical fiber construction shown in Figure is fragile. Thus, for most applications, the fiber must be made into a cable. There are many ways to construct a cable (tight buffer, loose tube, gel filled, distribution, breakout, etc). However, in our single fiber cable example, the 250 micron coating is jacketed with a 900 micron buffer and built into a 3.0mm outer sheath cable with aramid yarn as a strength member. As a typical example, figure portrays a cable with multiple optical fibers.

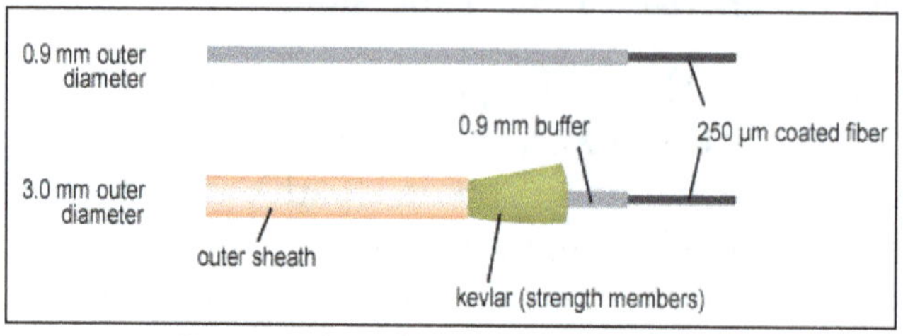

Construction of a Single Fiber Cable.

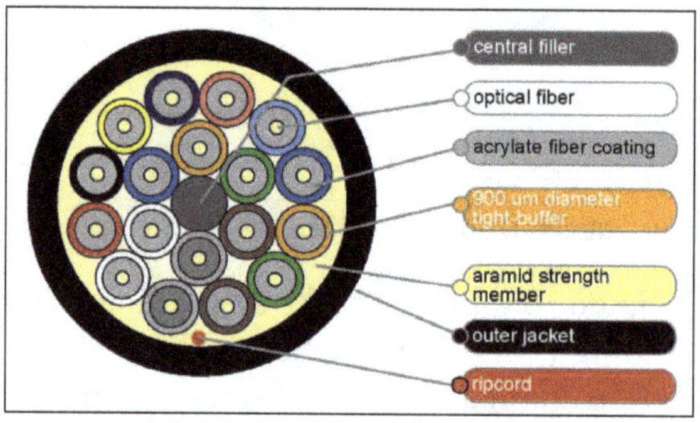

Example of the Construction of a Multi-Fiber Cable.

Connectivity

Fiber optic links require a method to connect the transmitter to the fiber optic cable and the fiber optic cable to the receiver. In general, there are two methods to link optical fibers together.

Fusion Splice

The first method is called a fusion splice. This operation consists of directly linking two fibers by welding with an electric arc or a fusion splicer. The advantages of this approach are that the linking method is fast and simple and there is very little insertion loss (the loss of light generated by a connection is called Insertion Loss [IL]). The disadvantages are that the link is relatively fragile, is permanent, and the initial cost (of the fusion splicer) is high.

Fiber-optic Fusion Splicer.

Connectors

The second method involves the uses of fiber optic connectors. A connector terminates the optical fiber inside a ceramic ferrule, using epoxy to hold the fiber in place. The connectors can be mated and unmated at any time. The advantages of this approach are that the connection is robust, the connector can be chosen according to the application, and the connector can be connected and disconnected hundreds or even thousands of time without damaging the connectors. The disadvantages of this approach are that the connectorization takes longer than fusion splicing, requires special tools, and the insertion loss can be higher when compared with fusion splicing.

Physical Contact Connectors

Physical contact connectors utilize fiber in a tightly tolerance ceramic ferrule. This allows easy handling of the fiber and protects it from damage. The principle of physical contact connectors involves the direct contact of polished fibers within two ceramic ferrules. The ferrules are aligned using a ceramic alignment sleeve. Insertion loss is a function of the alignment accuracy and the polish quality. There are springs behind the ferrule to ensure that the two ferrules are in constant contact even in high vibration and shock environments.

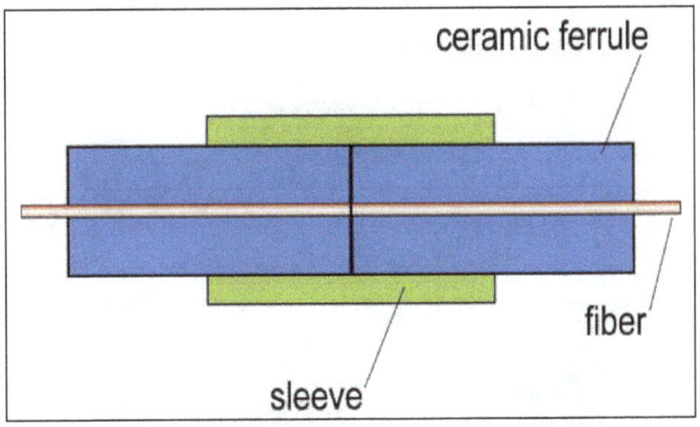

Physical contact connectors are the most common type of fiber optic connection. They are rugged, repeatable, easy to clean, cost-effective, and perform well. In addition, for physical contact connectors, the insertion loss is generally low (approximately 0.3dB). There are many types of fiber optic connectors used in various applications. The most popular single fiber connectors are:

- FC-Ferrule Connector: Although the FC connector is being replaced in many applications (telecom and datacom) by LC and SC connectors, it is still used in measurement equipment. The connector has a screw threading and is keyed allowing the ferrule to be angle polished providing low back reflection (light is reflected back to the transmitter, most often at the connector interface due to an index of refraction change).
- LC-Lucent Connector: LC connectors are supplanting SC connectors because of their smaller size and excellent panel packing density and push-pull design. They are also used extensively on small form-factor pluggable transceivers.
- SC-Subscriber Connector: SC connectors also offer a push-pull design (which reduces the possibility of end-face damage when connecting) and provide good packing density. They are still used in datacom and telecom applications.

- ST-Straight Tip Connector: ST connectors are engaged with a bayonet lock which is engaged by pushing and twisting the connector. The bayonet interlock maintains the spring-loaded force between the two fiber cores.

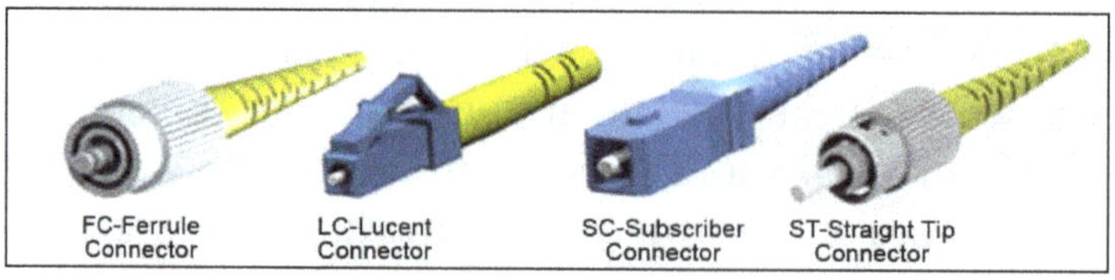

Popular Single Fiber Connectors.

These physical contact connectors perform well against particle contaminates (dust, mud, etc.) and are usually less sensitive to liquid contaminates (water or oil). The physical contact pushes liquid out of the way and the liquid does not degrade the connection. Physical contact connectors are cleaned by wiping the ferrule with a clean cloth or wipe, spraying with a cleaner or washing with water.

It is also common to provide multiple fibers in a single connector. An example is the MPO connector which supports 12 fibers in a single ferrule. Another example is the TFOCA-II® connector which provides 4 or 12 fiber optic channels for harsh environment fiber-optic applications.

MPO Multi-fiber Connector.

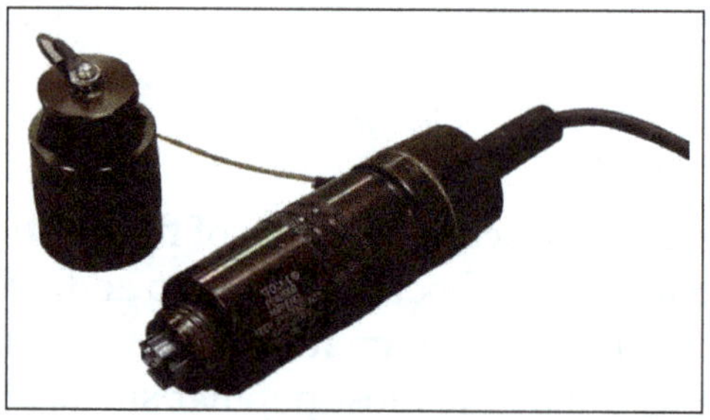

AFSI TFOCA-II Connector.

Expanded Beam Technology

The other connector technology is expanded beam, which consists of placing a lens at the exit of each fiber to widen and collimate the light. In this configuration, there is an air gap between the two optical fibers/lens assemblies. The mechanical interface between the connectors must be precise. Dust and dirt must not interfere with the alignment of the optical elements. Expanded beam connectors are less susceptible to particle contaminates such as dirt and dust but they perform poorly with liquids or film on the lenses. They can also be very difficult to terminate in the field.

The loss generated by an expanded beam connection is more than that of a physical contact connector due to the lenses, mechanical alignment and sometimes protective windows (approximately 0.8 to 2.5dB typical).

This type of connector performs well against particle contamination on the lens because the beam is expanded to a much larger size than the beam that comes directly from a fiber. However, any liquid or film (such as a fingerprint) on a lens creates significant loss in an expanded beam connector. Expanded beam connectors are also very sensitive to alignment of the lenses. Connectors must always be tightly coupled and kept clean on the mating surface in order to work properly. Cleaning an expanded beam connector must be done with care because any liquid (water, alcohol or another cleaner) that is trapped inside the connector may migrate to the surface of the lens, causing an unacceptable increase in insertion loss.

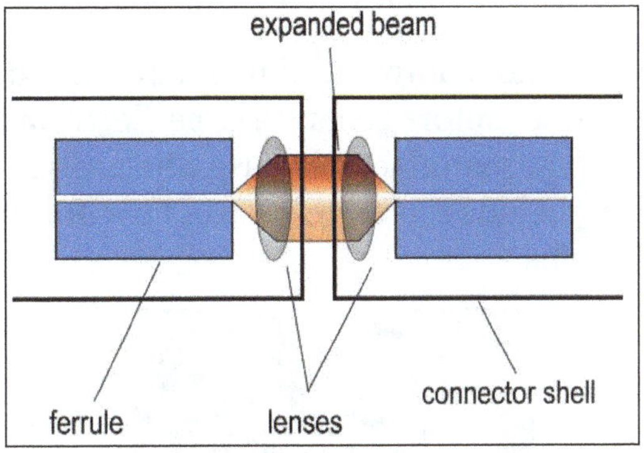

Expanded Beam Technology.

Receiver

The last component of the fiber optic link is the optical receiver, which uses a photodiode to convert the optical signals into electrical. The two types of photodiodes used are: Positive Intrinsic Negative (PIN) and the Avalanche PhotoDiode (APD). In a similar fashion as that of the laser transmitter, the photodiode will receive wavelengths depending on material composition.

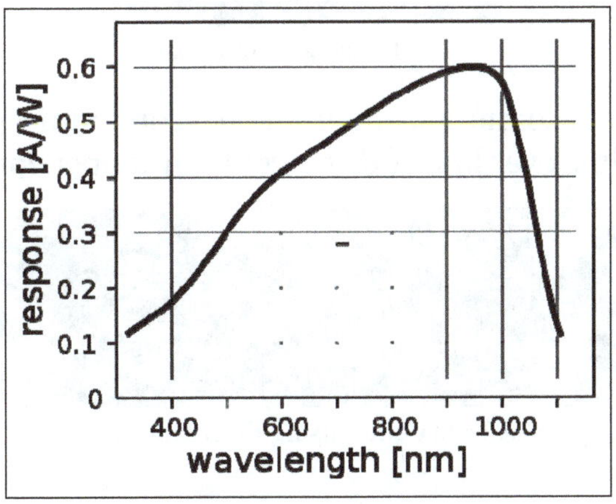

Responsivity of a Silicon Photodiode.

Fiber Optic Applications

Fiber optics is used in myriad applications. Due to its low weight, high bandwidth capacity and immunity to electromagnetic and RF interference, fiber optics is used extensively in avionics on both military and commercial aircraft systems. Applications include radar links, video systems, sensor networks, and in-flight entertainment systems.

Military Aircraft

Fiber optics is also used for command, control, and telemetry in industrial applications wherever there are large electric motors. These motors generate large electromagnetic fields. Instead of using heavily shielded copper conductors, manufacturers use fiber optics to eliminate electromagnetic interference concerns. Examples include top drive control links for drilling rigs and command and control for long wall mining systems.

Top Drive Control Link.

Fiber optics is also used in data communications and telecommunications systems due to its ability to transmit high bandwidths over longer distances than copper conductors.

A Data Center using Fiber Optics.

Applications of Optoelectronics

Lasers Applications

There are different types of lasers, which differ by the nature of the lasing medium employed which can be a solid, gas, liquid or semiconductor. They also differ on the emitted wavelength. In the table below you will find some typical lasers, their emission wavelengths and some of the corresponding applications.

Table: Laser types, materials and wavelengths.

Laser	Laser type	Wavelength (nm)	Application
Argon fluoride	Excimer laser	193 (UV)	Eye surgery
Krypton fluoride	Excimer laser	248 (UV)	Eye surgery
Diode lasers	Semiconductor lasers	630 nm - 680 nm	Laser pointers, low power applications, laser printers, CD players
Helium neon (green)	Gas laser	543 (visible)	Several uses
Helium neon (red)	Gas laser	633 (visible)	Several uses
Rhodamine 6G (organic)	Dye laser	570-650	Tunable over a broad frequency range
Ruby (red)	Solid-state laser	694	
Nd:Yag (neodymium:yttrium-aluminum garnet)	Solid-state laser	1064 (IR)	
CO_2 lasers	Gas laser	10600 (FIR)	Cutting of hard materials such as steel

The ruby laser is a solid-state laser and emits at a wavelength of 694 nm. If other emission wavelength is needed, other lasing materials can be selected. The CO_2 laser is a gas laser that emits light in the far-infrared region of the spectrum with very high power density which is used to cut materials such as steel. Infrared radiation is heat, and this laser beam melts the cutting materials. Diode lasers use low power and have small dimensions. They are used in laser pointers, CD players and laser printers.

Laser pointers, using diode lasers.

Excimer lasers use gases, such as chlorine and fluorine mixed with argon, krypton or xenon. When electrically stimulated, a pseudo molecule is produced, which produces light in the ultraviolet range. These lasers are used in refractive eye surgery to correct malformations or vision defects.

Applications in Medicine

- Refractive eye surgery is a recent surgical procedure that is used to correct vision defects such as myopia (short sightedness), astigmatism and hyperopia (far sightedness).
- Eye operations using lasers are the most used technique to correct eye problems.

- Before the operation, the patient's eye is measured to build a map of the eye's surface and determine the degree of visual correction needed. The required change in the cornea surface is calculated based on this information, and fed into the laser's computer.
- The surgery consists in reshaping the cornea surface by removing microscopic amounts of tissue from the outer surface with a cool, ultraviolet beam of light. An excimer laser which operates in this frequency range is normally used. The beam is controlled by computer and its precision is of the order of some nanometers (1 nm = 1 x 10^{-9} m). The surgery lasts only a few minutes, and patients can resume their daily activities are in two to three days.
- Other applications can be found in the field of dermatology where lasers are used to remove scars, resulting from acne, accidents, and burning scars. The laser performs a local abrasion which restores the skin. Lasers can also be used to remove tattoos.
- Laser skin resurfacing, using either the CO_2 or Erbium laser, is one of the latest developments in the war against wrinkles. This technique produces a controlled tissue injury by a process known as vaporization, in which the laser energy is absorbed by the water in the skin cells.

Applications in Communications

CDs and DVDs can store large amounts of information. Nowadays they are commonly used to store data, music, films or computer software. They are cheap to produce and to retrieve the information stored as bumps in the CD, all you need is a CD drive or player.

CD Drive

The CD drive consists of three main components:

- A drive motor which rotates between 200 and 500 rpm, spins the disc. This drive motor is controlled depending on which track is being read.
- A laser together with a lens system focus the laser beam on the disc surface tracks to read the bumps.
- A tracking system moves the laser, with a micron resolution, so that the laser's beam can follow the spiral track.

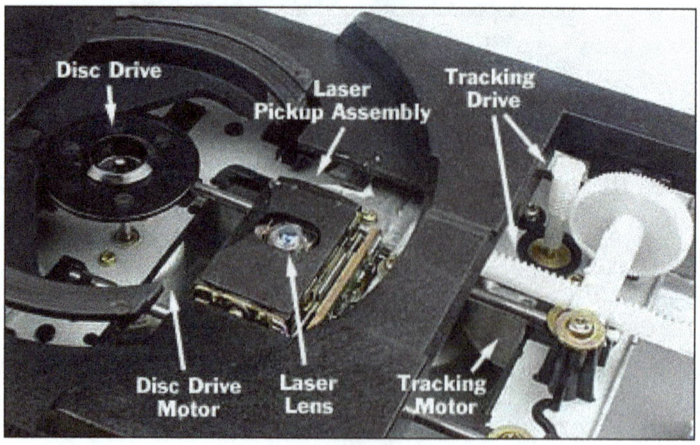

A CD player.

The laser beam is focused on the track on the CD, and reflects the Aluminium layer hitting an optoelectronic device that detects changes in light intensity. The bumps reflect light differently than the rest of the Aluminium layer (valleys), and the optoelectronic sensor detects the changes in reflectivity. The electronic part of the CD drive or player converts the changes in reflectivity into bits.

Laser Printers

The core component of this system is the photoreceptor, which has the form of a rotating cylinder. It is made of photoconductive material. A schematic representation of a laser printer is given in the figure below. Initially, the cylinder is given a positive charge by the corona wire, a wire with an electrical current running through it. A laser beam (diode laser usually) is focused on the cylinder surface to enable photons to discharge some areas of the surface, which represent the letters or images to be printed. Thus, an electrostatic image is created.

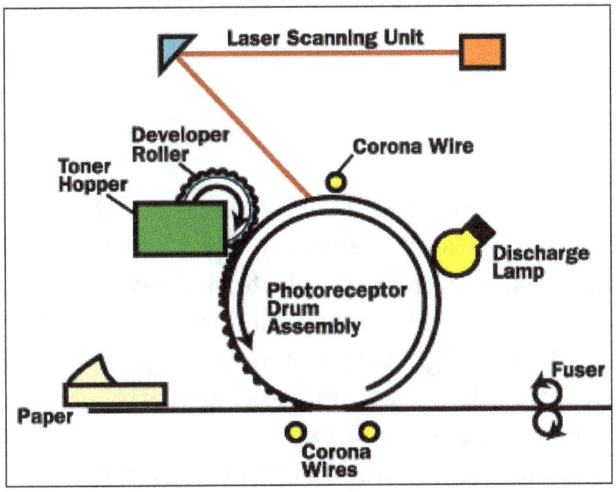

Schematic representation of a laser printer.

Then the cylinder is covered with black powder (the toner), which sticks to the discharged areas of the cylinder. Then the cylinder (the drum) passes on a sheet of paper which was given a negative charge, stronger than the negative charge of the electrostatic image, so that the paper can pull the toner powder away. Finally, the printer passes the paper through a pair of heated rollers melting the toner powder into the paper fibres and sends it out as a finished page. That is the reason why pages are always hot when they come out of a laser printer.

Optical Fibre Applications

- They are the most cost-effective way to move huge amounts of information (voice, data) quickly and reliably.
- Optical Fibres are insensitive to electrical interference, so they can be used in noisy, harsh environments.
- Fibre optic cables have less weight and consume less space than equivalent electrical links.

Recent and future applications of optical fibres include:

- High speed, long-haul trunks, bit rate: several Gbit/s.

- Extension of LAN's to broadband integrated services.
- Increased distance between repeaters of long-haul links.
- Interconnection of supercomputers networks.
- Medicine: Tele-surgery.
- CATV: Video on demand.
- Tele-learning: streaming video.
- Video-conferencing.
- Optical amplification.
- Photonic switching.

Photonics

Photonics is essentially the science that involves generation of a photon (light), its detection, as well as manipulation via transmission, emission, signals processing, modulation, switching, amplification, and sensing. Most importantly, photonics involves the proper use of light as a tool for the benefit of humans. Most photonics applications, even though they cover all technical applications over the entire electromagnetic spectrum, range from near-infrared light to visible region. The term "photonics" was derived from the root word "photon", which connotes the tiniest entity of light analogous to an electron in electricity. Just as the electronics revolutionized the 20th century, photonics is doing the same in the 21st century. Photonics is made up of many different technologies including optical fibers, lasers, detectors, quantum electronics, fibers, and materials.

The term photonics was first used to designate a field of research area responsible for utilizing light to perform tasks that are conventionally related to the traditional sphere of electronics, like telecommunications, information processing, and so on. Studies in the field of photonics began in 1960 after the discovery of lasers. Other progress followed including optical fibers for transmitting information, the laser diode in the 1970s, as well as erbium fiber amplifiers. These developments made the foundation for the industrial revolution in the telecommunications sector during the late 20th century and supplied the internet infrastructure. Although created before the 1980s, the word photonics was used commonly for the first time in the 1980s as network operators of telecommunications embraced fiber-optic data transmission. Through the years, until 2001 with the dot-com crash, research was primarily focused on optical fiber telecommunication. Nevertheless, the field of photonics has a broad range of scientific and technological applications. These include chemical and biological sensors, laser manufacturing, medical therapy and medical diagnosis, optical computing and displaying technology.

Advancement of photonics is possible due to the current success recorded concerning the development of silicon photonics. Photonics is related to opto-mechanics, electro-optics, quantum electronics and quantum optics. Nevertheless, these fields mean different things to both the scientific as well as the business community. Quantum optics is often concerned with fundamental theoretical research areas. Photonics, on the other hand, deals with applied research and progress. Optoelectronics is used to refer to the circuits or devices consisting of both electrical and optical components. The word "electro-optics" was utilized in the past to specifically relate to nonlinear interactions between electrical and optical devices. These devices include bulk crystal modulators and later include advanced imaging sensors that are typically employed by both government and private individuals in surveillance activities.

Photonics is said to be an "All-Pervasive" technology because it allows unlimited light to travel faster than the electrons that are used in electronic computer chips, which means that optical computers will compute thousands of times faster than any electronic computers because of the physical limitations of electronic conduction. More wavelengths can be packed into an optical fiber to allow an increase in the transmission bandwidth that can be in conventional copper wires. There is no electromagnetic interference in light compared to electrons in copper wires.

Applications of Photonics

Photonics have uses in almost every aspect of our life, ranging from daily life to highly innovative science. For instance, information processing, telecommunications, light detection, metrology, lighting, spectroscopy, photonic computing, holography, medical field (surgery, vision correction, health monitoring and endoscopy), fighting machinery, visual art, agriculture, laser material processing, robotics, and biophotonics. Similar to the way electronics have been used extensively since the creation of earlier transistors of 1948, the exceptional use of photonics continuously increases. Economically significant uses of photonic devices include fiber optic telecommunications, optical data storage, displays, optical pumping of high-power lasers and laser printing. Prospective applications of photonics are practically limitless and include medical diagnostics, organic synthesis, information, and communication, as well as fusion energy:

- Telecommunication: Optical down-converter to microwave, and optical fiber communications.
- Medical applications: Laser surgery, poor eyesight correction, tattoo removal and surgical endoscopy.
- Manufacturing processes in industries: Involves the use of laser in welding, cutting, drilling, and many surface modification techniques.
- Building and construction: Smart structures, laser range finding, and laser leveling.
- Space exploration and aviation: Including astronomical telescopes.
- Military operations: Command and control, IR sensors, navigation, mine laying, hunt and salvage, and discovery.
- Metrology: Range finding, frequency and time measurements.
- Photonic computing: Printed circuit boards, and quantum computing.
- Micro-photonics and nanophotonics.

These typically include solid-state devices and photonic crystals. In simple terms, photonics is currently solving and addressing the challenges of a modern world. Photonics enhances the quality of life; it safeguards our health, security, and safety, it drives our economic growth, and it creates jobs as well as global effectiveness. Photonics technology enhances the quality of life in many areas. Specific areas are communications and information technology, advanced manufacturing, defense, health and medicine, and energy. Photo detectors are used to detect light. They can be very slow, as in the case of solar cells that are used in harvesting sunlight energy, or very fast like photodiodes that are very fast and are employed in communications in conjunction with digital cameras. Numerous others centered on quantum, thermal, photoelectric and chemical areas also exist. Photonics likewise involves research on photonic systems. The term photonics system has found its application in optical communication systems.

Advances in Photonics

There has been an exponential growth in the research activities in the field of photonics and optics over the years. Photonic networks are the backbone of data dissemination, specifically in the modern and upcoming wireless communication systems. Photonic networks continue to gain interest for distribution of data from, say, central location to a remote antenna unit at base stations. While the demand for wireless photonic systems continues to rise, there is a need for implementation of low-cost systems.

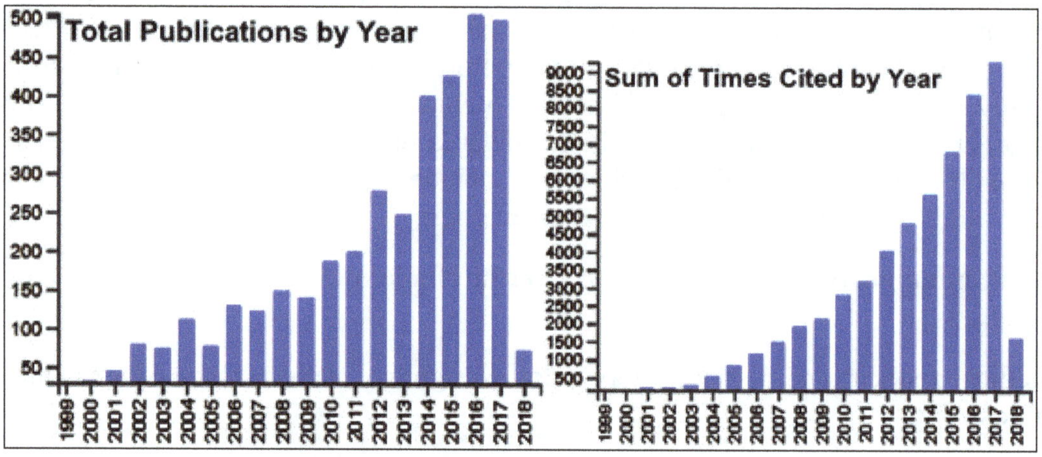

Publications and citation trends in Photonics.

Two of the most popular data transmission methods in wireless photonic systems are digital baseband and RoF (Radio-over-Fiber) optical communication. In addition, further emerging fields are opto-atomics, in which there is an integration of both atomic and photonic devices. Opto-atomics applications include precise time-keeping. Opto-mechanics, metrology, and navigation, as well as polaritonics, are different from photonics due to the presence of polarization as the primary carrier of information.

Microwave photonics is considered to be an emerging research field. Microwave photonics is an enabling technology for the generation, control, distribution, measurement, and detection of microwave signals. It also deals with the operation of new systems and devices. Part of the various functionalities facilitated by photonics, microwave measurements centered on photonics can offer greater performance regarding broad frequency coverage, significant direct bandwidth, high immunity to electromagnetic interference (EMI) and low frequency-dependent loss. Photonic microwave measurements therefore have been widely investigated in recent times.

Moreover, several new methodologies have been offered to address the challenges confronting electronic solutions. Plasmon lasers are among the categories of optical frequency amplifiers that send strong, penetrating, and guiding superficial plasmons underneath the diffraction walls. The interactions between light energy and matter can be intensely improved by the tightly held electric fields in plasmon lasers. This will also bring substantial innovative possibilities to data storage, bio-sensing, optical communications and photolithography. Because they can generate high-intensity nano-scale electromagnetic radiation in a fraction of a second, the modern development of plasmon lasers today has sparked the investigation of nanoscience and technology. This would enable more feature sizes than the conventional lasers. They could also be used to package additional information onto storage media such as hard disks or DVDs. The mid-IR spectroscopy offers a principal means for biological structure analysis as well as nonintrusive measurements. For instance, the broad cross-section for absorption allows for the detection of traces of vapors at the order of parts-per-trillion (ppt) as well as parts per-billion (ppb).

References

- An-introduction-to-optoelectronics, technical-articles: allaboutcircuits.com, Retrieved 18, February 2020
- Basics-of-Fiber-Optics: fibersystems.com, Retrieved 30, March 2020
- Electromagnetic-spectrum, science: britannica.com, Retrieved 24, June 2020

2

Optical Fiber: An Integrated Study

Optical fiber refers to a flexible and transparent fiber made by silica or plastic that is used to transmit light between two ends of the fibre. A few of the important aspects of optical fibres are propagation of light through optical fibre, characteristics of optical fibres and comparison of optical fibers. This chapter has been carefully written to provide an easy understanding of these varied facets of optical fibres.

An optical fiber is a piece of very thin and almost absolutely pure glass. It is thin as human hair. Its outside is made of a cladding of glass which is also another type of glass, with slightly different chemical properties and composition. Hence it has different refractive index compared to inner core material. No single fiber design meets all application requirements mainly due to many economic reasons.

Advantages and Disadvantages

The advantages of optical fiber include the following:

- Bandwidth is higher than copper cables.
- Less power loss and allows data transmission for longer distances.
- The optical cable is resistance for electromagnetic interference.
- The size of the fiber cable is 4.5 times better than copper wires.
- These cables are lighter, thinner, and occupy less area compare with metal wires.
- Installation is very easy due to less weight.
- The optical fiber cable is very hard to tap because they don't produce electromagnetic energy. These cables are very secure while carrying or transmitting data.
- A fiber optic cable is very flexible, easily bends, and opposes most acidic elements that hit the copper wire.

The disadvantages of optical fiber include the following:

- The optical fiber cables are very difficult to merge & there will be a loss of the beam within the cable while scattering.
- The Installation of these cables is cost-effective. They are not as robust as the wires. Special test equipment is often required to the optical fiber.
- Fiber optic cables are compact and highly vulnerable while fitting.
- These cables are more delicate than copper wires.
- Special devices are needed to check the transmission of fiber cable.

Propagation of Light through Optical Fiber

Propagation of light energy in the form of optical signals inside the core-cladding arrangement and throughout the length of the fiber takes place by a phenomenon called the Total Internal Reflection (TIR) of light. This phenomenon occurs only when the refractive index of core is greater than the refractive index of cladding and so the cladding is made from glass of lower refractive index. By multiple total internal reflections at the core-cladding interface the light propagates throughout the fiber over very long distances with low attenuation.

The essential requirements of the propagation of light through an optical fiber, over long distances with minimum loss, are discussed in detail. Figure shows a section of the core of an optical fiber. If a ray of light is incident on the core of an optical fiber from the side, the ray of light simply refracts out from the fiber on the other side. The ray shown in figure (in blue) demonstrates the situation.

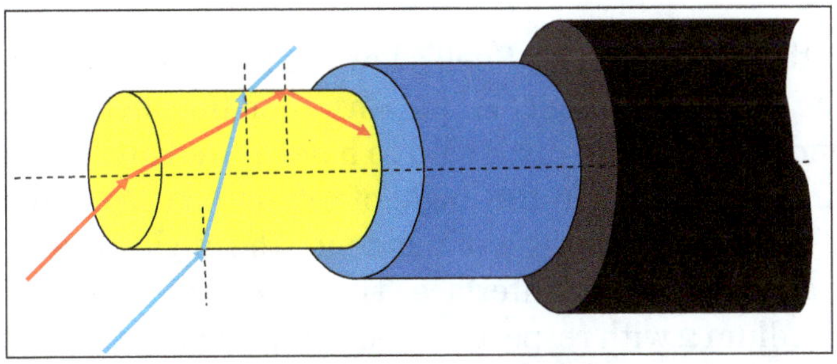

Launching light into optical fiber.

Any light that enters the optical fiber from the side does not propagate along the fiber. Therefore, one has to launch the light through the tip of the fiber. That is, in order to guide light along the fiber, the light must be incident from the tip of the optical fiber. The red ray of light in figure explains this situation. In other words, if the tip of the optical fiber is not exposed to light, no light will enter the fiber. Although there may be ambient light, as long as the tip is protected, no light from the sides propagates along the fiber. Equivalently, if there was propagation of light through the fiber, no light would emerge from the sides of the fiber. This characteristic of the optical fiber imparts the advantage of information security to the Optical Fiber Communication Technology.

Partial reflection at the core-cladding interface does not suffice the propagation of light along the fiber over long distances. The reason is that, at each reflection a part of the optical energy launched into the optical fiber would be lost and after a certain distance along the length of the fiber the optical power would be negligibly low to be of any use. Thus total internal reflection is an absolute necessity at each reflection for a sustained propagation of optical energy over long distance along the optical fiber. This precisely is the sole reason of launching light into the fiber at particular angles so that light energy propagates along the fiber by multiple total internal reflections at the core-cladding interface. The Ray-Model of light obeys the Snell's laws. Figure depicts a situation of a typical refraction phenomenon taking place at the interface of two optically different media having refractive indices n1 and n2:

$$n_1 \sin\theta_1 = n_2 \sin\theta_2 \text{ (Snell's law)}$$

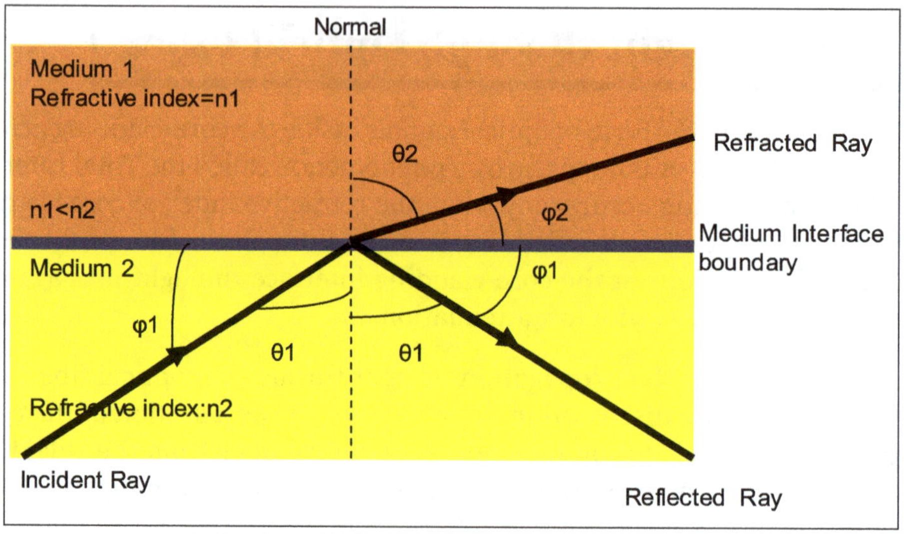

Refraction of light at medium interface boundary.

The angles measured in the expression for Snell's law are measured with respect to the normal to the media interface at the point of incidence. If n2 > n1, then the angle of refraction is greater than the angle of incidence and the refracted ray is said to have moved away from the normal. If the angle of incidence (θ1) is increased further, the angle of refraction (θ2) also increases in accordance with the Snell's law and at a particular angle of incidence the angle of refraction becomes 90° and the refracted ray grazes along the media interface. This angle of incidence is called the critical angle of incidence (θc) of medium 2 with respect to medium 1. That means, the same optically denser medium may have different critical angles with respect to different optically rarer media. If θ1 is increased beyond the critical angle, there exists no refracted ray and the incident light ray is then reflected back into the same medium. This phenomenon is called the total internal reflection of light.

The word 'total' signifies that the entire light energy that was incident on the media interface is reflected back into the same medium. Total Internal Reflection (TIR) obeys the laws of reflection of light. This phenomenon shows that light energy can be made to remain confined in the same medium when the angle of incidence is greater than the angle of reflection. There are two basic requirements for a TIR to occur:

- The medium from which light is incident, must be optically denser than the medium to which it is incident. In figure, n2 > n1.
- The angle of incidence in the denser medium must be greater than the critical angle of the denser medium with respect to the rarer medium.

Thus one can say that light propagates inside an optical fiber because of multiple rays undergoing TIRs at the core-cladding interface. In order to occur this, refractive index of the core glass must be greater than that of the cladding. This meets the first condition for a TIR. All the light energy that is launched into the optical fiber through its tip does not get guided along the fiber. Only those light rays propagate through the fiber that are launched into the fiber at such an angle that the refracted ray inside the core of the optical fiber is incident on the core-cladding interface at an angle greater than the critical angle of the core with respect to the cladding.

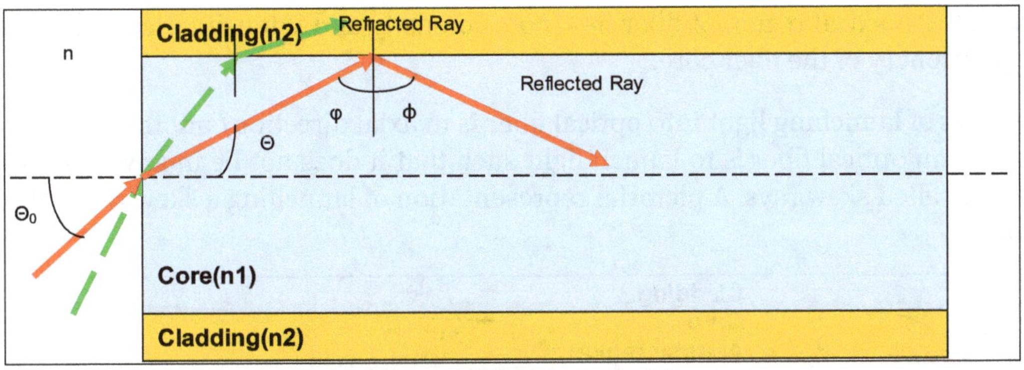

Launching the light into optical fiber.

Figure shows how to launch light into an optical fiber. The light ray lies in a plane containing the axis of the optical fiber. Such planes which contain the fiber axis are called meridional-planes and consequently the rays lying in a meridional-plane are called meridional-rays. Meridional rays always remain in the respective meridional plane. Infinite number of planes can pass through the axis of the fiber and an infinite number of meridional planes. This proves that there are an infinite number of meridional rays too, which incident on the tip of the fiber is making an angle with the fiber-axis as shown in the above figure.

These meridional rays after total internal reflection at the core-cladding boundary meet again at the axis of the optical fiber as shown in the figure. Consider the meridional plane is the plane of the paper which passes through the axis of the fiber and the incident rays, refracted rays and the reflected rays lie on the plane of the paper. For the sake of simplicity and clarity only two rays are shown in the figure. However in practice there would be a number of rays that would be convergent at the same point. One can classify these Meridional rays as bound and unbound rays. The rays that undergo TIR inside the fiber core remain inside the core at all times along the propagation and are called as bound rays. The red ray in the figure above is bound ray. The rays that fail to undergo TIR inside the core are lost into the cladding and are called unbound rays. The dotted green ray shown in figure above is an unbound meridional ray.

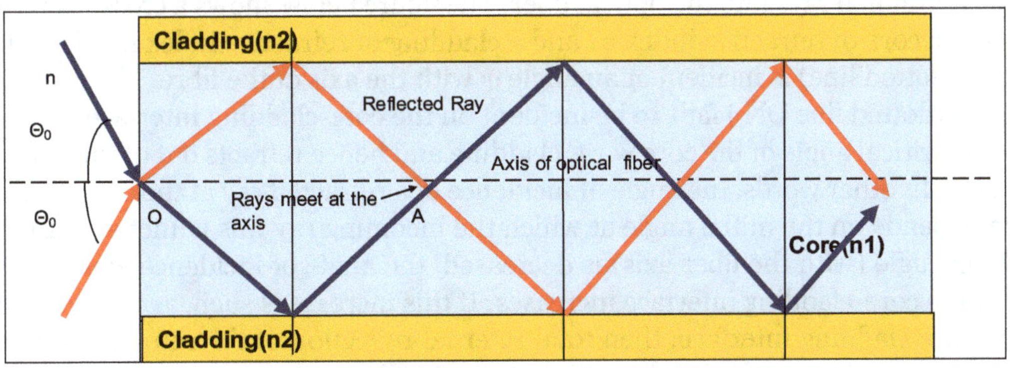

Propagation of meridional rays along the fiber axis.

Since all the reflected rays meet at the same point a region of high optical intensity is generated at that point. Due to multiple TIR at the core-cladding boundary, these rays meet repeatedly at the axis at regular intervals along the fiber. This causes multiple regions of maximum intensity along the axis of the fiber. Also, different sets incident rays would have different angles of reflection at the core-cladding boundary and consequently have different points of maximum intensities along the fiber axis. Thus axisal region of fiber has more intensity and intensity gradually decreases towards the pheriphery of the fiber core.

One of the ways of launching light into optical fiber is in axial direction/ another way of launching a light ray into an optical fiber is to launch light such that it does not lie in any meridional plane. These rays are called skew rays. A pictorial representation of launching a skew ray is shown in the figure below.

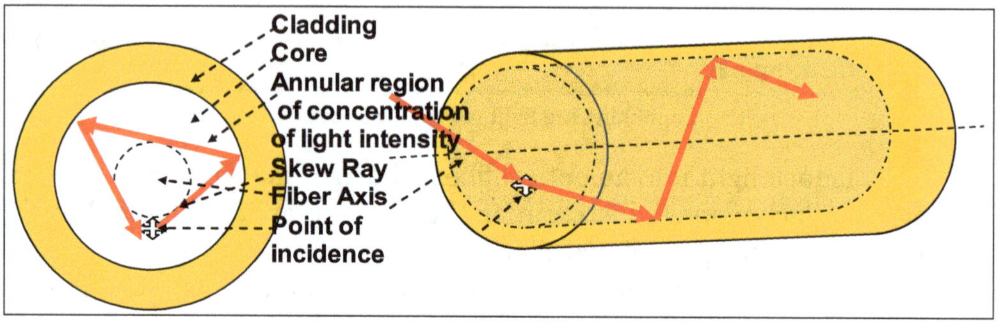

Launching light using skew rays into optical fiber.

Skew rays propagate without passing along the axis of the fiber. In fact the skew rays go on spiralling around the axis of the optical fiber. The light energy carried by them is effectively confined to an annular region around the axis as shown in figure. Therefore intensity at the output of optical fiber due to skew rays will have minimum energy at the axis of the optical fiber and it will gradually increase towards the periphery of the core. Hence one can conclude that when light energy is launched into an optical fiber, there arise two possible energy distributions; one, which has maximum intensity at the axis due to meridional rays and the other, which has minimum intensity at the axis due to the skew rays. Thus, on the whole, there are two ways of launching light into an optical fiber; light can be launched either as meridional or as skew rays.

One can simply analyze the propagation of light through optical fiber assuming that light is launched as meridional rays into the optical fiber. The figure below shows a cross-section of an optical fiber with a core of refractive index n1 and a cladding of refractive index n2. The incident ray AO (shown by dotted line) is incident at an angle φ with the axis of the fibre. The refracted ray for AO in the core (dotted line ON_1) fails to be incident on the core-cladding interface at angle greater or equal to the critical angle of the core w.r.t. cladding and hence refracts out of the core and is lost to the cladding. In other words, the angle of incidence of a refracted ray at the core-cladding interface in turn depends on the initial angle at which the incoming ray was launched into the fiber. If this launching angle (with the fiber axis) is decreased, the angle of incidence which the refracted ray makes at the core-cladding interface increases. If this increase is such, as to exceed the critical angle of the core-cladding interface, then total internal reflection of the refracted ray takes place and the light remains in the core and is guided along the fiber.

The ray CO is launched into the fiber at such an angle 'α' that its refracted ray is incident at the core-cladding boundary at its critical angle 'θc'. If any light ray is launched at an angle more than α then the refracted ray just refracts out to the cladding because the angle of incidence of its refracted ray at the core-cladding interface is less than the critical angle. Thus the angle α is indicative of the maximum possible angle of launching of a light ray that is accepted by the fiber.

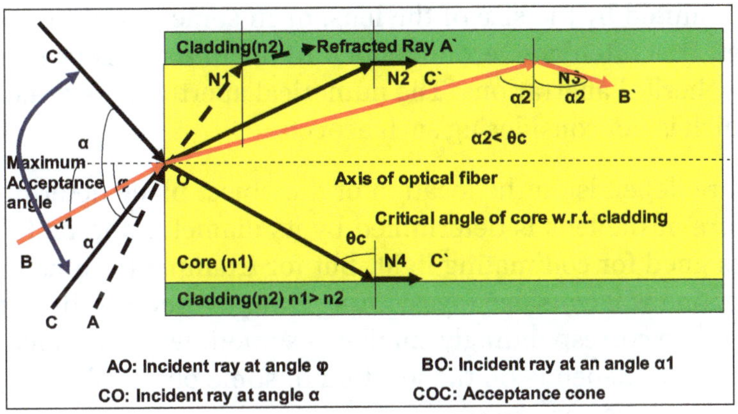

Launching light into optical fiber.

Consequently, the angle α is called the angle of acceptance of the fiber core. Since the optical fiber is symmetrical about its axis, it is very clear that all the launched rays, which make an angle α with the axis and form a sort of a cone. This cone is called the acceptance cone of the fiber as shown in the figure. Any launched ray that lies within this cone is accepted by the fiber and the light of this ray is guided along the fiber by virtue of multiple TIRs as shown by the red ray BO in the figure.

Numerical Aperture

The numerical aperture (NA) of an optical system (e.g. an imaging system) is a measure for its angular acceptance for incoming light. It is defined based on geometrical considerations and is thus a theoretical parameter which is calculated from the optical design. It cannot be directly measured, except in limiting cases with rather large apertures and negligible diffraction effects.

Numerical Aperture of an Optical System

The numerical aperture of an optical system is defined as the product of the refractive index of the beam from which the light input is received and the sine of the maximum ray angle against the axis, for which light can be transmitted through the system based on purely geometric considerations (ray optics):

$$NA = n \sin \theta_{max}$$

For the maximum incidence angle, it is demanded that the light can get through the whole system and not only through an entrance aperture.

NA of a Lens

A simple case is that of a collimating lens:

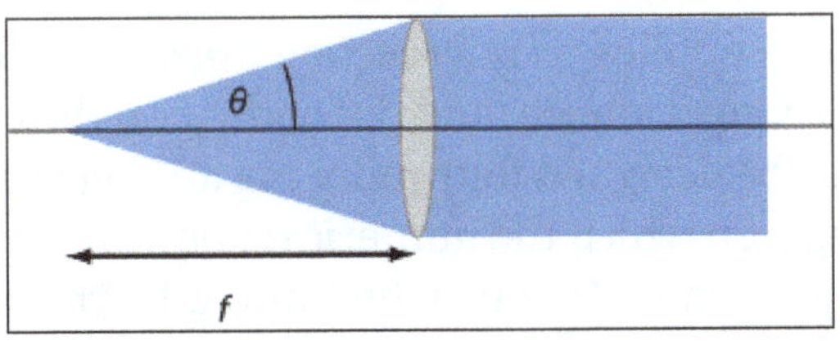

A collimating lens can theoretically accept light from a cone, the opening angle of which is limited by its size.

The extreme rays are limited by the size of the lens, or in some cases somewhat less if there is a non-transparent facet. It is often not recommended to operate a lens or its full area, since there could be substantial spherical aberrations. The numerical aperture, however, is a completely geometrical measure, which is not considering such aspects.

The numerical aperture depends on the location of the object plane determined by the designer. The numerical aperture of the lens is determined by its diameter and its focal length. However, a lens may not be designed for collimating light, but for example for imaging objects in a larger distance. In that case, one will consider rays coming from that object distance, and the obtained numerical aperture will be correspondingly smaller – sometimes even much smaller. This shows that the numerical aperture depends on the location of some object plane determined by the designer according to the intended use.

Some lenses are used for focusing collimated laser beams to small spots. The numerical aperture of such a lens depends on its aperture and focal length, just as for the collimation lens. The beam radius w_{lens} at the lens must be small enough to avoid truncation or excessive spherical aberrations. Typically, it will be of the order of half the aperture radius of the lens (or perhaps slight larger), and in that case ($w_{lens} = D/4 = NA \cdot f/2$, with the beam divergence angle being only half the NA) the achievable beam radius in the focus is,

$$W_f = \frac{\lambda f}{\pi W_o} = \frac{4\lambda f}{\pi D} \approx \frac{2\lambda}{\pi NA}$$

where D is the aperture diameter, f the focal length and λ the optical wavelength. The calculation is based on the paraxial approximation and therefore not accurate for cases with very high NA.

A somewhat smaller spot size may be possible with correspondingly larger input beam radius, if the performance is not spoiled by aberrations. High-NA lenses (e.g. with NA above 0.6 or even 0.8) are required for various applications:

- In players and recorders of optical data storage media such as CDs, DVDs and Blu-ray Discs it is important to focus laser light to a small spot (pit) and receive light from such a spot.
- Lenses with high NA are also required for collimating laser beams which originate from small apertures. For example, this is the case for low-power single-mode laser diodes. When a lens with too low NA is used, the resulting collimated beam can be distorted (aberrated) or even truncated.

NA of a Microscope Objective

The same kind of considerations applies to microscope objectives. Such an objective is designed for operation with a certain working distance, and depending on the type of microscope with which it is supposed to be used; it may be designed for producing an image at a finite distance or at infinity. In any case, the opening angle on which the numerical aperture definition is based is taken from the center of the intended object plane. It is usually limited by the optical aperture on the object side, i.e., at the light entrance.

In many cases, the light input comes from air, where the refractive index is close to 1. The numerical aperture is then necessarily smaller than 1, but for some microscope objectives it is at least not much lower, for example 0.9. Other microscope objectives for particularly high image resolution are designed for the use of some immersion oil between the object and the entrance aperture. Due to its higher refractive index (often somewhat above 1.5), the numerical aperture can then be significantly larger than 1 (for example, 1.3).

The NA of a microscope objective is important particularly concerning the following aspects:

- It determines how bright the observed image can be for a given illumination intensity. Obviously, a high-NA objective can collect more light than one with a low numerical aperture.
- More importantly, the NA sets a limit to the obtained spatial resolution: the finest resolvable details have a diameter of approximately $\lambda/(2\,\mathrm{NA})$, assuming that the objective does not produce excessive image aberrations.
- A high NA leads to a small depth of field: only objects within a small range of distances from the objective can be seen with a sharp image.

Photographic Objectives

In photography, it is not common to specify the numerical aperture of an objective, because such objectives are not thought to be used with a fixed working distance. Instead, one often specifies the aperture size with the so-called f-number, which is the focal length divided by the diameter of the entrance pupil. Usually, such an objective allows the adjustment of the f-number in a certain range.

Numerical Aperture of an Optical Fiber or Waveguide

Although an optical fiber or other kind of waveguide can be seen as a special kind of optical system, there are some special aspects of the term numerical aperture in such cases. In case of a step-index fiber, one can define the numerical aperture based on the input ray with the maximum angle for which total internal reflection is possible at the core–cladding interface:

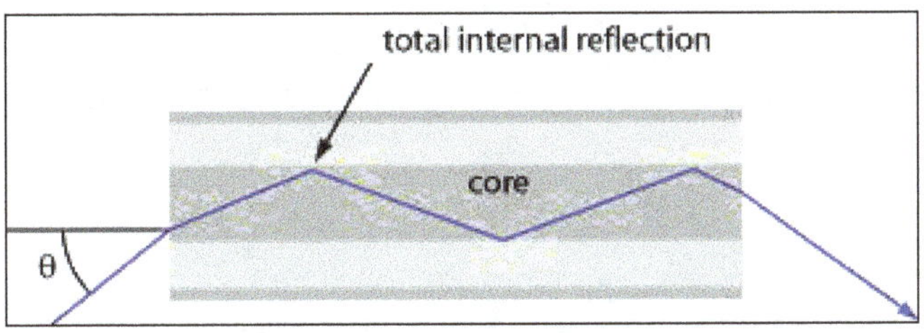

An incident light ray is first refracted and then undergoes total internal reflection at the core–cladding interface. However, that works only if the incidence angle is not too large. The numerical aperture (NA) of the fiber is the sine of that maximum angle of an incident ray with respect to the fiber axis. It can be calculated from the refractive index difference between core and cladding, more precisely with the following relation:

$$\mathrm{NA} = \sqrt{n_{core}^2 - n_{cladding}^2}$$

The NA is independent of the refractive index of the medium around the fiber. For an input medium with higher refractive index, for example, the maximum input angle will be smaller, but the numerical aperture remains unchanged.

The equation given above holds only for straight fibers. For bent fibers, one may use an approximate modified equation, which also contains the bend radius R and the core radius:

$$NA = \sqrt{n_{core}^2 - n_{cladding}^2\left(1 + \frac{r_{co}}{2R}\right)}$$

For fibers or other waveguides not having a step-index profile, the concept of the numerical aperture becomes questionable. The maximum input ray angle then generally depends on the position of the input surface. Some researchers calculate the numerical aperture of a graded-index fiber based on the maximum refractive index difference between core and cladding, using the equation derived for step-index fibers. However, some common formula in fiber optics involving the NA can then not be applied.

Light propagation in most optical fibers, and particularly in single-mode fibers, cannot be properly described based on a purely geometrical picture (with geometrical optics), because the wave nature of light is very important; diffraction becomes strong for tightly confined light. Therefore, there is no close relation between properties of fiber modes and the numerical aperture. Only, high-NA fibers tend to have modes with larger divergence of the light exiting the fiber. However, that beam divergence also depends on the core diameter. As an example, Figure shows how the mode radius and mode divergence of a fiber depend on the core radius for fixed value of the numerical aperture. The mode divergence stays well below the numerical aperture.

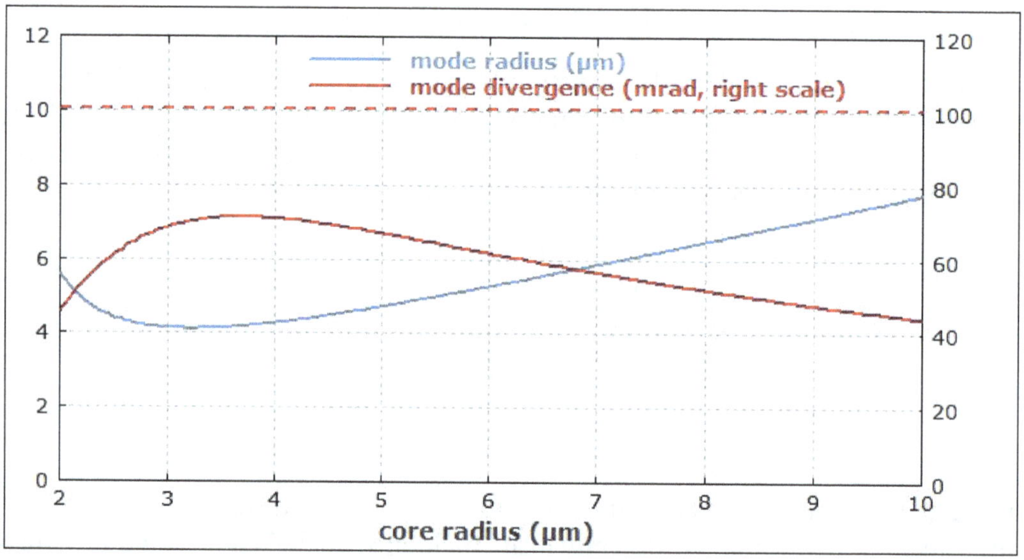

Mode radius and divergence angle for the fundamental mode of a step-index fiber as functions of the core radius for a fixed numerical aperture of 0.1 and the wavelength of 1000 nm. In Figure below one can see that the angular intensity distribution somewhat extends beyond the value corresponding to the numerical aperture. This demonstrates that the angular limit from the purely geometrical consideration is not a strict limit for waves.

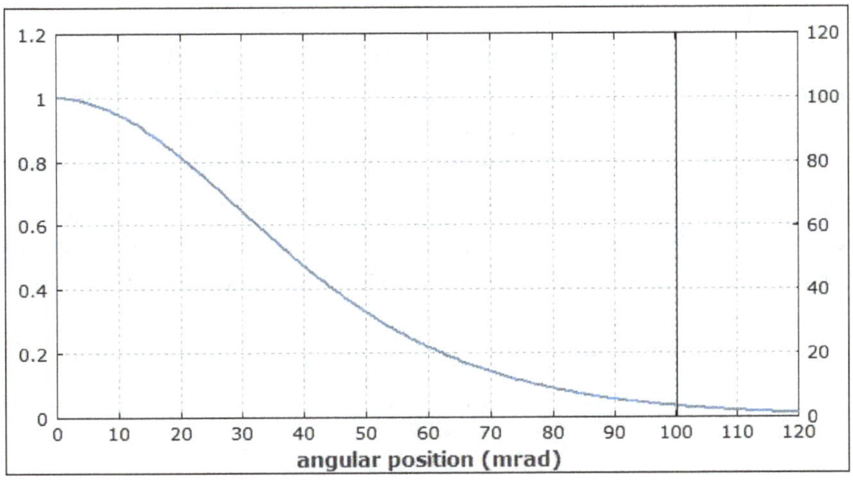

Far field intensity distribution of a fiber mode at 1000 nm for a core radius of 3.5 μm, a numerical aperture of all 0.1. The intensity distribution somewhat extends beyond the value corresponding to the numerical aperture.

For a single-mode fiber, the NA is typically of the order of 0.1, but can vary roughly between 0.05 and 0.4. Because wave optics effects are particularly important for single-mode fibers, but ignored in considerations based on the NA, those values are not particularly relevant for calculations. Multimode fibers typically have a higher numerical aperture of e.g. 0.3. Very high values are possible for photonic crystal fibers. A higher NA has the following consequences:

- For a given mode area, a fiber with higher NA is more strongly guiding, i.e. it will generally support a larger number of modes.
- Single-mode guidance requires a smaller core diameter. The corresponding mode area is smaller, and the divergence of the beam exiting the fiber is larger. The fiber nonlinearity is correspondingly increased. Conversely, a large mode area single-mode fiber must have a low NA.
- Bend losses are reduced; the fiber can be more strongly bent before bend losses become significant.
- The sensitivity of the guidance to random refractive index fluctuations is reduced. (For large mode area low-NA single-mode fibers, it can be a problem).
- The higher doping concentration of the fiber core (e.g. with germanium), as required for increasing the refractive index difference, may increase scattering losses. The same can be caused by irregularities of the core/cladding interface, which are more important for a larger index difference.
- A low NA increases the influence of random refractive index variations. Therefore, fibers with very low NA may exhibit increased propagation losses.

Pulse Broadening in Optical Fiber

When optical fiber is used for transmission of information, light signal launched into it cannot be of continuous nature. For a carrier signal to carry information, one or more of its characteristics has to be altered in accordance to the data signal. In an optical fiber light is launched in the form of optical pulses to transmit the required information.

Light energy launched into the fiber may be considered to travel in the form of numerous rays in accordance to the Ray-Model. These rays travel different paths inside the core of an optical fiber because different light rays are incident on the tip of the optical fiber at different angles within the acceptance cone itself. This causes different light rays in the acceptance cone to travel along different paths in the core of the optical fiber and accordingly take different time intervals to travel a given distance too, which leads to a phenomenon of pulse broadening inside the core of the optical fiber. Thus the pulse of light which might originally be of width T seconds now might be of T+ΔT seconds inside the fiber core. The figure below depicts a pictorial description of how light pulse broadens inside the core of the fiber.

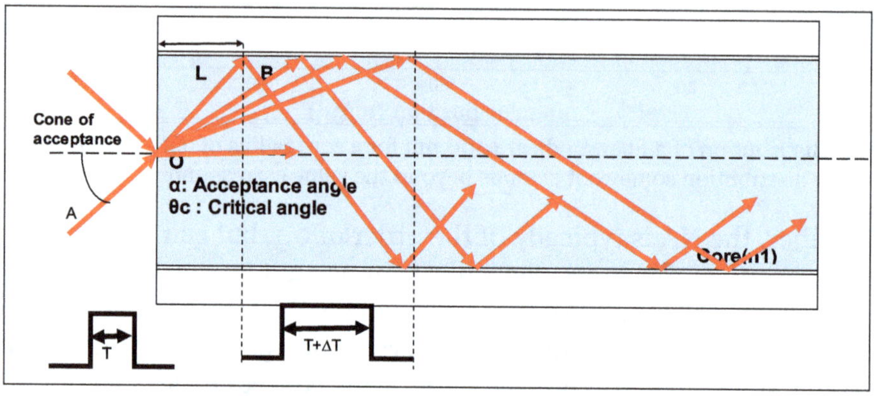

Pulse broadening mechanism in optical fiber.

Any incident ray that lies within the acceptance cone gets guided inside the optical fiber by virtue of multiple total internal reflections. Since the angle of refraction different incident rays are different, they travel along different paths in the optical fiber as shown in the above figure. This causes the initially launched narrow light pulse to broaden as shown. The amount of broadening is measured in terms of the increase in the pulse time width and is denoted by ΔT the value of ΔT is given by:

$$\Delta T = \frac{L}{C}\frac{n_1(n_1 - n_2)}{n_2}$$

Where, ΔT= Pulse Broadening; c = velocity of light in free space; n_1 = refractive index of core and n_2 = refractive index of the cladding.

The quantity L is the horizontal distance travelled before suffering the first total internal reflection by the refracted ray OB which corresponds to the incident ray AO, incident at the acceptance angle as shown in the figure. The amount of pulse broadening is effectively the difference in time of travel between the ray travelling along the axis and the incident ray AO. This pulse broadening effect signifies that if a second pulse is now launched into the fiber within the time interval T+ΔT, the two pulses will overlap and no identifiable data would be obtained on the output. Thus for a given length L, there would be a corresponding value of ΔT which would limit the rate at which light pulses can be launched into the optical fiber. In other words, it limits the rate at which data can be transmitted along the fiber. This indirectly limits the bandwidth available on the fiber. Thus we can say that more the pulse broadening lower the bandwidth. That is:

$$\text{Bandwidth} = \text{BW} = \frac{1}{\Delta T}$$

Equation above suggests that for higher bandwidth of transmission the pulse broadening, ΔT should be as low as possible. In equation 10, we see that the value of ΔT is dependent on the value of L, the difference (n1 – n2) as well as the value of n1/n2. But reducing the value of L would signify the reduction in the length of the optical fiber, which is not desirable. As 1<n2, the ratio, n1 / n2 is very close to 1. Thus for low ΔT values, the only option available with us is to decrease the value (n1–n2) or in other words, to increase the refractive index of the cladding n2.

One can now notice that a contradictory situation has been generated as to whether the cladding should be removed for high NA or to use a cladding of large refractive index value for higher bandwidth? The answer to this query is purely application specific. That means if an optical fiber is used as a sensor (say), where lowest possible light has to be accepted, we use fiber with low n2 values. When the optical fiber is used for data communication, fibers with high values of n2 are used. For practical communication purposes the value of (n1 – n2) is made of the order of about 10^{-3} to 10^{-4}. If the cladding is removed, the value of n2 becomes 1 and the value of the above difference becomes about 0.5. The bandwidth corresponding to this value of n1-n2 is of the order of few Kilohertz, which is far worse than that of a normal twisted pair of wires. Thus cladding is an extremely important requirement for optical fiber when the bandwidth is the prime concern of the application and its refractive index is made as close to that of the core as the available technology permits, but not made equal. This is brought about by varying the amount of doping in a single glass rod. The differently doped regions have different refractive indices and serve as core and cladding of the optical fiber.

Study of Total Internal Reflection: Phase Front Aspect

Let us now have a study of the phenomenon of total internal reflection at the core-cladding interface on a backdrop of the wave-fronts of the incident and the reflected light. Wave-fronts are nothing but the constant phase planes of the light wave and are also called as phase-fronts. They are perpendicular to the direction of propagation of the wave at every point. Any light ray launched meridional within the acceptance cone will propagate along the fiber core by virtue of multiple total internal reflections at the core-cladding interface. Thus in accordance to the ray-model of light we may visualise a solid cone of light (having angle = double the acceptance angle) that enters an optical fiber and propagates through the fiber by TIRs. Figure below shows the phenomenon of total internal reflection of a ray of light at the core-cladding boundary along with the wave-fronts of the incident and the reflected rays. The red and green coloured dotted lines represent the wave fronts of the light rays which are perpendicular to their direction of propagation. The light rays, actually, are fictitious lines which, in reality, represent the direction of propagation of these wave-fronts.

The distance between a red and a green wave-front corresponds to a phase difference of 180° (π radians). The similar coloured wave-fronts have either 0° or 360° phase difference between them. Thus, when two similar coloured wave-fronts meet, they interfere constructively and dissimilar coloured wave-fronts interfere destructively. This is evident from the interference pattern that sets up in the core as shown in the above figure. In the core, the interference between the incident and the reflected wave-fronts constitutes a standing wave pattern of varying light intensity with discrete maxima and minima in a direction normal to the core-cladding interface.

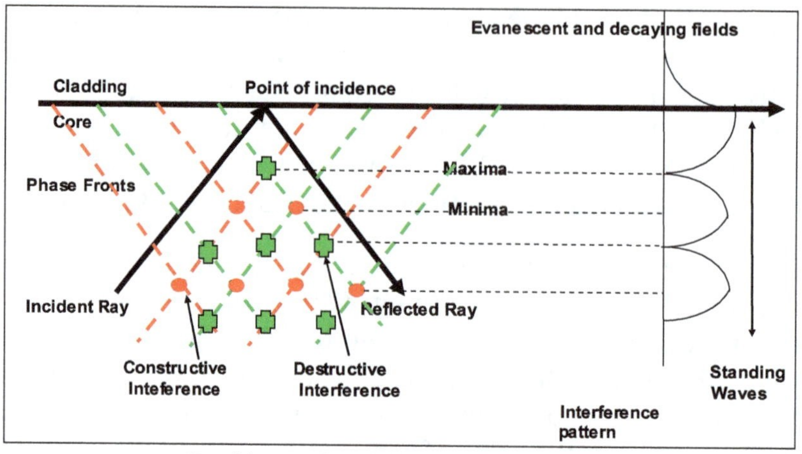

Total internal Reflection in optical fiber.

Total internal reflection is also accompanied by an abrupt phase change between the incident and the reflected rays at the core-cladding boundary. This phase change depends on the angle of incidence of the incident ray at the core-cladding boundary, the refractive index or the core and cladding and various other parameters.

If we refer to the electromagnetic wave theory of light, it shows that at total internal reflection, the light intensity inside the cladding is not completely zero. Instead, there exist some decaying fields in the cladding, which do not carry any power but support the total internal reflection phenomenon by satisfying the boundary conditions at the core-cladding interface. These fields are called as evanescent fields. The Ray-model of light does not offer any explanation about the evanescent fields, which indeed are as equally important as the fields in the core for total internal reflection to occur.

The importance of these evanescent fields in the TIR can be clearly ascertained from the fact that even the slightest disturbance to these fields in the cladding could lead to the failure of the TIR at the corecladding boundary accompanied by leakage of optical power to the cladding. This is one of the instances when the ray-model of light becomes inadequate in explaining the phenomena exhibited by light. Though the evanescent fields are decaying fields, they never become zero, at least theoretically. In other words, they remain present up to infinite distance from the core-cladding boundary. But in practice, these fields decay down to a negligibly small value as we move away from the core-cladding boundary deeper into the cladding. Larger the value of the angle of incidence of the incident ray at the core-cladding boundary, sharper is the decay of the evanescent fields. Thus there must me a sufficient thickness of cladding provided for these evanescent fields to be accommodated so that they decay to a negligibly small value in the cladding and cannot be disturbed by external sources.

With this backdrop of knowledge about the propagation of light in an optical fiber, let us now look into certain finer aspects of such propagation and investigate the condition required for successful propagation of light in the fiber. Figure below shows two parallel rays that are launched into an optical fiber and they propagate as shown. The dotted lines represent the wave-fronts of the rays. The refractive indices of core and cladding are n1 and n2 respectively. The diameter of the core is 'd'. The phase-front AE is common to both Ray 1 and Ray 2. The phase-front DB is common to Ray 2 and BF. The Ray 2 is thus common to both the phase-fronts. Hence for a sustained constructive interference, the distance between these two phase-fronts must be multiples of 2π. In other words, it can be said that the phase difference between the phase change undergone by Ray 1 in travelling distance s1 and the Ray 2 in travelling s2 must be 0 or integral multiples of 2π.

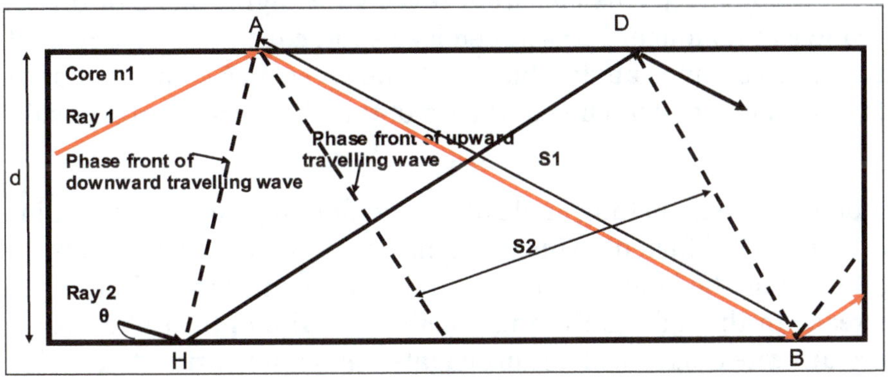

Phase front representation of propagation of light through optical fiber.

Mathematically,

$$s1 = \frac{d}{\sin\theta}$$

$$s2 = AD\cos\theta = \frac{\left(\cos^2\theta - \sin^2\theta\right)d}{\sin\theta}$$

If δ is the phase change undergone in each TIR of Ray 1, then the total phase change undergone by Ray 1 in travelling S1 is given by,

$$\phi 1 = \frac{2\pi nl}{\lambda} s1 + 2\delta$$

Where n1 = refractive index of core; λ=Wavelength of the light in the core. The phase change undergone by Ray 2 in travelling s2 is given by,

$$\phi 2 = \frac{2\pi nl}{\lambda} s2$$

For a sustained constructive interference, both φ1 and φ2 must have a phase difference of either 0 or integral multiples of 2π. That is, for an integer m (=0,1,2,3,...) the following condition must be satisfied:

$$\phi 1 - \phi 2 = 2m\pi$$

$$\frac{2\pi nl}{\lambda}\left(s1 - s2\right) + 2\delta = 2m\pi$$

$$\frac{2\pi nl \sin\theta}{\lambda} + \delta = \pi m$$

The significance of the above equation is that only those rays, which are incident on the tip of the fiber at angles such that their angle of refraction in the core satisfies above equation, can successfully travel along the fiber. For this equation, we find that since 'm' can take only discrete integral values, the value of angle θ is also discrete. This suggests that there are only some discrete launching angles within the acceptance cone (N.A. cone) for which the rays can propagate inside the fiber core. A 3D visualisation reveals the significance of this observation, i.e. the acceptance cone can no longer be assumed as a solid cone of rays, launched at all possible angles (smaller than acceptance angle), but has to be viewed as composed of discrete annular conical rings of rays which are launched at the tip of the fiber core at angles which satisfy above equation.

Thus the condition that the launching angle of the incident ray should be within the acceptance cone is necessary but not sufficient. This angle has to be such that the above equation is satisfied. Thus light can only be launched at certain discrete angles within the N.A. cone leading to a further decrease in the light gathering efficiency of the optical fiber. Any ray that is not launched at these discrete angles will not propagate inside the optical fiber. This discretization in the values of launching angles lead to formation of what are called as modes in an optical fiber, which are nothing but different patterns of light intensity distribution around the axis of the core. In the subsequent sections, we shall see the modal propagation of light in an optical fiber, in detail, and perform some rigorous mathematical analysis of the light propagation in an optical fiber.

Characteristics of Optical Fibers

The Structure and Physics of an Optical Fiber

The optical fibers used in communications have a very simple structure. They consist of two sections: the glass core and the cladding layer. The core is a cylindrical structure, and the cladding is a cylinder without a core. Core and cladding have different refractive indices, with the core having a refractive index, n_1, which is slightly higher than that of the cladding, n_2.

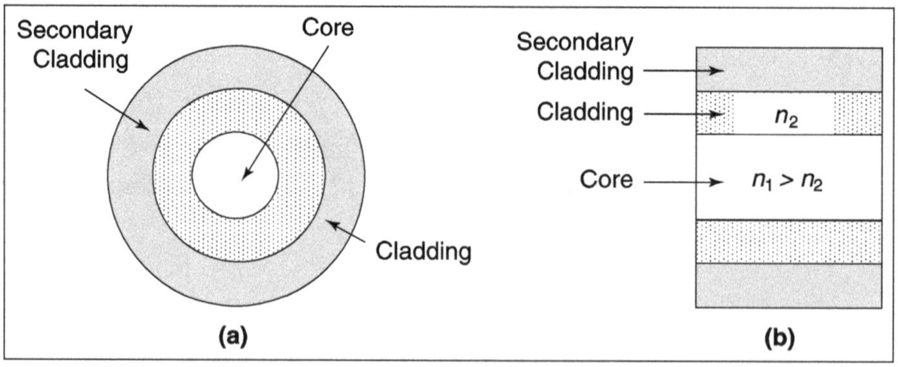

(a) Cross section and (b) longitudinal cross section of a typical optical fiber.

It is this difference in refractive indices that enables the fiber to guide the light. Because of this guiding property, the fiber is also referred to as an "optical waveguide." As a minimum there is also a further layer known as the secondary cladding that does not participate in the propagation but gives the fiber a minimum level of protection. This second layer is referred to as a coating.

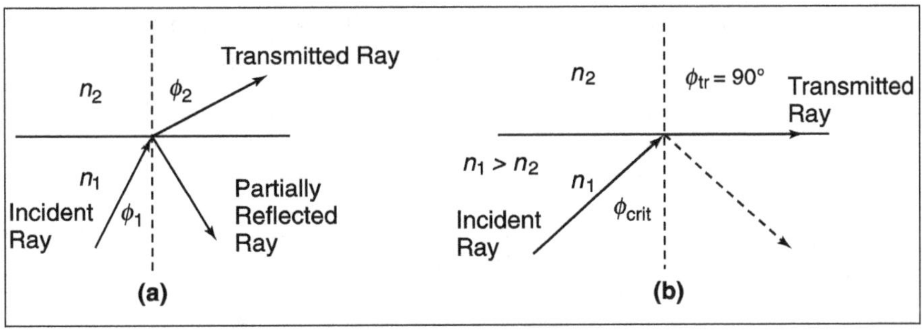

Snell's law

The basics of light propagation can be discussed with the use of geometric optics. The basic law of light guidance is Snell's law. Consider two dielectric media with different refractive indices and with $n_1 > n_2$ and that are in perfect contact. At the interface between the two dielectrics, the incident and refracted rays satisfy Snell's law of refraction—that is,

$$n_1 \sin\phi_1 = n_2 \sin\phi_2$$

or

$$\frac{\sin\phi_1}{\sin\phi_2} = \frac{n_2}{n_1}$$

In addition to the refracted ray there is a small amount of reflected light in the medium with refractive index n_1. Because $n_1 > n_2$ then always $\varphi_2 > \varphi_1$. As the angle of the incident ray increases there is an angle at which the refracted ray emerges parallel to the interface between the two dielectrics. This angle is referred to as the critical angle, φ_{crit}, and from Snell's law is given by,

$$\sin\phi_{crit} = \frac{n_2}{n_1}$$

If the angle of the incident ray is greater than the critical angle, the ray is reflected back into the medium with refractive index n_1. This basic idea can be used to propagate a light ray in a structure with $n_1 > n_2$.

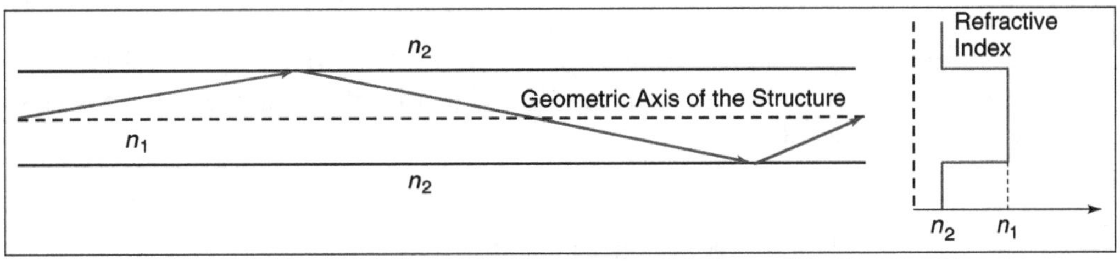

Light guidance using Snell's law.

The light ray incident at an angle greater than the critical angle can propagate down the waveguide through a series of total reflections at the interface between the two dielectrics. The ray is referred to as a "meridional ray" because it passes through the axis of the fiber. For rays not satisfying this condition. Clearly, the picture of total internal reflection assumes an ideal situation without imperfections, discontinuities, or losses. In a real-world fiber, imperfections will introduce light that is refracted as well as reflected at the boundary. In arriving at the basic idea of light propagation, it was assumed that somehow the ray has been launched into the fiber.

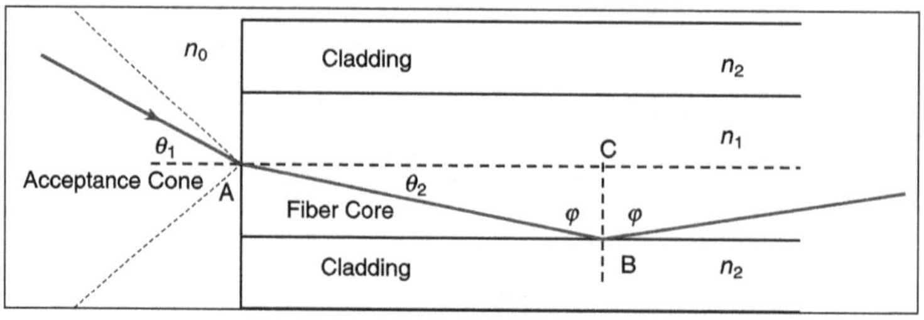

Geometry for the derivation of the acceptance angle.

For a ray to be launched into the fiber and propagated it must arrive at the interface between the two media (with different refractive indices) at an angle that is at minimum equal to θ_{crit} and in general less than that. Figure above illustrates the geometry for the derivation of the acceptance angle. To satisfy the condition for total internal reflection, the ray arriving at the interface, between the fiber and outside medium, say air, must have an angle of incidence less than θ_{acc}, otherwise the internal angle will not satisfy the condition for total reflection, and the energy of the ray will be lost in the cladding.

Consider that a ray with an incident angle less than the θ_{acc}, say θ_1, enters the fiber at the interface of the core (n_1) and the outside medium, say air (n_o), and the ray lies in the meridional plane. From Snell's law at the interface we obtain,

$$n_o \sin\theta_1 = n_1 \sin\theta_2$$

From the right triangle ABC, the angle φ is given by,

$$\phi = \frac{\pi}{2} - \theta_2$$

where the angle φ is greater than the critical angle. Substituting above equation into equation $n_o \sin\theta_1 = n_1 \sin\theta_2$ we obtain,

$$n_o \sin\theta_1 = n_1 \cos\phi$$

In the limit as the incident angle, θ_1, approaches θ_{acc}, the internal angle approaches the critical angle for total reflection, φ_{crit}. Then, by introducing the trigonometric relation $\sin^2\varphi + \cos^2\varphi = 1$ into equation $n_o \sin\theta_1 = n_1 \sin\theta_2$ we obtain,

$$n_o \sin\theta_1 = n_1 \cos\phi = n_1\left(1 - \sin^2\phi\right)^{1/2} = n_1\left(1 - \left(\frac{n_2}{n_1}\right)^2\right)^{1/2} = \left(n_1^2 - n_2^2\right)^{1/2}$$

This equation defines the angle within which the fiber can accept and propagate light and is referred to as the "Numerical Aperture" (NA).

$$NA = n_o \sin\theta_{acc} = \left(n_1^2 - n_2^2\right)^{1/2}$$

When the medium with refractive index n_o is air, the equation for the NA of the glass fiber simplifies to,

$$NA = \sin\theta_{acc} = \left(n_1^2 - n_2^2\right)^{1/2}$$

This equation states that for all angles of incident where the inequality $0 \le \theta_1 \le \theta_{acc}$ is satisfied the incident ray will propagate within the fiber. The parameter NA expresses the propensity of the fiber to accept and propagate light within the solid cone defined by an angle, $2\theta_{acc}$. The equation for the NA can be also expressed in terms of the difference between the refractive indices of core and cladding—that is,

$$\Delta = \frac{n_1^2 - n_2^2}{2n_1^2} \approx \frac{n_1 - n_2}{n_1}$$

With these simplifications the NA can now be written as,

$$NA = n_1 (2\Delta)^{1/2}$$

To appreciate the numbers involved, consider a fiber made of silica glass whose core refractive index is 1.5 and that of the cladding is 1.46. The φ_{crit} and the NA of the fiber are calculated to be,

$$\phi_{crit} = \sin^{-1}\left(\frac{n_2}{n_1}\right) = \sin^{-1}\left(\frac{1.46}{1.50}\right) = 74.73^\circ$$

$$NA = \left(n_1^2 - n_2^2\right)^{1/2} = \left(1.50^2 - 1.46^2\right)^{1/2} = (2.25 - 2.13)^{1/2} = 0.346$$

$$\theta_{acc} = \sin^{-1} NA = \sin^{-1} 0.346 = 20.24^\circ$$

If one uses the equation for the approximation of NA, equation $NA = n_1 (2\Delta)^{1/2}$ the result is,

$$NA \approx n_1 (2\Delta)^{1/2} = 1.5 \times \left(2 \times \left(\frac{1.50 - 1.46}{1.5}\right)\right)^{1/2} = 1.5 \times (2 \times 0.0266)^{1/2} = 0.346$$

If the ray does not lie in the meridional plane, the analysis for obtaining the NA is slightly more complex. The discussion up to now has been based on the launching and propagation of a single ray. Because all the rays with a reflection angle less that the critical angle can propagate, it is expected that a number of rays will propagate provided they can be launched into the fiber. The class of fiber that can support the simultaneous propagation of a number of rays is known as multimode fiber. The term "mode" is used in a more sophisticated analysis of light propagation using Maxwell's equations, and it corresponds to one solution of the equations.

A fiber that allows the propagation of one mode only is called a single-mode fiber. A fiber can be multimode or single mode, and the behavior depends on the relative dimensions of the core and the wavelength of the propagating light. Fibers with a core diameter much larger than the wavelength of the launched light will support many propagating modes, and the propagating conditions can be analyzed with geometric optics. A fiber with a core diameter similar to that of the light wavelength supports only one propagation mode.

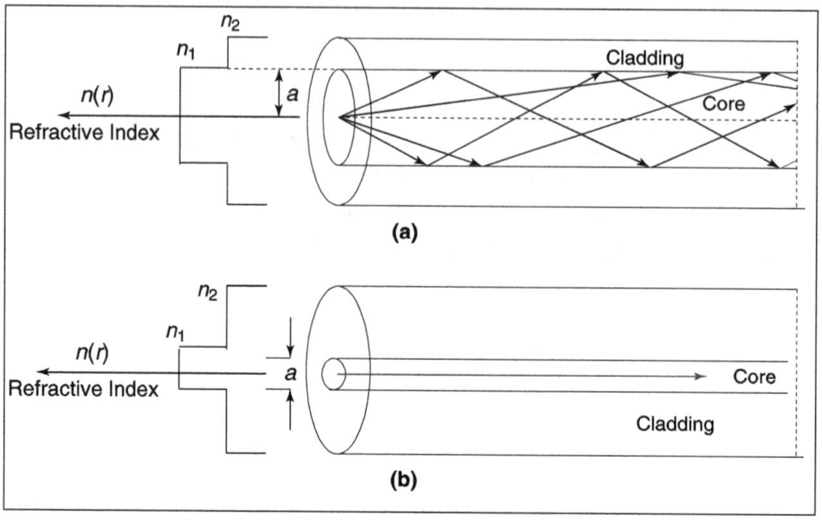

(a) Multimode and (b) single-mode propagation for step index fiber.

Material Characteristics of Fibers—Losses

The basic material used in the manufacture of optical fiber for optical transmission is silica glass. A large number of glasses have been developed and studied with the objective of improving fiber transmission properties. There are two parameters of glass that have a substantial impact on its performance: the losses and the changes of refractive index with wavelength. The basic material used in the manufacture of optical fibers is vitreous silica dioxide (SiO_2), but to achieve the properties required from a fiber, various dopants are also used, (Al_2O_3, B_2O_3, GeO_2, P_2O_5). Their task is to slightly increase and decrease the refractive index of pure silica (SiO_2). Initially the fiber losses were high, but through improvements in the quality of the materials and the actual production process, the losses have been reduced so as to be close to the theoretical expected losses.

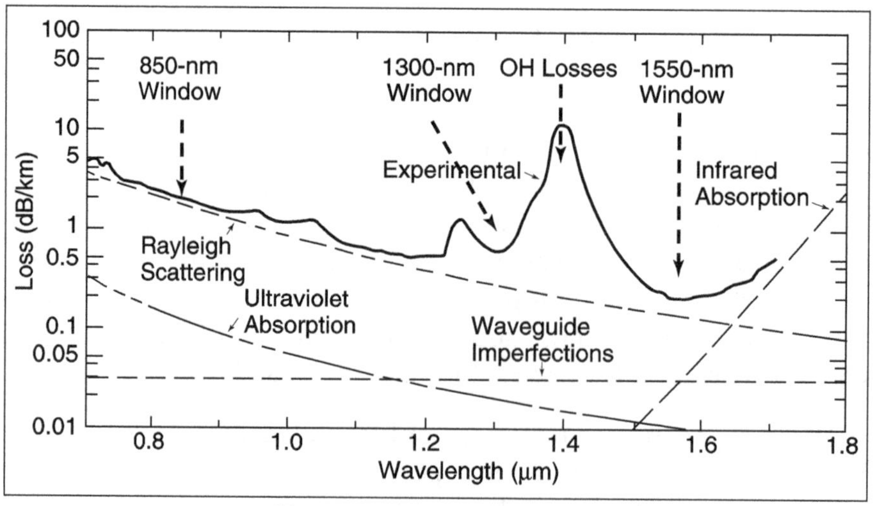

Measured loss versus wavelength for SiO_2 fibers.

In the part of the electromagnetic spectrum where optical fiber transmission takes place, the losses are bracketed between two asymptotes: ultraviolet absorption and infrared absorption. The measured loss–versus–wavelength curve is shown in figure below. The actual loss in the window between 0.8 and 1.6 μm is dictated by the Rayleigh scattering losses. The physical origin of the Rayleigh scattering losses is the excitation and re-radiation of the incident light by atomic dipoles whose dimensions are much less than the wavelength of the light. The loss can be expressed in terms of decibels per kilometer by the expression,

$$\alpha_{scat} = \frac{A_R}{\lambda^4}$$

Where the constant A_R reflects the details of the doping process and the fiber fabrication, and the value of A_R is around 0.95 dB/km – μm⁴. In fiber with a pure silica core, the value of A_R can be as low as 0.75 dB/km – μm4. The actual scattering within a fiber is higher than in that of bulk material because of additional scattering from interfaces and in homogeneities in the fiber structure. The losses discussed up to now are usually referred to as intrinsic because their origin is in the physical properties of the material. The total intrinsic loss can be expressed as the sum of the three—that is,

$$\alpha_{intrinsic} = \alpha_{ultraviolet} + \alpha_{infrared} + \alpha_{scat}$$

The theoretical intrinsic loss has a minimum value of 0.185 dB/km at a wavelength close to 1550 nm. Figure below explains the intrinsic theoretical losses of a silica glass fiber. In addition to the intrinsic losses, the fiber has additional losses that are referred to as extrinsic. These are associated with the presence of various substances in the glass, the quality of the glass, the processing, and the mechanical imperfections associated with the fiber structure.

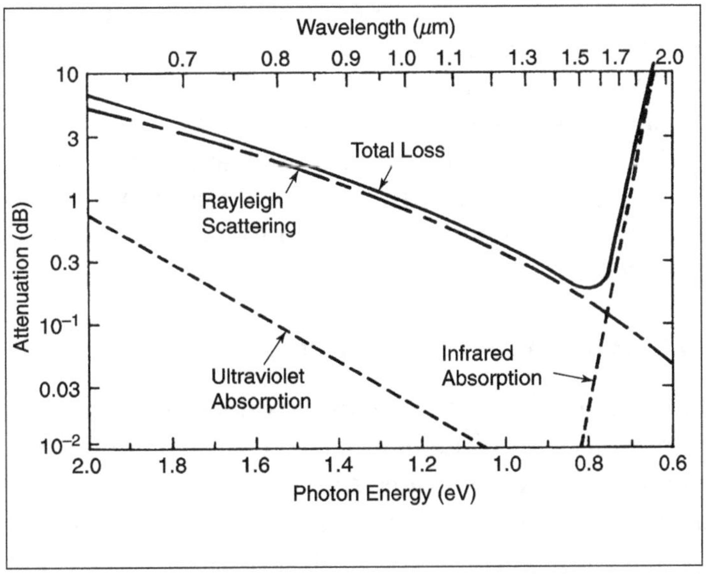

Theoretical intrinsic loss characteristics of germania doped silica (GeO_2–SiO_2) glass.

These losses can be removed with refinements in the fabrication process and in the quality of glass. The presence of metallic and rare-earth impurities contributes to the extrinsic fiber loss as does the presence of the hydroxyl group OH that enters the glass through water vapors. These contributors to the extrinsic loss are the most difficult to remove. Another contributor to the extrinsic losses is the micro-bending and macro-bending of the fiber that arises from periodic microbends as a result of the spooling or cabling of the fiber and the bending of the fiber for cabling and deployment.

When the intrinsic and extrinsic losses are combined, the total fiber loss is as shown in figure. The experimental results shown there are from the early 1980s, and the fiber of today has a loss that is very close to the intrinsic loss of the material. This was achieved by removing the OH extrinsic loss and improving the quality of the glass. The total loss of a modern fiber is shown in figure.

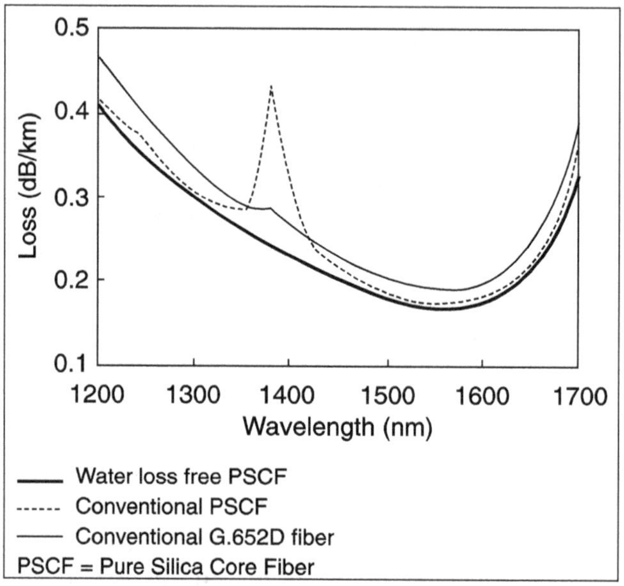

Spectral loss of water loss free PSCF.

The evolution of fiber in terms of loss generates three optical windows. The first window was around 850 nm with multimode fiber. The reason for operating at 850 nm was the availability of semiconductor lasers at this wavelength. When the silica zero-dispersion wavelength at 1300 nm was identified, the first single-mode systems operated at 1300 nm, the second window. When it was possible to reduce the losses at the minimum-loss wavelength, 1550 nm, then single-mode systems started operating at 1550 nm, the third window. Long-haul systems operate in the region of 1550 nm either with a single wavelength or with multiple wavelengths. The 1300-nm window is still used for short-haul systems.

Material Characteristics of Fibers—Dispersion

The variation of refractive index of a material with wavelength is known in optics as dispersion, and it is responsible for resolution of white light into its constituent colors. In the context of fiber light propagation, the dispersion can be divided into two parts. The first part is the dispersion induced on the light by the material used in the waveguide, and this is known as material dispersion. The second part is the impact of the actual waveguide structure, and it is known as waveguide dispersion. The speed of propagation of monochromatic light in an optical fiber is given by the simple equation,

$$u_{phase} = \frac{c}{n_1(\lambda)}$$

This speed is the phase velocity of the light wave, and it is different for each wavelength. In transmitting a pulse of light through the fiber, the pulse can be expressed as the summation of a number of sine and cosine functions, which is known as the spectrum of the pulse. If the spectrum is centered on a frequency, ω, and has a small spectral width around ω, then a velocity can be associated with this group of frequencies, and this is known as the group velocity. In conformity with the idea of velocity of propagation for monochromatic radiation, a group index can be defined as corresponding to the group of frequencies around ω. From a detailed analysis of propagation, group velocity is defined by,

$$u_{group} = \frac{c}{n_1 - \lambda(dn_1/d\lambda)} = \frac{c}{N}$$

where the group index, N, of the material is defined as,

$$N \equiv n_1 - \lambda\frac{dn_1}{d\lambda}$$

The packet of frequencies corresponding to the pulse will arrive at the output of the fiber sometime after the pulse is launched. This delay is the group delay, and it is defined as,

$$\tau_g = \frac{L}{c}\left(n_1 - \lambda\frac{dn_1}{d\lambda}\right) = \frac{L}{c}\bullet N$$

where L is the fiber length. Figure above shows the refractive index and group index of SiO_2 glass fiber.

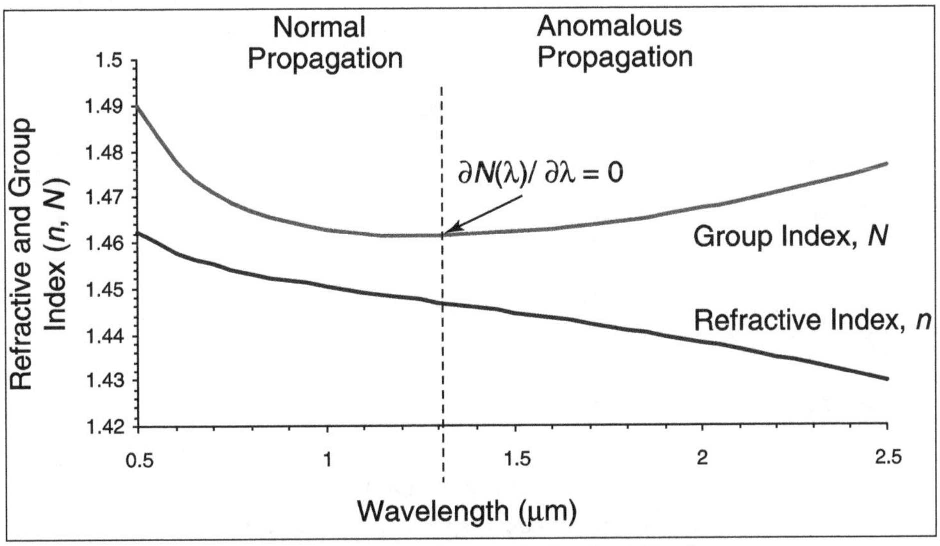

Refractive index and group index for SiO_2 glass.

For any material, the zero group dispersion is the wavelength at which the curve of the refractive index has an inflection point. For silica this wavelength is at 1300 nm, and it is known as λ_o. From Figure above one can distinguish two regions associated with the group index curve. The first region is the region with wavelengths less than λ_o. In this region the group index decreases as the wavelength increases. This means that the spectral components of longer wavelength of the pulse travel faster than spectral components of shorter wavelengths. This regime is identified as the normal group dispersion regime and imposes a positive chirp. For wavelengths greater than lo the opposite behavior is observed, and the region is identified as the anomalous group dispersion regime and a negative chirp is imposed on the pulse.

When measurements of these parameters are considered, it is useful to define a new parameter known as the material dispersion parameter, $D_m(\lambda)$. This parameter is defined as,

$$D_m(\lambda) \equiv \frac{d(\tau_g / L)}{d\lambda} = \frac{1}{c} \bullet \frac{dN}{d\lambda}$$

where $D_m(\lambda)$ is expressed as picoseconds per nanometer of source bandwidth per kilometer of distance (ps/nm-km). Figure below shows the material dispersion of a pure silica glass fiber.

There are two ways to write in the time domain the spread of a pulse whose spectrum is centered at λ_s. For pulses without initial chirp, the change in pulse width is,

$$\Delta\tau_m = -\Delta\lambda_{width} D_m(\lambda_s) L$$

and in terms of the r.m.s. (Root Mean Square) width,

$$\sigma_m = -\sigma_\lambda D_m(\lambda_s) L$$

where σ_λ is the r.m.s. width of the optical source. The value of λ_o depends weakly on the dopants used in the fiber. For example, the use of 13.5 percent of GeO_2 shifts the λ_o by 0.1 nm with respect to the λ_o of pure SiO_2. The material dispersion is also referred to as the chromatic dispersion.

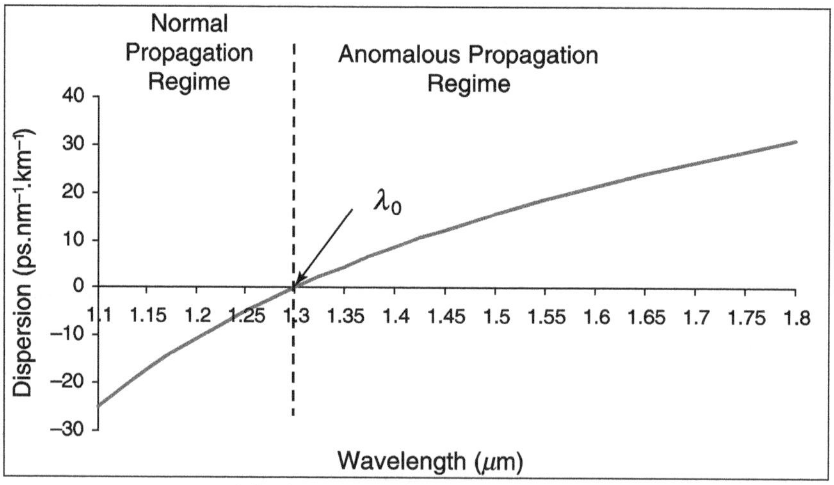

Material (chromatic) dispersion of pure silica.

Multimode Fibers

A multimode fiber can support the propagation of a number of modes (rays). Each of these modes carries the signal imposed on the optical wave. When the modes arrive at the receiver, they create a multi-image of the pulse launched in the waveguide. This multi-image can force the receiver to make a wrong decision regarding the transmitted bit of information.

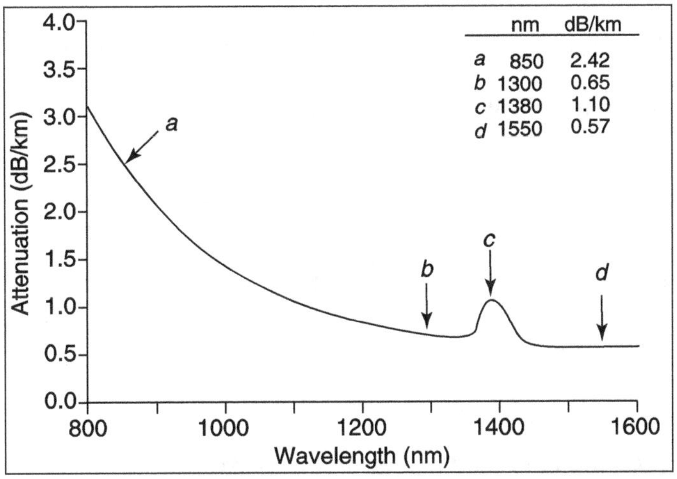

Typical spectral loss for multimode fiber. Notice the low loss at the OH wavelength.

The loss of multimode fibers should not be different from the spectral loss because it is a question of material and processing. However, some manufacturers make available fiber that still has a higher loss, around 1380 nm. The reason for this is the high cost of completely eradicating the residual loss at the OH wavelength. In above figure the spectral loss of a multimode fiber is shown; it is important to notice that the loss at the OH wavelength is only around 0.5 dB, compared to 10 dB.

Light propagating in a multimode fiber is subject to the material dispersion of the fiber. However, there is another form of dispersion that is specific to multimode fibers. Consider a step index multimode fiber with two propagating modes.

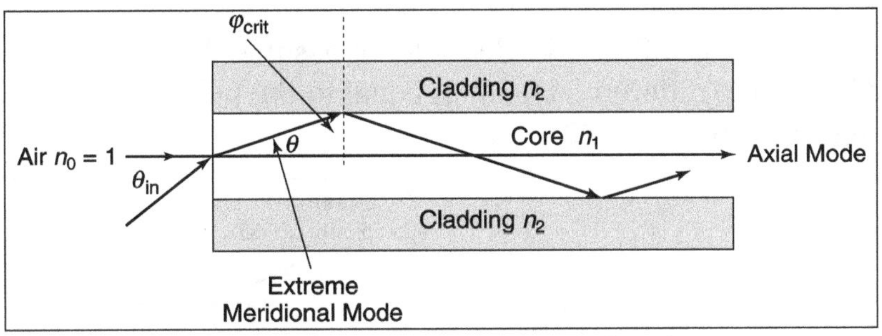

Optical paths of meridional and axial modes.

The first is the axial mode that propagates along the geometric axis of the fiber. The time taken by this mode to reach the end of the fiber is the minimum possible and the output is delayed by that time, which is given by Senior and Buck:

$$\tau_{min} = n_1 \frac{L}{c}$$

where the symbols have their usual meaning. Now the extreme meridional mode will reach the output of the fiber after a delay that in this particular geometric arrangement will be the maximum and given by,

$$\tau_{max} = \frac{n_1}{c} \bullet \frac{L}{\cos\theta}$$

From Snell's law (equations $\sin\phi_{crit} = \frac{n_2}{n_1}$ and $n_0 \sin\theta_1 = n_1 \cos\phi$), we have for the critical angle at the core–cladding interface the following equation:

$$\cos\theta = \sin\phi_{crit} = \frac{n_2}{n_1}$$

Then, substituting into equation $\tau_{max} = \frac{n_1}{c} \bullet \frac{L}{\cos\theta}$ for cos θ,

$$\tau_{max} = \frac{n_1^2}{n_2} \bullet \frac{L}{c}$$

The difference in the delay between the two modes is given by,

$$\Delta\tau_S \approx \Delta\frac{Ln_1}{c} \approx (NA)^2 \bullet \frac{L}{2cn_1}$$

where NA stands for the numerical aperture of the fiber and Δ is given by equation $\Delta = \frac{n_1^2 - n_2^2}{2n_1^2} \approx \frac{n_1 - n_2}{n_1}$. The r.m.s. broadening of the pulse is given by,

$$\sigma_s|_{step} \approx \Delta\frac{Ln_1}{2\sqrt{3}c} \approx (NA)^2 \bullet \frac{L}{4\sqrt{3}cn_1}$$

To appreciate the impact of this differential delay, let us assume that a pulse of nominal width T is launched into the fiber. If the differential delay is equal to the pulse width, the output consists of two pulses occupying a total width of 2T.

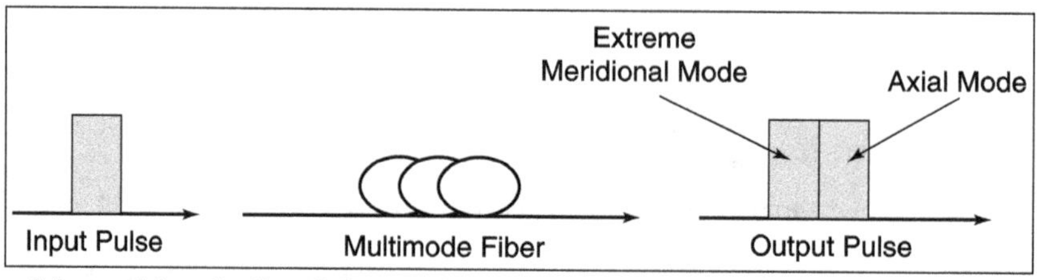

Impact of the multimode fiber delay on the pulse output for a delay equal to the pulse width.

The receiver will therefore detect two pulses when only one was sent. This effect is called the intermodal dispersion of the fiber, and it is an additional dispersion imparted on the pulse. There are a number of reasons why one has to be cautious in using equation $\Delta\tau_S \approx \Delta\frac{Ln_1}{c} \approx (NA)^2 \bullet \frac{L}{2cn_1}$ for long-haul communications. The first is that equation $\Delta\tau_S \approx \Delta\frac{Ln_1}{c} \approx (NA)^2 \bullet \frac{L}{2cn_1}$ is a worst-case scenario. In an actual fiber, the power between adjacent modes is coupled back and forth, which leads to a partial mode delay equalization. The second is related to the physics of propagation. Modes that are weakly confined in the core spread out in the cladding, and if they meet higher loss than in the core, then the number of modes propagating is limited to those strongly confined in the core. The additional loss might be present in the cladding because the material used has a higher intrinsic loss or some of the modes might be radiated into the secondary cladding. These two effects limit the spread of the delay per mode, reducing the intermodal dispersion. In addition to pure fiber effects, there is another source of unpredictability in the mode structure of the light source and its statistics, leading to modal noise.

To appreciate the quantities involved, consider a multimode step index fiber of 10 km length with a core refractive index of 1.5 and Δ = 2%. Then from above equation the r.m.s. pulse broadening is,

$$\sigma_s|_{step} = \Delta\frac{L\times n_1}{2\sqrt{3}\times c} = 0.02\frac{10\times10^3\times1.5}{2\sqrt{3}\times2.998\times10^8} = 2.88\times10^{-2}\frac{ns}{m}\times10\times10^3 m = 288ns$$

The maximum transmission bitrate in terms of the pulse r.m.s. width is given by,

$$B_T\ max = \frac{0.25}{\sigma_{pulse}}$$

Therefore, for $\sigma_s = \sigma = 288$ ns, the maximum bitrate is 868 Kbit/s, which is not a useful value for most modern applications. The key question now is whether the differential delay for a multimode fiber can be improved. The reason the differential delay between the axial mode and the extreme meridional mode is high is that the meridional mode has to reach the boundary between core and cladding before it is reflected back into the core. If the flight time of a meridional mode is reduced, then the differential delay will also be reduced. This can be achieved with the use of graded index fiber.

The basic concept here is to vary the refractive index from a maximum at the center of the core to a minimum at the core–cladding interface. The general equation for the variation of refractive index with radial distance is,

$$n(r) = \begin{cases} n_1\left(1-2\Delta(r/a)^{\alpha}\right)^{1/2} & r<a \text{ core} \\ n_1(1-2\Delta)^{1/2} = n_2 & r\geq a \text{ cladding} \end{cases}$$

where Δ is the relative refractive index difference, r is the axial distance, and a is the profile parameter that gives the refractive index profile. For a = ∞ the representation corresponds to the step index profile. Figure below illustrates the fiber refractive index for various values of the profile parameter a. The step index profile is obtained by setting a = ∞.

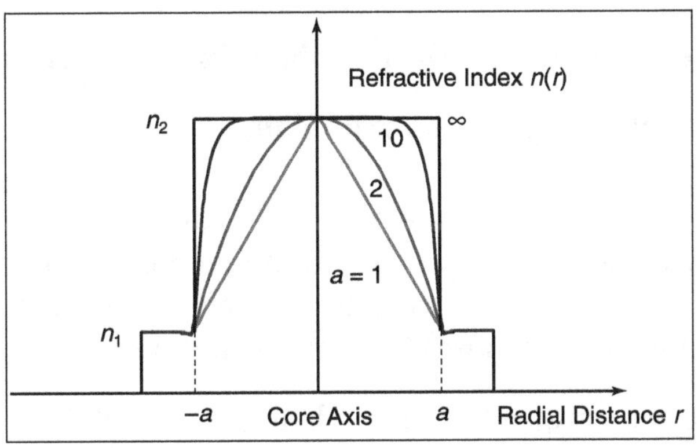

Fiber refractive index for different values of the profile parameter a.

The improvement in differential delay can be observed by considering the modes of a multimode fiber with the profile parameter a set to 2. Two effects may be observed. First, the axial mode propagates through the section of the fiber core where the refractive index has its maximum value, which implies that the axial mode is slowed down. The meridional modes are bent toward the axis of the fiber, reducing their flight time. Together these two effects reduce the differential delay.

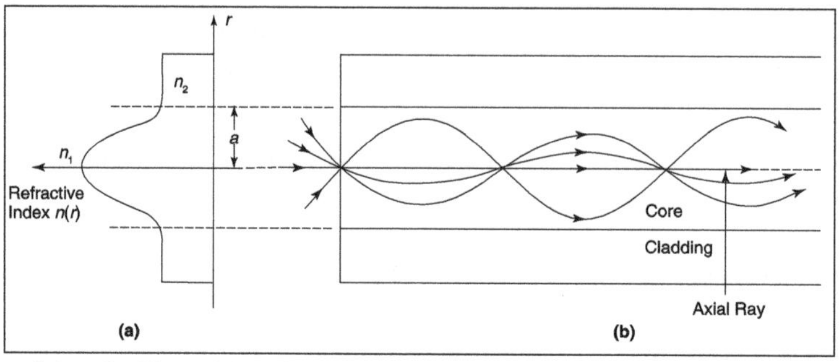

Graded index multimode fiber: (a) Refractive index profile; (b) Meridional modes within the fiber.

If electromagnetic theory is employed to analyze the differential delay, the value obtained is,

$$\Delta\tau_s|_{graded} = \frac{n_1 L}{c} \times \frac{\Delta^2}{8} = \Delta\tau_s|_{step} \times \frac{\Delta}{8}$$

The implication of this equation is to highlight the fact that the reduction of the differential delay for step index fiber is Δ/8. The r.m.s. pulse broadening is now given by,

$$\sigma_s|_{index} = \frac{n_1 L\Delta}{2 \times \sqrt{3} \times c} \times \frac{\Delta}{10} = \sigma_s|_{step} \times \frac{\Delta}{10}$$

and there is a Δ/10 reduction in the r.m.s. pulse broadening. It is now straightforward to compare the r.m.s. pulse broadening for step and graded index fibers. From the example studied before, we have a 10-km graded index fiber with a core refractive index of 1.5 and Δ = 2%. Then, from above equation, the r.m.s. pulse broadening is,

$$\sigma_s|_{index} = \frac{n_1 L\Delta}{2 \times \sqrt{3} \times c} \times \frac{\Delta}{10} = \sigma_s|_{step} \times \frac{\Delta}{10} = 288\,\text{ns} \times \frac{0.02}{10} = 0.576\,\text{ns}$$

and the maximum bitrate that can be used with this graded index fiber is now 434 Mbit/s.

Comparing the r.m.s. pulse broadening of the two fibers for a length of 1 km, one obtains,

$$\sigma_s|_{step} = 28.8\,\text{ns/km} \text{ and } \sigma_s|_{graded} = 0.057\,\text{ns/km}$$

Because of its substantially improved performance, graded index multimode fiber is the clear choice when one wants to exploit the advantages of the multimode fiber with low intermodal dispersion. Multimode fibers have been fabricated in various core–cladding dimensions addressing specific requirements. For communication purposes two core–cladding sizes have been standardized.

The two fibers differ in the core size, with the first being 50 μm (ITU-I G standard G.651) and the second being 62.5 μm. The cladding for both fibers is 125 μm. At the moment there are no standards for the 62.5-μm fiber. The major advantage of the 50/125 μm fiber is the higher bandwidth it offers compared to that of 62.5/125 μm at 850 nm. An advantage of the 62.5/125 μm is that more power can be coupled into the fiber from an optical device such as a Light Emitting Diode (LED) or laser because of the larger NA (NA = 0.275).

One of the important questions related to the performance of a fiber is that of the bandwidth. In general time and frequency domain techniques can be used to establish the performance, but the industry decided to adopt the bandwidth × length product as the performance index of a multimode-fiber bandwidth. A simple equation relating the bandwidth, BW, and the length of the fiber is,

$$\frac{BW_L}{BW_S} = \left[\frac{L_L}{L_S}\right]^{\gamma}$$

where the subscripts L and S correspond to the bandwidth and length for long and short lengths of fiber, and the exponent γ is the length ratio of interest. Usually $\gamma = 1$, but there has long been debate about the value of γ because it reflects the conditions under which the measurements are taken. The value is calculated from the solution of above equation — that is,

$$\gamma = \frac{\log\left[\frac{BW_L}{BW_S}\right]}{\log\left[\frac{L_L}{L_S}\right]}$$

The value of $\gamma = 1$ will be assumed. The first-order estimate of the system length that a fiber can support is calculated from the given bandwidth × length product. For example, for a bandwidth × length product of 1000 MHz × km the system length for a 2.4 Gbit/s system is,

$$\text{System length} = \frac{\text{Bandwidth} \times \text{Length}}{\text{Bitrate}} = \frac{1000\,\text{MHz} \times \text{km}}{2.4 \times 10^3\,\text{MHz}} = 401\,\text{m}$$

When this approach is used to estimate the system length, one should understand the technique or techniques used to estimate the bandwidth × length product of the fiber. Some of the key parameters of the Corning InfiniCor graded index fiber are summarized in Tables below. Similar fibers are available from other manufacturers.

Table: Attenuation of Corning InfiniCor Fiber.

Wavelength (nm)	Maximum value (dB/km)
850	≤2.3
1300	≤0.6

Table: Bandwidth × km Performance of Corning InfiniCor Fiber.

Corning Optical Fiber	High-performance 850 nm only (MHz × km)	Legacy performance (MHz × km)	
Infi niCor eSX+	4700	1500	500
Infi niCor SX+	2000	1550	500

Infi niCor SXi	850	700	500
Infi niCor 600	510	500	500

The multimode fiber was introduced and discussed up to now with the tacit assumption that there are a large number of modes, but their number was not stated. This question will be addressed now. To derive the number of modes supported by a given multimode fiber, the Maxwell equations must be solved in the context of the fiber structure (cylinder) and the material (dielectric, glass).

From such an analysis a very useful quantity emerges called normalized frequency, V. It combines the wavelength, the physical dimensions of the fiber, and the properties of the dielectric (glass), and is given by,

$$V = \frac{2\pi}{\lambda} a \bullet (NA) = \frac{2\pi}{\lambda} \bullet a \bullet n_1 (2\Delta)^{1/2}$$

where a is the core radius, Δ is the relative refractive index difference, and λ is the wavelength.

It can be shown that the number of modes supported by a step index multimode fiber, M_{step}, is,

$$M_{step} \approx \frac{V^2}{2}$$

and by a graded index multimode fiber,

$$M_{graded} \approx \left(\frac{\alpha}{\alpha+2}\right) \bullet \left(\frac{V^2}{2}\right)$$

where α is the refractive index profile parameter. For a parabolic profile (α = 2) the number of modes is $M_{graded} \approx V^2/4 = M_{step}/2$. Consider a multimode fiber operating at 1300 nm with radius 25 μm, $n_1 = 1.5$, and Δ = 2%. Then the step index fiber supports,

$$V \approx \frac{2\pi}{\lambda} \bullet a \bullet n_1 (2\Delta)^{1/2} = \frac{2 \times \pi}{1300 \times 10^{-9}} \times 25 \times 10^{-6} \times (2 \times 0.02)^{1/2} = 24$$

and

$$M_{step} \approx \frac{V^2}{2} = \frac{(24^2)}{2} = 288 \text{ guided modes}$$

With a parabolic refractive index profile the same fiber will support 156 guided modes. The number of modes and their fluctuations can impair the performance of a multimode fiber system because of modal noise.

After a pulse propagates in a fiber the r.m.s. pulse broadening s given by,

$$\sigma_{total} = \left(\sigma_{mat}^2 + \sigma_w^2 + \sigma_{mod}^2\right)^{1/2}$$

where σ_{mat} is the material (chromatic) dispersion, σ_w is the waveguide dispersion, and σ_{mod} is the intermodal dispersion (for step index or graded index fiber). For multimode transmission the material and waveguide dispersion terms are negligible compared to the intermodal dispersion, and they are usually ignored. Therefore, the pulse in the output of the fiber has an r.m.s. width given by,

$$\sigma_{out} = \left(\sigma_{inp}^2 + \sigma_{mod}^2\right)^{1/2}$$

where σ_{inp} is the r.m.s. width of the input pulse.

Single-mode Fibers

The concept of the single-mode fiber was introduced, and the most important physical feature of a single-mode fiber is its small core. For a useful image of how a single-mode fiber operates, one can use the idea that only the axial mode propagates in the fiber. The bandwidth of single-mode fiber is so large compared to that of the multimode fiber that single-mode fiber is used in all long-haul communications today, terrestrial and submarine. Concepts used to arrive at a reasonable understanding of the propagation features of multimode fiber cannot be used in single-mode fibers, and the solution of Maxwell equations in the context of the geometry and materials is required.

Single-mode fibers are available with these loss characteristics as products for terrestrial and submarine applications. The most striking difference between multimode and single-mode fibers is in the dispersion. In multimode fibers the total dispersion is dominated by the intermodal dispersion. In single-mode fibers only one propagation mode is supported, so the total dispersion consists of three components: material dispersion, waveguide dispersion, and polarization mode dispersion.

The reason waveguide dispersion and polarization mode dispersion became important is the low value of the material dispersion. The normalized frequency parameter, V, plays an important role in single-mode fiber. The parameter V is given by,

$$V = \frac{2\pi a}{\lambda} \bullet n_1 (2\Delta)^{1/2}$$

where the symbols have their usual meaning. Single-mode operation takes place above a theoretical cutoff wavelength λ_c given from above equation as,

$$\lambda_c = \frac{2\pi a}{V_c} \bullet n_1 (2\Delta)^{1/2}$$

where V_c is the cutoff normalized frequency. With the help of equation $V = \frac{2\pi a}{\lambda} \bullet n_1 (2\Delta)^{1/2}$, above equation can be written as,

$$\frac{\lambda_c}{\lambda} = \frac{V}{V_c}$$

For a step index single-mode fiber—that is,

$$n(r) = \begin{cases} n_1 & r < a \text{ core} \\ n_2 & r \geq a \text{ cladding} \end{cases}$$

$V_c = 2.405$ and the cutoff wavelength is given by,

$$\lambda_c = \frac{V\lambda}{2.405}$$

For operation at 1300 nm the recommended cutoff wavelength ranges from 1100 to 1280 nm. With this choice for the cutoff wavelength, operation at 1300 nm is free of modal noise and intermodal dispersion. The material dispersion for silica glass with a step index single-mode fiber is given by,

$$D_m(\lambda) \approx \frac{\lambda}{c} \bullet \frac{d^2 n_1}{d^2 \lambda}$$

where $d^2n_1/d^2\lambda$ is the second derivative of the refractive index with respect to the wavelength.

The value of $D_m(\lambda)$ is usually estimated from fiber measurements, but for theoretical work a simple equation with two parameters offers a reasonable approximation. The equation is,

$$D_m(\lambda) = \frac{S_0 \lambda}{4} \bullet \left[1 - \left(\frac{\lambda_0}{\lambda}\right)^4\right]$$

where S_0 and λ_0 are the slope at λ_0 and the zero material dispersion wavelength, respectively. For standard single-mode fiber the slope at λ_0 is in the range 0.085 to 0.095 ps/nm² × km.

Table: Attenuation of Corning SMF-28e+Single-Mode Fiber.

Wavelength (nm)	Maximum values (dB/km)
1310	0.33–0.35
1383	0.31–0.33
1490	0.21–0.24
1550	0.19–0.20
1625	0.20–0.23

Table: Single-mode Fiber Dispersion of Corning SMFe+ in the Third Optical Window.

Wavelength (nm)	Dispersion value [ps/(nm × km)]
1550	≤18.0
1625	≤22.0

The Corning single-mode fiber SMF-28e+ is a typical example of high-quality modern fiber. For this Corning fiber, above equation is valid for the range $1200 \leq \lambda \leq 1625$ nm. Single-mode fibers of similar performance are also available from other fiber manufacturers. After material dispersion the next important dispersion component is waveguide dispersion. The physical origin of waveguide dispersion is the wave-guiding effects, and even without material dispersion, this term is present. A typical waveguide dispersion curve is shown in figure.

Table: Zero-Dispersion Wavelength and Slope of Corning SMF-28e+ Single-Mode Fiber.

Zero-dispersion wavelength (nm)	$1310 \leq \lambda_0 \leq 1324$
Zero-dispersion slope [ps/(nm² × km)]	$S_0 \leq 0.092$
λ_0 typical value (nm)	1317
S_0 typical value [ps/(nm² × km)]	0.088

The importance of this term is not so much in the value of the dispersion but in the sign. In silica fiber the waveguide dispersion is negative, and because the material dispersion is positive above λ_0, the value of the total dispersion is reduced. The immediate impact is to shift λ_0 to longer wavelengths. By itself this small shift in λ_0 is of marginal importance, but if the value of the waveguide dispersion is designed to take much higher values, then the shift is substantial and gives rise to two new classes of single-mode fiber: dispersion shifted and dispersion flattened.

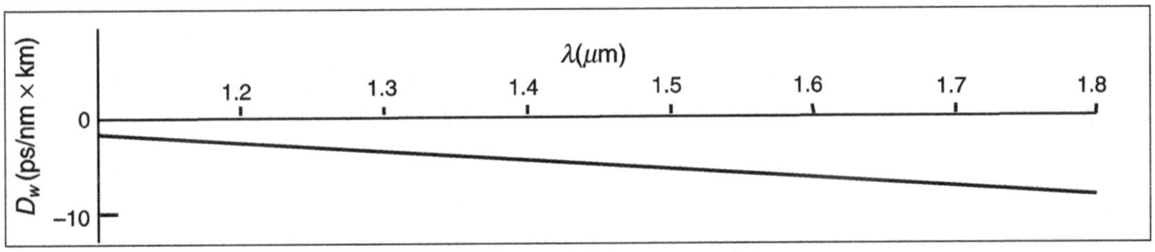

Waveguide dispersion with $\lambda_c = 1.2$ and $\Delta = 0.003$.

The third component of dispersion in single-mode fiber is polarization mode dispersion, or PMD. For ease of study it has been assumed tacitly that the fiber is a perfect cylindrical structure. In reality the structure is only nominally cylindrical around the core axis. The core can be slightly elliptical. Also, the cabling process can impose strain on the fiber, or the fiber might be bent as it is installed. In these situations the fiber supports the propagation of two nearly degenerative modes with orthogonal polarizations. When the optical wave is modulated by the carried information, the two waves corresponding to the two states of polarization are modulated. Therefore, two modes orthogonal to each other propagate in the waveguide. In systems where the receiver can detect both states of polarization, the receiver will function as expected if both modes arrive at the same time; that is, there is no differential delay between the two modes. However, the refractive index of the silica glass can be slightly different along the two orthogonal axes, and consequently there is differential delay. The average Differential Group Delay (DGD) is the PMD coefficient of the fiber.

$$\text{PMD} = \langle \text{DGD} \rangle$$

To compound the difficulties, the DGD is not the same along the length of the fiber but varies from section to section. Because of this feature, the PMD is not described by a single value but rather by a probability distribution function. The impact of PMD on pulse width is illustrated in Figure.

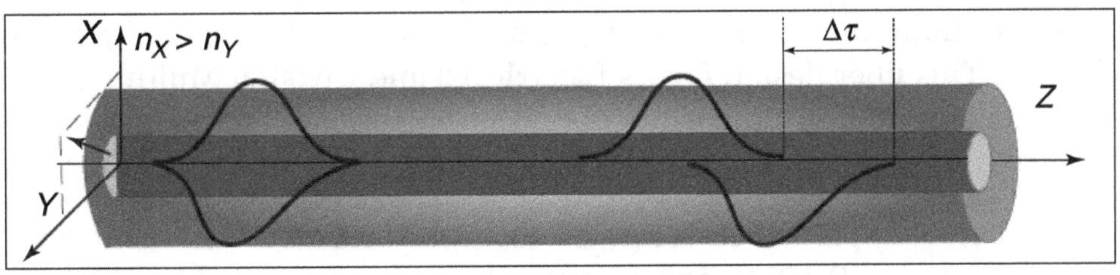

Effect of PMD on a pulse.

The PMD is a complex effect, and its study and mitigation have attracted a very substantial effort. Theoretical studies and experiments have demonstrated that the PMD is proportional to the square root of the fiber length—that is,

$$D_{PMD} = PMD_{coeff}\sqrt{L}\ \ ps\sqrt{km}$$

Because of its importance for high-speed systems, the PMD fiber coefficient is part of the fiber specifications. For example, the Corning fiber STM-28e+ has a PMD coefficient of $\leq 0.06\ ps/\sqrt{km}$. The calculation of the amount of PMD a system can tolerate is an involved process and is part of the system design. An easily remembered guideline is that the PMD of the link satisfies the inequality,

$$D_{PMD} \leq \frac{\text{Pulse width}}{10}$$

The impact of PMD on high-speed, systems is quite severe. From this table it is clear that long-haul high-capacity systems are very vulnerable to PMD, and the selection of the fiber type is a very important decision in system design. The ITU-T has issued standards for the conventional fiber that are described by the G.652 recommendation. The fiber covered by this recommendation is the most widely installed in the worldwide telecommunications network.

Table: Impact of PMD on System Length for Typical Bitrates.

			System length (km)	
Bitrate (Gbit/s)	Timeslot (ps)	Link PMD (ps)	PMD_{coeff} 0.02 ps/ $\sqrt{km}$	PMD_{coeff} 0.05 ps/ $\sqrt{km}$
2.5	400	40.0	4.0×10^6	6.4×10^3
10.0	100	10.0	2.5×10^5	4.0×10^2
40.0	25	2.5	15.6×10^3	25.0
100.0	10	1.0	2.5×10^3	4.0

Until now the discussion has been limited to what may be called the conventional single-mode fiber. That is a single-mode fiber with $\lambda_0 \approx 1300$ nm, low waveguide dispersion ($D_m \geq D_w$), and a step index profile. Since the introduction of the conventional single-mode fiber, a number of applications have arisen where a fiber optimized for the application can deliver better performance.

Special Single-mode Fibers

Waveguide dispersion can be used to shift the λ_0 of a fiber to a direction that the dispersion of the fiber will be substantially altered. Initially, two classes of fiber emerged to exploit this property of waveguide dispersion. The first was the dispersion-shifted fiber in which the value of the waveguide dispersion was large and the λ_0 shifted to 1550 nm. At this wavelength the zero dispersion and the minimum loss value of silica glass coincided, offering the best of both worlds: dispersion and loss. This fiber design favors Electrical Time Division Multiplexed (ETDM) systems, and it was introduced at a time when the only practical way to increase the capacity appeared to be ETDM.

Later, however, a better approach in terms of cost–performance was identified with Wavelength Division Multiplexing (WDM), and another fiber was introduced in which the dispersion presented a plateau enabling the WDM channels to face similar dispersion. Unfortunately, again the fiber could not be employed in large quantities because of the nonlinear behavior of the silica fiber when optical amplifiers are used demands the presence of substantial dispersion between channels if the nonlinear effects are to be mitigated. Therefore, the two first attempts to break the confines of the conventional single-mode fiber were not very successful. The dispersion characteristics of standard fiber, dispersion-shifted fiber, and dispersion-flattened fiber are shown in Figure.

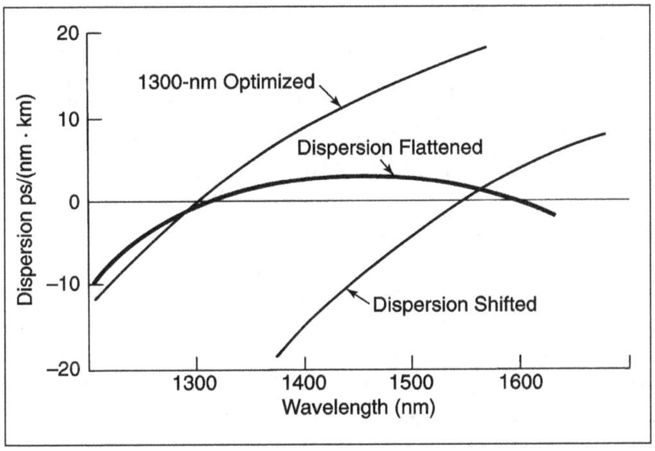

Dispersion characteristics of standard, dispersion-flattened, and dispersion-shifted fibers.

The dispersion-shifted fiber standards are addressed in ITU-T recommendation G.653. The introduction of the erbium-doped silica optical amplifier operating over the band 1530 to 1565 nm and the suppression of the OH loss peak prompted the division of the available low loss window of the silica fiber into a number of bands whose exploitation will depend on the availability of the relevant technology.

Table: Optical Bands Available for Fiber Communications with SiO_2 Fibers.

Band	Description	Wavelength range
O-band	Original band	1260–1360 nm
E-band	Extended band	1360–1460 nm
S-band	Short-wavelengths band	1460–1530 nm
C-band	Conventional (erbium window)	1530–1565 nm
L-band	Long-wavelengths band	1565–1625 nm
U-band	Ultralong-wavelengths band	1625–1675 nm

A number of fibers have been introduced to address the use of these optical bands. The aim is to reduce the amount of dispersion in the intended band of operation and also to minimize the impact of fiber nonlinearities. According to the ITU-T G.655 recommendations, the dispersion over the C-band should lie in the range 1 to 10 ps/(nm × km). Then the operation can be extended into the L-band because the dispersion will still be low there. Both C- and L-bands are used for long haul applications. ITU-T recommendation G.656 is designed to extend operation into the S-band. With chromatic dispersion in the range of 2 to 14 ps/(nm × km) from 1460 to 1625 nm, this fiber can be used for Coarse WDM (CWDM) and Dense WDM (DWDM) throughout the S-, C-, and L-bands. These fibers are referred to as Non-Zero Dispersion-Shifted Fibers (NZ-DSF).

Table: Attenuation of Nonzero Dispersion-Shifted Fiber.

Wavelength (nm)	Loss (dB/km) Maximum	Loss (dB/km) Typical	The maximum attenuation in the range 1525– 1625 nm is no more than 0.05 dB/km greater than the attenuation at 1550 nm.
1310 nm	≤0.40	≤0.35	
1383 nm	≤0.40	≤0.25	
1450 nm	≤0.26	≤0.25	
1550 nm	≤0.22	≤0.20	
1625 nm	≤0.24	≤0.21	

The dispersion of the NZ-DSF at 1550 nm is nearly four times less than that of the standard single-mode fiber, reducing the amount of dispersion compensation needed in long-haul systems. The reader should be aware that all leading fiber manufacturers market NZ-DSF fiber.

Table: Dispersion of Nonzero Dispersion-Shifted Fiber.

Optical band	Dispersion ps/(nm × km)	Zero dispersion wavelength: ≤1405 nm Dispersion slope at 1550 nm: ≤0.045 ps/nm²-km Mode field diameter: 8.6 ± 0.4 mm at 1550 nm
C-band 1530–1565 nm	5.5–8.9 ps/nm-km	
L-band 1565–1625 nm	6.9–11.4 ps/nm-km	
S- and L-bands 1460–1625 nm	2.0–11.4 ps/nm-km	

A significant part of the cost of DWDM systems is the cost of the narrowband active and passive optical components required by the specifications. The availability of the NZDSF fibers has enabled a substantial reduction of the cost of WDM by spacing the channels at 20 nm. This channel spacing accommodates 16 channels from 1310 to 1625 nm and is supported by ITU-T recommendation G.652C. A schematic of the spectral loss and the CWDM channel scheme is shown in Figure.

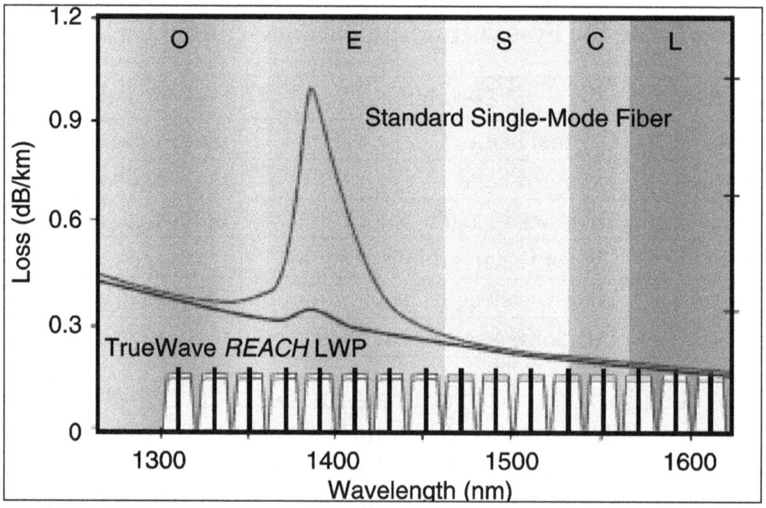

Use of NZ-DSF in a coarse WDM channel allocation scheme.

Special Optical Fibers

No single fiber design meets all application requirements due to economic reasons. Therefore some special types of fibers are manufactured for specialized uses. These are special optical fibers.

High Purity Silica Fiber (HPSUV)

This type of fiber is suitable for transmission of light in the range of 180 to 800 nm. It is good and cheap. It is coated with aluminum which gives very high mechanical strength and extra power handling capacity as aluminum dissipates heat more quickly. These fibers operate at temperatures up to 400 degree celcius and vacuum.

Characteristics of High Purity Silica Fiber (HPSUV)

- Fiber type: Step index multimode.
- Core material: High purity synthetic silica.
- Cladding material: Doped silica.
- Primary coating: Aluminum or polymer.
- Secondary coating: Polymer.
- NA: 0.24.
- Tensile strength: 7 G Pa.
- Minimum bend radius: 40 times fiber radius.
- Temperature range: -169 to 400 degree celcius.
- Humidity: 100%.
- Attenuation values:
  - 248 nm (KrF laser) <1.2dB/m.
  - 308 nm (XeCl laser) <0.26dB/m.
- Radiation resistance: Good.
- Maximum intensity of the transmitted power:
  - CW upto 100KW/cm^2.
  - Pulse upto 500 KW/cm^2.

High Purity Silica (HPSIR)

It is similar to HPSUV fiber with slightly different dopants to give it a longer wavelength capability in the near IR from 500 nm to 2600 nm.

Chalcogenide Fiber

This type of fiber is used for transmission of light from 1 to 0.6 µm. They have extremely low losses. There are two types of chalcogenide fibers i.e. A and B. The loss in A type is down to 0.1dB/m. Hence they are particularly suitable for medical applications. The characteristics are:

- Fiber type: Step index multimode.

- Core material: As_2S_3.
- Cladding material: As_xS_{1-x}.
- Primary coating: PTFE.
- Secondary coating: PTFE or PVC.
- NA: 0.3 to 0.5.
- Minimum bend radius: < 10 nm.
- Temperature range: < 200 < T < 100 degree celcius.
- Attenuation values:
    - A type minimum 0.1dB/m.
    - B type minimum 0.6dB/m.
- Radiation resistance: Good.
- Maximum intensity of the transmitted power: 10W (CO laser).
- Core diameter:
    - CHAL 100: 100μm.
    - CHAL 200: 200μm.
    - CHAL 300: 300μm.

Halide Fibers

These fibers are extended to 15 μm region. The most common application is for use with the CO_2 laser in medicine to replace bulk optics delivery systems or in industry. Silver halide fibers are used for transmission of light from 3 to 15 μm. They have low losses and are currently the only known optical fibers for transmission of light from high power, long wavelength lasers like the CO_2 laser. They are very flexible and much more convenient than normal mechanical delivery systems for these long wavelengths.

The fibers comprise a core polycrystalline silver halide surrounded by an opaque tube. Because of high refractive index (2.2) of the material the total internal reflection takes place at the boundary with air. The tube is necessary to prevent UV light from reaching the core which causes premature failure. The normal life time of these fibers is 6 to 12 months depending upon how they are used.

Characteristics of Halide Fibers

- Numerical Aperture: < 0.7.
- Outer diameter with protective covering: 0.36 to 1.5 mm.
- Core diameter: 0.1 to 1.2 mm.

- Attenuation at 10.6µm i.e. CO_2 laser: 0.5 to 1.5 dB/m.
- Attenuation at 5 to 6µm i.e. CO_2 laser: < 2dB.
- Usable wavelength range: 3-15mm.
- Maximum length available: < 15m.
- Yield strength: 150-170MPa.
- Radius of elastic bending: >0.4 cm.
- Operating maximum temperature: > 100 degree celcius.

Tapered Optical Fibers

Tapered fibers are useful for getting maximum amount of power from a poor quality laser spot into a fiber. The use of tapered optical fiber is an efficient low cost method of transforming a poor quality laser beam into a uniform output spot. The concept of conversion of brightness states that the spatial and angular parameters of light anywhere within or at either end of the taper are mutually connected by the formula,

$$A_i n_i^2 \sin^2\theta_i = A_o n_o^2 \sin^2\theta_o$$

Where n's and θ's represent refractive indices and angles respectively. Subscripts 'I' and 'o' refer to input and output.

$$\frac{A_i}{A_o} = R^2$$

Where R is taper diameter ration and it follows that $\frac{(NA)}{(NA)_o} i = R$

It should be noted that if the product of the input NA and the taper diameter ratio R exceeds the NA of the taper's pigtail, the light will escape into the cladding and will be lost. In that case the above relation is no longer true. To avoid light losses the NA of the light beam at the taper's input should be less than the practical value. Figure shows structure of tapered optical fiber.

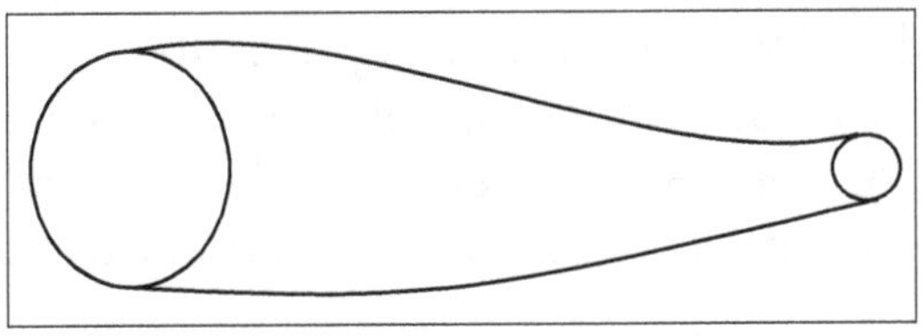

Characteristics of Tapered Optical Fibers

- Input to output ratios: Upto 5:1.
- Input core diameter: 100µm to 4.0 mm.
- Output core diameter: 50 µm to 1.5 mm.

- Taper length: 1-3m.
- Total length: 3-10m.
- Core material: Pure synthetic silica.
- Primary coating: Aluminum or acrylic.
- Optional secondary coating: Epoxy acrylate, flouropolymers.
- Operating temperature: 196-400 degree celcius (Al coated).
- Humidity: 100%.
- Radiation resistance: Good.
- Power transmission:
    - CW upto 100KW/cm^2.
    - Pulsed 145, 500KW/cm^2.
- NA: 0.24.

Applications of Optical Fibers (Wavelength Wise)

Table: Use of different types of optical fibers according to operating wavelength.

Type	Range of operating wavelength
HPSUV	180nm to 800nm
HPSIR	500 nm to 2.6 μm
Taper	180nm to 2.6 μm
Chalcogenide	1 μm to 6 μm
Silver halide	3.0 μm to 15 μm

Nonlinear Optical Effects in Fibers

The discussion of fiber performance was based implicitly on a linear model. That is, when the input is doubled the output is also doubled. However, the optical fiber is essentially a waveguide loaded with a dielectric material, the glass, and under suitable conditions the behavior of the glass departs from a linear characteristic. The study of nonlinear optical effects belongs to the field of nonlinear optics, but because the effects on optical fiber communications are serious.

The origin of nonlinear optical effects lies in the response of the atoms and molecules of the glass dielectric in high electric fields. One might be surprised that in spite of the low optical powers used in fiber communications, the nonlinear behavior of the fiber impairs system performance. The electric field in a medium with effective area A_{eff} and power P is given by,

$$|E| = \sqrt{\frac{2\times 377}{n}} \times \sqrt{\frac{P}{A_{eff}}}$$

With a semiconductor laser of 10 mW and a fiber with an effective area of 53 μm², the electric field in the fiber is,

$$|E| = \sqrt{\frac{2\times377}{n}} \times \sqrt{\frac{P}{A_{eff}}} = \sqrt{\frac{2\times377}{1.5} \times \frac{10\times10^{-3}}{53\times10^{-12}}} = 3.0\times10^{5}\,V/m$$

The electric field experienced by the electron in the hydrogen atom is of the order of 5×10^{11} V/m, so one could think that the field generated by a laser beam of 10 mW would not give rise to nonlinear optical effects. Although the optical power is low, the interaction length between the optical radiation and the glass dielectric can be very long when the fiber operates with optical amplifiers. The effect is cumulative, and system performance is impaired.

The channel power used in this simple example refers to one laser, but in a WDM system the number of channels can exceed 100, so the total power can be very high. In nonlinear effects associated with the propagation of light in fibers there are two fiber parameters that dimension the impact of the effects. The first is the effective area of the radiation mode, A_{eff}. It is defined as shown in equation $\alpha = \frac{\alpha_{dB}}{4.343}$:

$$A_{eff} = \pi r_o^2$$

where r_o is the mode field radius.

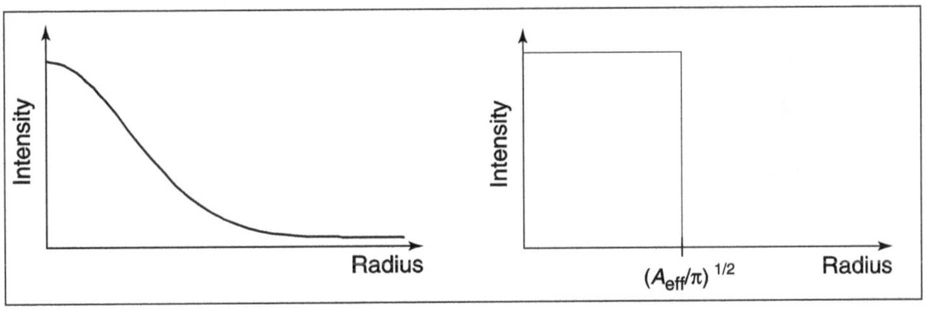

Concept of the effective area of A_{eff}.

The second is the effective length, L_{eff}, and it is defined by,

$$L_{eff} = \int_0^L \exp(-\alpha z)\,dz = \frac{1-\exp(-\alpha L)}{\alpha}$$

where L is the fiber length and α is the attenuation coefficient of the fiber, which here is not in dBs. The attenuation coefficient of the fiber in this equation is derived from the loss in decibels through the equation,

$$\alpha = \frac{\alpha_{dB}}{4.343}$$

For low-loss fiber $L_{eff} \approx L$ and for high loss fiber, $L_{eff} \approx 1/\alpha$. Because the fiber for optical transmission has low loss, the first approximation is usually used. The physical meaning of L_{eff} is simple.

We know that for low power the nonlinear effects are negligible, so there is a "threshold" of power above which the propagation is nonlinear and below which it is linear.

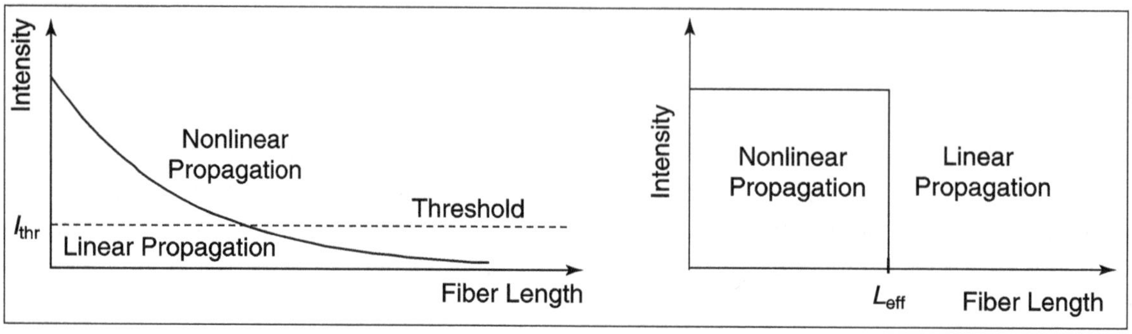

Concept of the effective length, L_{eff}, in nonlinear optical effects.

For example, consider a fiber with a loss of 1.8 dB/km and length of 25 km. Using the approximation for low loss, $L_{eff} \approx L$. If we use equation 5.50 the result is 15.5 km. In a system with optical amplifiers, the effective length of the system is given by,

$$L_{eff} = \frac{1-\exp(-\alpha l)}{\alpha} \times \frac{L_{total}}{\Delta l}$$

where L_{total} is now the total system length and Δl the spacing between amplifiers. It is clear that to reduce the system effective length it is better to use as few amplifiers as possible. The nonlinear effects relevant to fiber communications are:

- Single-channel systems:
  - Self-Phase Modulation (SPM).
  - Stimulated Brillouin Scattering (SBS).
- Multichannel systems:
  - Cross-Phase Modulation (XPM).
  - Stimulated Raman Scattering (SRS).
  - Four-Wave Mixing (FWM).

Self-phase Modulation

To understand the self-phase modulation effect, one has to understand the effect of high intensity on the refractive index of the material. This impact of optical power on the refractive index can be stated mathematically as,

$$n(t) = n_1 + \Delta n(t) = n_1 + n_2 E^2(t)$$

where n_1 is the conventional refractive index of the silica glass and n_2 is the nonlinear refractive index. The value of n_2 is 3.2×10^{-20} m²/W and it is therefore very small compared to the linear refractive index, n_1, which for silica glass at 1550 nm is 1.444. The difference is 20 orders of magnitude, but the impact of n_2 is still important because of the long length interaction between material and radiation and the high intensity of the radiation.

Over a fiber length L the total phase shift due to Δn(t) is given by,

$$\Delta\varphi(t)=\int_0^L \frac{\omega_o}{c}\Delta n(t)dz=\int_0^L \frac{2\pi}{\lambda}\Delta n(t)dz=\frac{\omega_o}{c}n_2E^2(t)L$$

$$\Delta\omega(t)=-\frac{d}{dt}\Delta\varphi(t)=-\frac{d}{dt}\left[\frac{\omega_o}{c}n_2E^2(t)\right]L$$

To proceed, one needs an analytical expression of the electric field. For convenience let us assume that the input light pulse has a Gaussian envelope:

$$E(t)=E_o\exp\left[-\frac{(t-t_o)^2}{\tau^2}\right]$$

where E_0 is the peak pulse amplitude, t_o is the time position of the pulse, and t is the half-width at the 1/e point of the peak amplitude. Substituting above equation into equation $\Delta\varphi(t)=\int_0^L \frac{\omega_o}{c}\Delta n(t)dz=\int_0^L \frac{2\pi}{\lambda}\Delta n(t)dz=\frac{\omega_o}{c}n_2E^2(t)L$, we obtain,

$$\Delta\varphi(t)=\Delta\varphi_{max}\exp\left[-\frac{2(t-t_o)^2}{\tau^2}\right]$$

$$\Delta\varphi_{max}=\frac{\omega_o}{c}n_2E_o^2L$$

where the length L corresponds to the nonlinear interaction length, L_{eff}. The maximum frequency change is obtained by substituting equation $\Delta\varphi(t)=\Delta\varphi_{max}\exp\left[-\frac{2(t-t_o)^2}{\tau^2}\right]$ into equation $\Delta\omega(t)=-\frac{d}{dt}\Delta\varphi(t)=-\frac{d}{dt}\left[\frac{\omega_o}{c}n_2E^2(t)\right]L$:

$$\Delta\omega(t)=\Delta\varphi_{max}\frac{4(t-t_o)}{\tau^2}\exp\left[-\frac{2(t-t_o)^2}{\tau^2}\right]$$

At the center of the pulse the frequency shift is zero. To find the position of the maximum frequency shift we calculate,

$$\frac{d}{dt}\left[\Delta\omega(t)=0\right]$$

One finds that the maximum frequency changes occur at $(t-t_o)=\pm\tau/2$. For example, consider the Gaussian pulse with $\tau = 1$ psec, $\Delta\varphi_{max}=8\pi$.

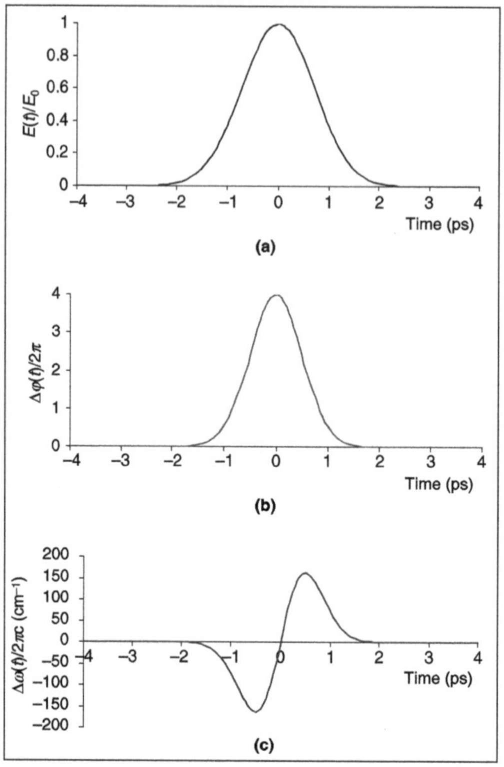

(a) Normalized input pulse amplitude; (b) phase shift due to nonlinearity; (c) frequency chirp.

For the central position of the pulse the frequency chirp is linear between $\pm\tau/2$ and is given by,

$$\omega(t) = \omega_0 + \left(\frac{4\pi n_2}{\lambda_0 \tau^2} L \frac{P}{A_{eff}} \right) \times t$$

The impact of SPM on the performance of an optical system depends on the dispersion. If there were no dispersion in the fiber, SPM would have no impact on performance. First consider operation at $\lambda > \lambda_0$. Then $D < 0$ and the pulse will broaden because the leading "red" frequencies travel faster and the "blue" frequencies slower. For $\lambda > \lambda_0$ the situation changes substantially. Now, the "red" frequencies in the rising edge of the pulse are delayed and the "blue" frequencies at the falling edge are speeded up; consequently, the pulse width contracts. However, as the blue frequencies catch up with the red ones and overtake them, the pulse starts broadening again. This feature of the interaction of SPM with anomalous dispersion is used in soliton transmission.

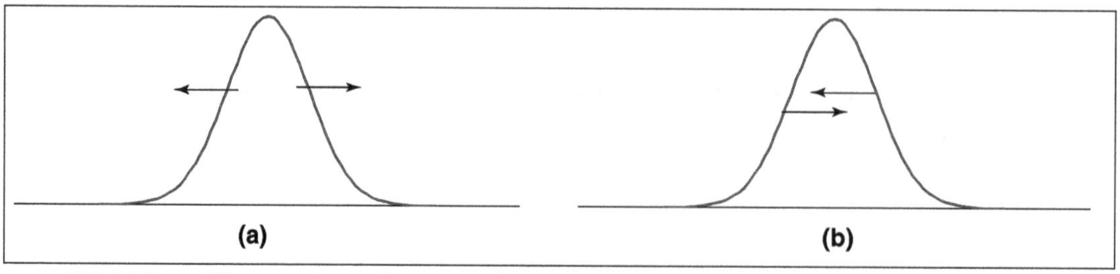

SPM followed by negative and positive dispersion: (a) $D < 0$, additional pulse broadening; (b) $D > 0$, pulse compression.

Stimulated Brillouin Scattering

The physical origin of Stimulated Brillouin Scattering (SBS) is the interaction between an incident optical wave and the elastic acoustic wave induced by the radiation. A threshold can be established for the onset of SBS, and it is given by,

$$P_{th}^{Brillouin} \approx 21 \times p_{Pol} \times \frac{A_{eff}}{g_B L_{eff}} \times \left(1 + \frac{\Delta f_{source}}{\Delta f_{Br}}\right)$$

where p_{Pol} takes the value 1.5 if the polarization is completely scrambled and 1.0 otherwise; A_{eff} is the effective area of the mode; g_B is the Brillouin (Br) gain, which is around 4×10^{-11} m/W at 1550 nm; L_{eff} is the effective fiber length; Δf_{source} is the bandwidth of the incident radiation; and DfBr is the bandwidth of the SBS effect, which is approximately 20 MHz.

In the worst case for $\Delta f_{source} << \Delta f_{Br}$ and $p_{Pol} = 1$ with $A_{eff} = 50\ \mu m^2$ and $L_{eff} = 20$ km the threshold power is 1.3 mW. This is very low and care must be taken to enhance the threshold. For high-speed modulation this is not a problem. In SBS the scattered wave propagates backward and the effect appears in the receiver as a reduction of the power of the wave carrying the information. The SBS is a narrowband effect and consequently does not interact with other waves unless they fall within its bandwidth. For this reason SBS once suppressed in individual channels is not an issue in WDM systems.

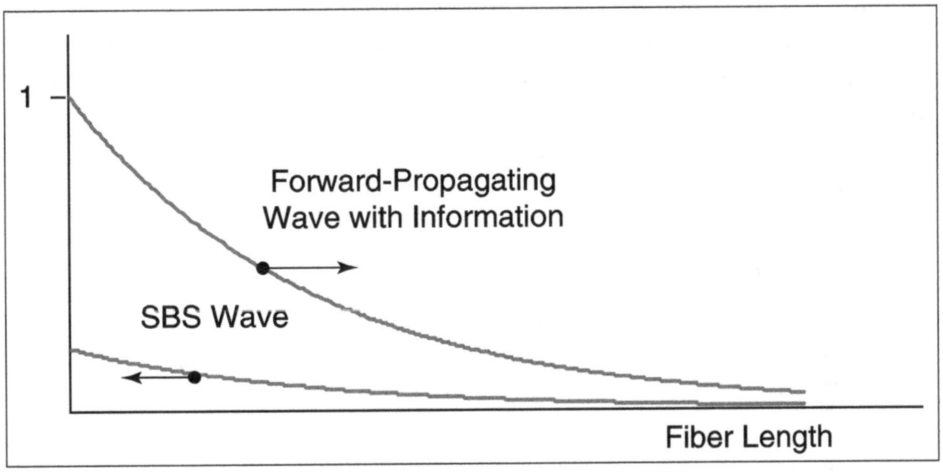

Stimulated Brillouin scattering (SBS) effect.

Cross-phase Modulation

The SPM explained earlier operates within each channel. However, if a channel has enough power to induce the effect on the refractive index of the material for another channel propagating through the same material (that is to say, in a WDM system) then impairment arises in this other channel. This effect is referred to as Cross-Phase Modulation (XPM). The total induced frequency chirp, $\Delta\omega_{NL}(t)$, is given by,

$$\Delta w_{NL}(t) = -\gamma \times L_{eff}^{System} \times \left(\frac{\partial P_n}{\partial t} + 2\sum_{j \neq n}^{N} \frac{\partial P_j}{\partial t}\right)$$

Where,

$$\gamma = \frac{2\pi n_2}{\lambda A_{eff}}$$

Because the XPM term is twice that of a single channel in the first-order design of WDM, only the XPM term is taken into consideration.

Stimulated Raman Scattering

With an very intense incident optical wave, Stimulated Raman Scattering (SRS) can occur in which a forward-propagating wave, referred to as the Stokes wave, grows rapidly in the fiber so that most of the energy of the incident wave is transferred into it. The incident radiation is also referred to as the pump. In a fiber the forward-scattered wave grows as,

$$I_s(L) = I_z(o) \times \exp\left(g_R I_p L_{eff} - \alpha_S L\right)$$

where g_R is the Raman gain coefficient; I_p is the intensity of the incident radiation; $I_S(o)$ is the initial value of the forward-scattered wave; L_{eff} is the effective length; α_S is the fiber loss at the wavelength of the Stokes wave; and L is the fiber length. The Raman gain, g_R, in silica fiber extends over a large frequency range (up to 40 THz) with a broad peak near 13 THz. This behavior is due to the non-crystalline nature of silica glass. In amorphous materials such as fused silica, molecular vibrational frequencies spread out into bands that overlap and become a continuum, unlike most materials where Raman gain occurs at specific frequencies. The Raman gain for silica is measured with a pump at 1.0 μm.

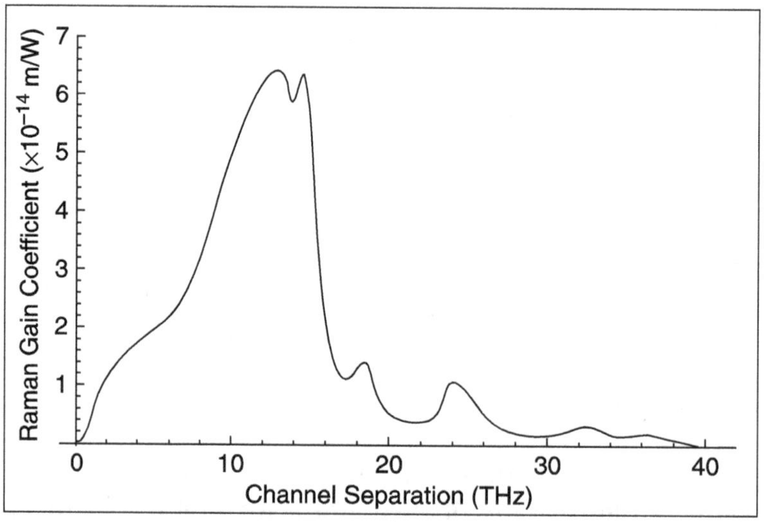

Raman gain of silica fiber.

The impact of the SRS on the performance of a single-channel optical system is negligible because the threshold power required is,

$$P_{th}^{Raman} \approx 16 \frac{A_{eff}}{g_R L_{eff}}$$

and for a fiber with $A_{eff} = 50\ \mu m^2$, $g_R = 6 \times 10^{-14}$ m/W and with $L_{eff} = 20$ km, the threshold is reached for 0.666 W.

Considering that optical systems are low-power systems in terms of optical power, single-channel systems are not affected by SBS. With optical amplifiers and single-channel systems one has to use equation 5.50 for L_{eff}. The threshold given by equation 5.66 is defined as the threshold at which the power of the forward-scattered wave, the Stokes wave, equals the incident radiation (pump).

The Raman gain coefficient at 1550 nm is approximately 6×10^{-14} m/W. This is of course much smaller than the gain coefficient of SBS, but the fundamental difference is in the bandwidth of the two processes. From above figure the gain extends up to 15 THz (125 nms), and two waves 125 nm apart will still be coupled. Although SRS is of no particular importance in the performance of single-channel systems, in WDM systems it becomes very important.

The reason for this is the wide bandwidth of the SRS. Let us consider a WDM system and one wavelength, say λ_o, that is at 10 THz in the Raman gain profile. All the channels with shorter wavelengths act as pumps and amplify the wavelength at λ_o, losing photons (power) in the process. The SRS process can be used for optical amplification, and the amplifiers using it are referred to as Raman amplifiers. In a WDM system the SRS can lead to performance impairments if care is not taken. Figure illustrates the effect of SRS in a four-channel WDM system.

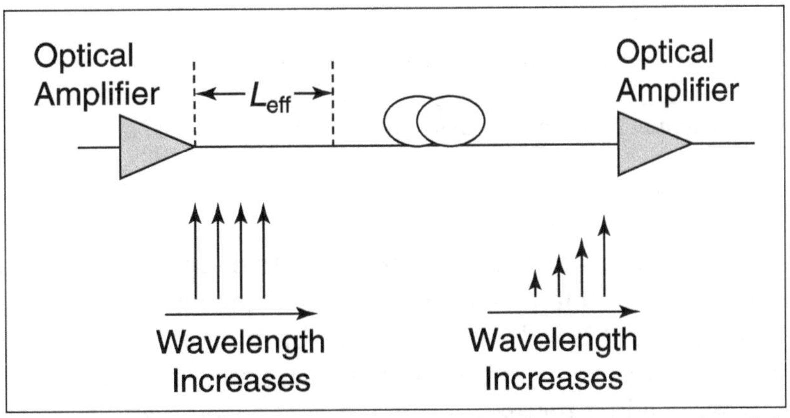

Impacts of SRS in a four-channel WDM system. The channel of the shortest wavelength is most depleted.

It is clear that SRS cannot be avoided in WDM systems, and the way to handle it is to accept impairment in system performance and design with this as a constraint. If this approach is acceptable, then the loss of power to the channel of the shortest wavelength is the constraint. After some assumptions, such as that the channel spacing is equal, the power in each channel is the same, and all channels fall within the Raman gain bandwidth, the power loss by the channel with the shortest wavelength, say channel 0, is given by,

$$P_o = \frac{[(N-1)\Delta f][NP_{ch}]}{6\times10^4} \times \frac{g_{Rpeak} \times L_{eff}}{A_{eff}} = \frac{B_{total} \times P_{total}}{6\times10^4} \times \frac{g_{Rpeak} \times L_{eff}}{A_{eff}}$$

The penalty for channel 0 is now given by,

$$\text{Penalty}\big|_o = -10\text{Log}[1-P_o]$$

If a penalty equal to or less than 0.5 dB is desired, then $P_0 \leq 0.1$ and after some changes in the units of parameters,

$$P_{total} \times B_{total} \times L_{eff} < 40 \times 10^3 \quad mW \times nm \times km$$

This equation is conservative because it ignores chromatic dispersion; with chromatic dispersion in the system the right side of above equation is nearly doubled. Using a channel spacing of 0.8 nm (10 GHz), an amplifier spacing of 80 km, and an L_{eff} of 20 km, the limitations imposed on channel power by SRS. It is clear that for a small number of channels SRS does not impose a strong upper bound on power per channel, but for a large number of channels care should be taken to control the impact of SRS.

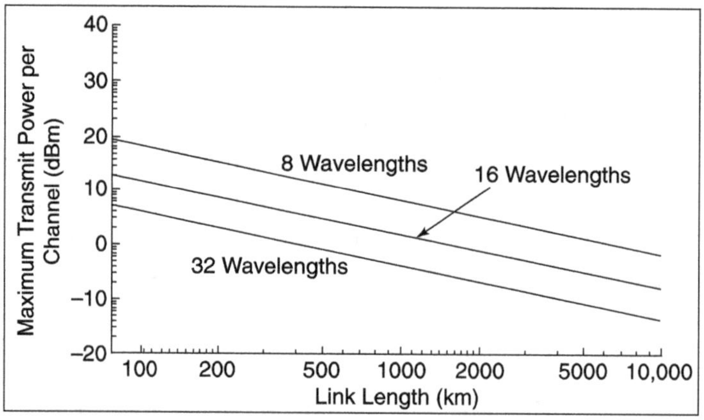

Maximum power limitations imposed by SRS in a WDM system.

Four-wave Mixing

The origin of four-wave mixing is the dependency of refractive index on optical power. This dependency not only gives rise to SPM but also induces new frequencies to appear. This effect is called Four-Wave Mixing (FWM) and it is independent of the bitrate but dependent on channel spacing and the chromatic dispersion present in the propagation path. It is a complex effect to analyze, but FWM efficiency is reduced by increasing the channel spacing and the chromatic dispersion between channels. New frequencies are generated by the process,

$$\omega_{ijk} = \omega_i + \omega_j - \omega_k \text{ with } i, j \neq k$$

where the indices i, j, and k refer to the frequencies present. Assuming that N frequencies are present, the number of new frequencies is given by,

$$\text{Number of new frequencies} = N(N-1)^2$$

As an example, consider a three-channel system. The number of new frequencies expected to be induced by the process is 12. The spectrum of the 12 new frequencies is shown in Figure. It clear that the penalty of FWM is twofold: First, there is loss of power to the new frequencies generated but falling outside the bandwidth of the original three-channel system; second, there is interference as some of the new frequencies coincide with the original frequencies. The assessment of the performance is complex, and we can only illustrate the impact of performance for a few cases.

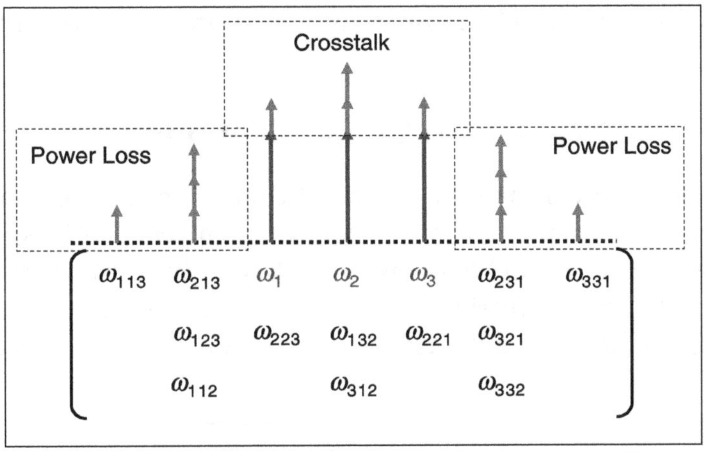

Spectrum generated from a FWM process for a three-channel system.

It should be clear that the impact increases as the dispersion between channels decreases (for conventional single-mode fiber and dispersion-shifted fiber) and the channel spacing is reduced. One might wonder what fiber parameters one should expect from a fiber suitable for long-haul multichannel operation. The key to answering this question is equation $P_o = \frac{[(N-1)\Delta f][NP_{ch}]}{6\times10^4}\times\frac{g_{Rpeak}\times L_{eff}}{A_{eff}} = \frac{B_{total}\times P_{total}}{6\times10^4}\times\frac{g_{Rpeak}\times L_{eff}}{A_{eff}}$. From that equation it is clear that the propensity of the fiber to induce nonlinear effects depends on the effective area. For this reason fiber manufacturers have introduced fibers with larger effective areas.

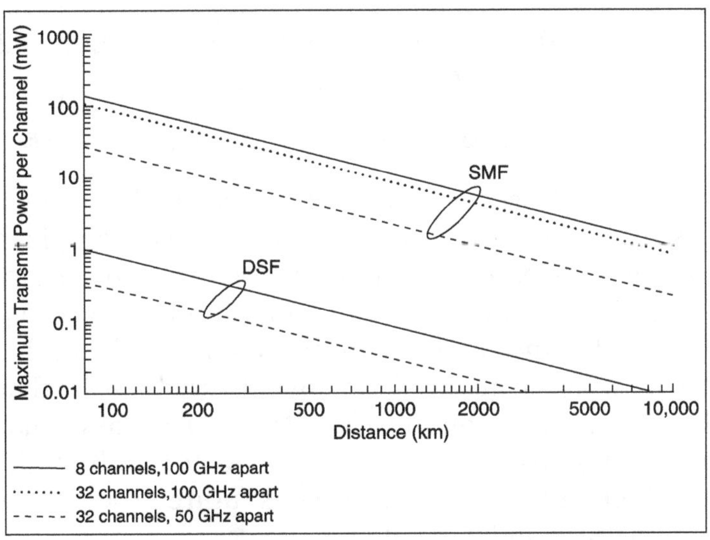

Performance limitation imposed by the FWM process for two types of fiber and two classes of WDM systems.

For example, the Large Effective Area Fiber (LEAF) manufactured by Corning, Inc. has an area of 72 μm²; the True wave XL Ocean fiber manufactured by OFS also offers the same effective area. The standard SMF has an effective area around 50 μm². The Corning Vascade L1000 fiber offers an effective area of 101 μm². A fiber design using a novel approach claims an effective area of 1470 μm². System design with optical nonlinearities present is a complex undertaking and after some preliminary calculations one has to use a simulation package that solves the nonlinear propagation equation (nonlinear Schrödinger equation) under the conditions imposed by the system details.

Future Evolution of Optical Fiber Communications

Silica glass proved to be the ideal material in its optical and mechanical properties and also for manufacturing cabling and reliability. The losses have been reduced to such an extent that any further progress in that direction will be met with the law of diminishing returns. There has been work on new glass materials whose intrinsic theoretical losses are less than that of silica. They are collectively referred to as telluride and fluoride glasses. Figure shows the theoretical loss spectrum of these glasses.

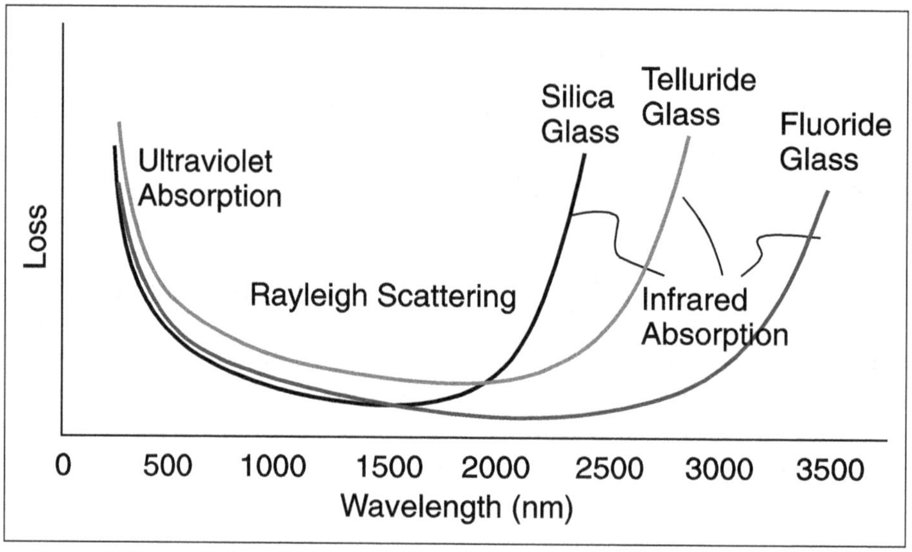

Theoretical performance of some potential new fiber materials.

The interesting aspect of these losses is that, at the wavelengths where optical fiber systems operate now, silica glass is still as good as or better than these new glasses. Therefore, to exploit the theoretical low fluoride losses, the optical window has to shift to longer wavelengths. Here semiconductor optical devices will have difficulties delivering the performance one has come to expect in the C- and L-bands and around 1300 nm. In spite of the theoretical low losses, experimental work with fluoride material has not delivered results that will give confidence for the future. Therefore, from the material perspective, silica glass will continue to dominate the telecommunications market as far as one can see. As for multimode and single-mode fibers, the trend is to use single-mode fiber because it is future proof for bandwidth and its cost is low because it is manufactured in large volume.

However, multimode fiber will keep a part of the market because of its special properties. There was a time when multimode fibers were considered obsolete on the grounds that operation in the fundamental waveguide mode provides better dispersion performance. However, it subsequently became apparent that there will be a huge market for optical LANs, data communications networks, and access networks. Multimode fibers offer greater ease of handling and splicing and therefore their role is being reevaluated, especially for cost-sensitive, short-span applications.

Other Types of Optical Fibers

Plastic Fibers

Plastic fibers are manufactured from a polymer perform drawing into a fiber. The losses associated with this type of fiber are about 100s of dB. The operate at low temperature i.e. upto 125 degree celcius while glass fibers can be used upto 1000 degree celcius. These are used in sensors, process control and short distance communication Features:

- High light gathering capacity.
- Large core area.
- Low cost components.
- Uses visible LEDs.
- Easy to connect or couple.

Table: Various plastic fibers with specification commercially available.

Type	Bandwidth	NA	Operating temperature range in degree celcius
SK or SH	50 MHz(50 m length)	0.50	-40 to +75
EK or EH	120MHz(50 m length)	0.47	-40 to 85
DK or DH	50MHz(50 m length)	0.54	-40 to 115
FH	100MHz(20 m length)	0.75	-40 to 125

Plastic fibers are not available for use at long wavelength, because fibers of that type are very difficult to fabricate and also very expensive. A fiber with glass core and plastic cladding is called as "plastic clad silica" or PCS fiber. They have following characteristics:

- High NA
- Large core diameter
- High attenuation
- Low bandwidth

Advantage of large core is the greater power coupling. The high value of NA permits use of less expensive surface emitting LED's. Along with high attenuation and low bandwidth plastic fibers have poor mechanical strength and low maximum operating temperature.

Comparison of Optical Fibers

- Bandwidth: A good quality step index fiber may have a bandwidth of 50MHz km whereas equivalent GRIN fiber can have 200,400 or 600 MHz km bandwidth. Theoretically, the correction gradation in the GRIN fiber will produce nearly an infinite bandwidth.

- Attenuation: GRIN fiber tends to be generally lower attenuation than the step index fiber. There is no fundamental reason why a step index fiber should have higher attenuation than the GRIN fiber. At least theoretically they are made of the exact same kind of glass, with same amount of energy absorption. But as a practical matter, in step index fiber there may be some irregularities at the interface between the core and the cladding which in turn will contribute to mode conversion and some reflection-loss. But in GRIN fiber there is no such interface. Hence the GRIN fiber has no real counterpart for this loss mechanism in the step index fiber. Certain amount of loss from microscopic bubbles and actual attenuation due to dielectric hysteresis etc. are present in both the fibers. Attenuation may also be caused by impurity absorption. But the prime culprit is the interface dissipation of the step index fiber.
- Mode Dispersion: There is an inherent advantage in mode dispersion for the graded index fiber compared to the step index fiber. For if a wave front had originally crossed the fiber axis with smaller inclinations it would have spent more time wading through glass with higher index and would have travelled more slowly. The rays having wide swinging actually travel farther but they spend more of their time travelling at a faster rate.
- Numerical aperture: It has been observed that for a given fiber diameter the numerical aperture of GRIN fiber is generally smaller than the one on a step index fiber. For a GRIN fiber with an attenuation between 5 and 10dB/km the NA will tend to run between 0.16 and 0.2 whereas a step index fiber of about the same physical size with a loss of the order of 12dB/km will tend to have NA if the order of 0.2 to 0.35.

Table: Comparison of single mode and multimode fibers.

Single mode fiber	Multimode fiber
Only a single mode propagates through fiber.	Many modes propagate through fiber.
Diameter is much less than multimode fiber < 10 μm.	Diameter much larger than single mode fiber.
Largest transmission bandwidth.	Transmission bandwidth is small compared to single mode fiber.
Exhibits low loss.	Comparatively more loss.
Superior transmission quality due to absence of modal noise.	Lesser transmission quality than single mode fiber.

Table: Comparison of step index and graded index optical fibers.

Step index optical fiber	Graded index optical fiber
Bandwidth: 50MHz.	Bandwidth: 200,600 MHz.
Modal dispersion is more.	Modal dispersion is lower.
NA: 0.2 to 0.5 (for 12dB/km loss).	NA: 0.16 to 0.2(for 5 to 12dB/km loss).

References

- Optical-fiber-working-and-its-applications: elprocus.com, Retrieved 13, July 2020
- Numerical-aperture: rp-photonics.com, Retrieved 08, April 2020
- The-Cable-and-Telecommunications-Handbook: ttgtmedia.com, Retrieved 10, August 2020

3

Fiber Optic Link Design

Fiber optic link refers to a part of the optical fiber communication systems that provides a point-to-point data connection. There are some significant aspects of fiber optic link design such as design criteria, design of point-to-point optical link, factors for evaluation of fiber optic link, calculations for fiber optic link, etc. These topics are crucial for a complete understanding of the subject.

Design Criteria

The design criteria for a fiber optic link design procedure is mainly divided into broad categories which can be further subdivided as shown by the tree diagram below:

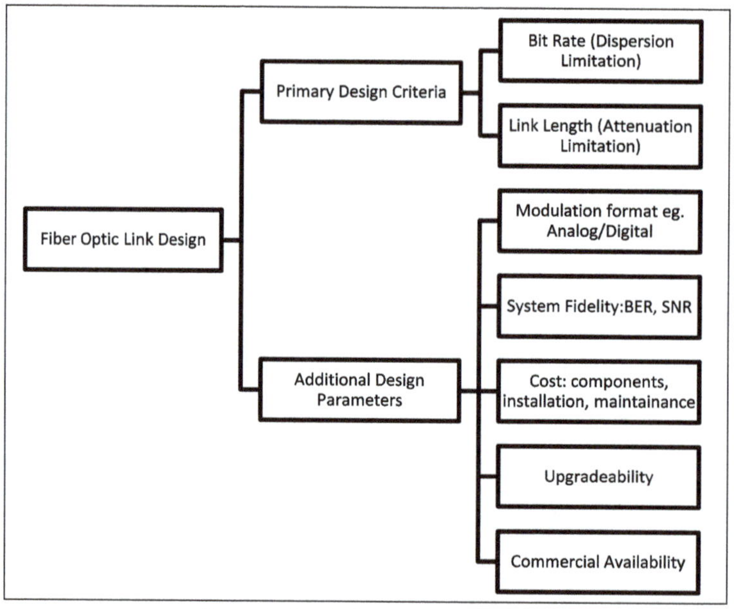

Fiber Optic Link Design Criteria.

The primary design criteria signify the most basic and fundamental information parameters to be made available by the user to the designer for designing a reliable fiber optic link. The first important information to be specified by the user is the desired bit rate of data transmission. However, the dispersion in the optical fiber exerts a limitation on the maximum achievable and realisable data rate of transmission. The next intricate information to be provided for the design process is the length of the optical link so as to enable the designer to ascertain the position of the optical repeaters along the link for a satisfactory optical data link.

Along with the primary design criteria, there are some additional parameters which facilitate better design and quality analysis of the optical link. These factors consist of the scheme of modulation, the system fidelity, cost, upgradeability, commercial availability etc. A fundamental and very simple point-to-point optical communication link can be schematically drawn as shown in the figure.

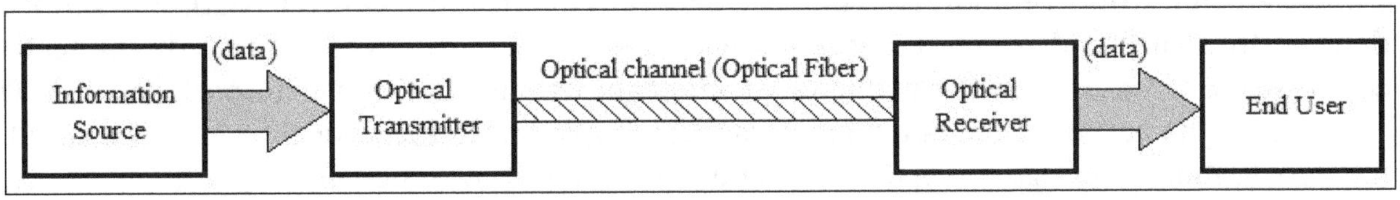

Basic Optical communication link.

The information source provides the data (analog/digital) at the modulating frequency which are to be transmitted to the end user. These data bits are then impressed on a carrier signal and then coupled onto the optical channel (optical fiber cable) for transmission. The optical fiber carries this signal and the receiver receives the transmitted signal at the output of the optical fiber. This received signal is then processed appropriately to retrieve the data in the signal and this data is then communicated to the intended user.

Although the above block diagram looks simple and practically inadequate for a reliable communication link, yet it provides the basic structure for designing of advanced optical communication links with diverse capabilities and better performance characteristics. In practical optical communication links, number of different factors need to be ascertained and monitored both online and offline to ensure satisfactory and reliable optical link. The design parameters mentioned above encompass only the three main modules of the optical communication link namely the optical transmitter, optical fiber and the optical receiver. A more appropriate model along with different important performance parameters and factors is shown in the figure below:

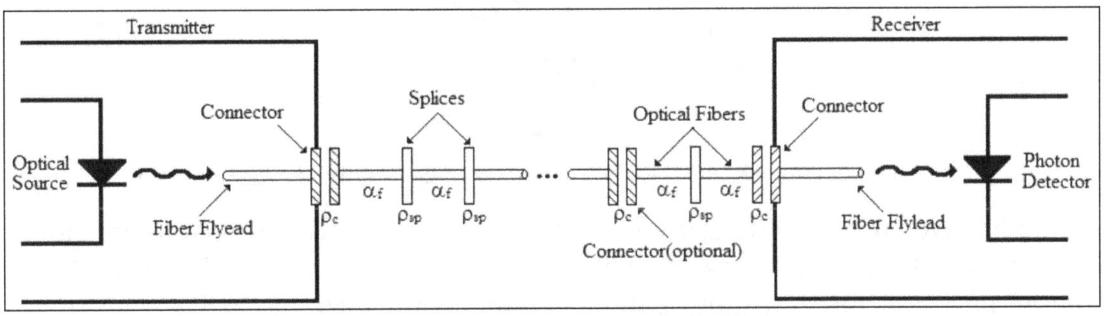

Practical Representation of an optical link.

The optical transmitter module consists of an optical source such as LASER diode, LED etc. and the light from this source is coupled onto the optical fiber via suitable connector as shown in the above figure. The quantities indicated below certain components in the above figure signify the amount of optical energy degradation caused in the corresponding component (per metre) when light energy propagates through it. The coupled light in the optical fiber then travels along the fiber to reach the receiving end. However, in practical links, the distance between the transmitter and the receiver modules may range from a few metres to tens of kilometres.

To realise an optical link over such long distances, optical fibers of such lengths are not commercially available. Generally optical fibers come in spools which may contain optical fibers up to lengths of about 3-5km and so to realise long distance links, often, spools of optical fibers have to be cascaded and fiber ends have to be joined together. This joint may either be permanent or temporary depending on the requirement. Permanent joints are done by a mechanism called splicing and the joint produced as a result of splicing is called as a splice. Temporary joints are generally done via appropriate connectors. Temporary joints are generally preferred wherever routine monitoring of fiber performance and measurements are necessary so that the fibers can be disconnected/ re-connected with ease without causing any significant change to the system performance. The light energy at the output of the optical fiber is then coupled onto the optical receiver via suitable connectors to minimize losses in optical power and to reduce power penalty.

The diagram below shows the performance of different optical sources and optical receivers with respect to the desired bit rate of data transfer which helps the designer to choose the appropriate optical source and the appropriate optical receiver which would successfully serve the desired application.

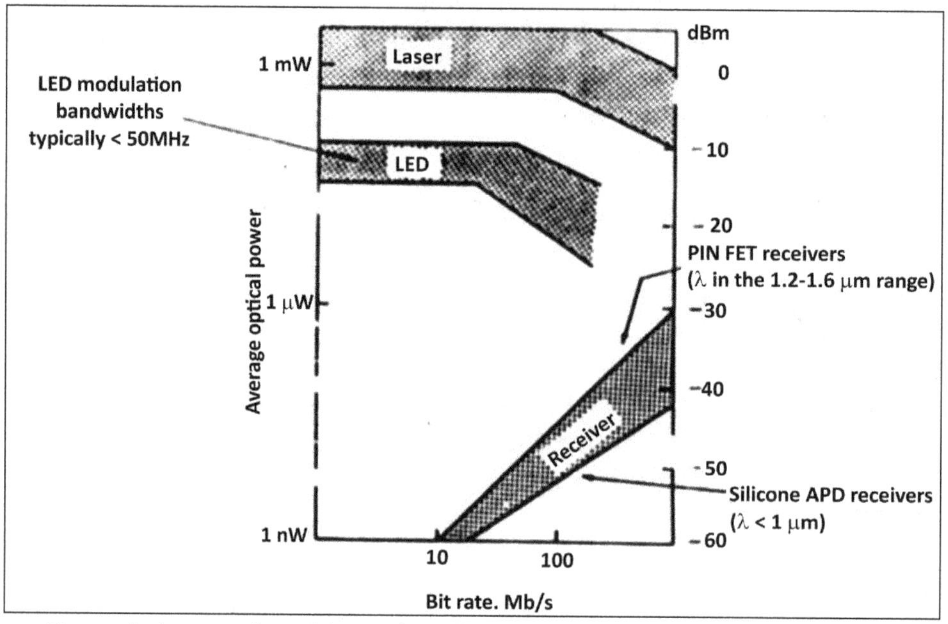

Transmission margin vs. bit rate for a digital fiber-optic communication system.

LEDs generally have an optical output power of about a few micro-watts to a few hundred micro-watts at modulation bandwidths of about 50MHz. However, the LASER diode has a much higher output power at the same input power than the LED which may range from 1mW to tens of miliwatts. One important observation that can be made from the above figure is the decrease in the optical output power from both the type of sources at higher data rates. This observation can be attributed to the fact that both the types of optical sources are based on p-n semiconductor junctions which, in principle, act as low pass filters and therefore the output power reduces at the high frequencies (or higher data rates).

The above figure also shows the performance of an optical receiver with respect to increasing data rate of transmission. It is clearly seen from the above figure that as the data rate of transmission increases, the average power required to be detected for an acceptable BER, also increases. This fact was also visible in our discussion of the average power required per bit and the discussions on power penalty that in the presence of noise, as the rate of bit transfer increases, the minimum detectable power required also increases (this increase is almost as the square root of the bandwidth as seen in the thermal noise dominated receiver).

In optical link design problems, the power of the optical source and the optical receiver are generally expressed in dBm. The dB equivalent power of 1mW power is taken as the reference i.e. 0 dBm and the increasing powers are expressed in their equivalent dBm values by normalizing them with respect to 1mW and for every 10 fold increase in the actual power, increases the dBm equivalent power by 10. That is:

$$1mW = 0dBm;$$

$$10mW = 10dBm;$$

$$100mW = 20dBm; \text{ and so on}.$$

For typical optical LASER type of sources the output power normally ranges between 3-5 dBm. And a typical optical receiver requires -30 to -40 dBm of detectable power for a BER of about 10^{-9}. The following figure shows a plot of the sensitivity of different materials that qualify to be used as material for construction of an optical receiver.

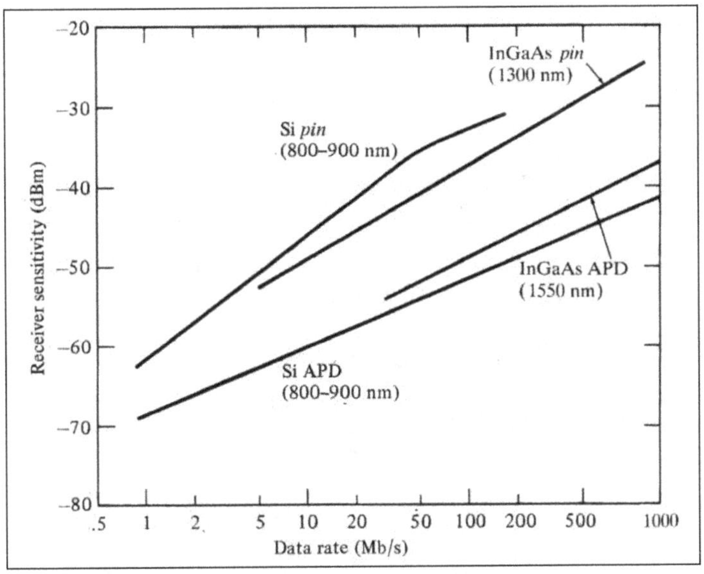

Receiver Sensitivities vs. Data rate.

The above figure enables a designer to choose the appropriate material based on the desired wavelength of operation and the data rate specified by the user and also helps to calculate the minimum required detectable power for the chosen material so that the system achieves the desired BER performance. The plots are for the variation in the caused in the receiver sensitivity as a result of the variation in the data rate at a particular wavelength. The choice of the wavelength depends upon the characteristic properties of the material.

After having chosen an appropriate type of source which emits a known amount of optical power and an equally suitable optical receiver material with known sensitivity at the desired data rate of operation, the next step in the design process is to estimate the maximum possible allowable losses and other performance parameters in the system during operation so that the designed system renders the desired SNR or BER. Such estimations are known as power budget calculations. Hence the two known quantities are:

- P_s = Optical output power from optical source (Transmitter) in dBm.

- P_r = Sensitivity of optical detector in dBm for given BER.

Having fixed the above two parameters, the maximum possible allowable loss that can occur is the difference between the transmitter and the received powers. The loss occurs in the different components connected in the system such as the connectors, splices, the optical fiber and also in the system itself which is known as the system margin. Generally a system margin of about 6dB is pre-set in practical systems. The total loss is, hence, the sum total of all the losses occurring in each of these components (calculated per unit length).

$$\text{Maximum allowable loss, } \alpha_{max}(\text{in dB}) = P_T - P_r$$

$$\alpha_{max} = \alpha_{fiber} + \alpha_{connectors} + \alpha_{splices} + \alpha_{system}$$

$$\Rightarrow \alpha_{fiber} = \alpha_{max} - (\alpha_{connectors} + \alpha_{splices} + \alpha_{system})$$

The maximum possible length (L_{Pmax}) of the optical fiber that can be used in the above design, without affecting the system BER, can then be determined as:

$$L_{Pmax} = \frac{\alpha_{fiber}}{\text{Loss / Km}}$$

The length of the fiber determined from the above expression is termed as the power budget limited link length of optical fiber. If an optical fiber of length beyond the power budget limited link length is used, the system BER deteriorates and the system performance degrades. So the power budget calculations are a first priority in optical link design.

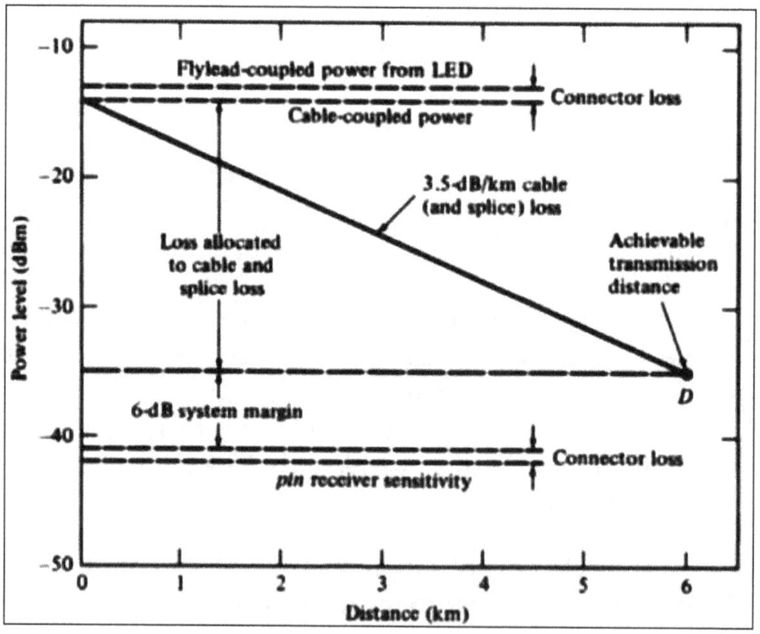

Power budget calculations.

The next step in the design of an optical link is the rise time budget calculations. Rise time of a system is the time taken by the system to attain 90% of the steady state response of the system (from the initial state) to particular input. The rise time of a system varies inversely with its bandwidth; a system with infinite bandwidth has zero rise time i.e. it can instantaneously respond to any instantaneous input. However, practical systems, including optical systems, have finite bandwidth and so have non-zero rise time value. That is, there is always a finite amount of time, however small, that the system requires, responding to a particular input or variations in the input. The rise time of the system determines the speed of the device and also indicates the maximum possible frequency of variations in the input signal that the system can be subjected to, so that it produces a reliable output.

Since, the optical transmitter, optical receiver and the optical channel may be assumed to be individual systems, they have individual rise times which may be different and there is always a need to synchronize the three so that the system as a whole operates on the desired data rate of transmission with best possible performance of the system as a whole. However, the aim of this section is to estimate the maximum allowable rise time of the system as whole to realise the required data rate. Such estimation is known as rise-time budget calculations. Rise time analysis gives the effective bandwidth of the optical link. The system rise time (t_{sys}) is calculated as a root mean square value of the transmitter rise time (t_{tx}), the receiver rise time (t_{rx}) and the rise time associated with the optical channel dispersion ($D_{\sigma\lambda}L$). 'D' is the dispersion parameter, 'σ_λ' is the spectral width of the optical source and 'L' is the length of the optical link. That is:

$$t_{sys} = \left\{t_{tx}^2 + D^2\sigma_\lambda^2 + t_{rx}^2\right\}^{\frac{1}{2}}$$

For a satisfactory operation of the optical link, the system rise time should be less than or equal to 70% of the bit duration (T_b) at the specified data rate. That is:

$$t_{sys} \leq 0.7 T_b$$

The above relationship is true only for NRZ kind of data bits. For RZ kind of data bits, the system rise time must be less than or equal to $0.35T_b$. If we now substitute equation $t_{sys} \leq 0.7 T_b$ into $t_{sys} = \left\{t_{tx}^2 + D^2\sigma_\lambda^2 + t_{rx}^2\right\}^{\frac{1}{2}}$ we can rearrange the resultant equation to find an expression for the maximum possible length of the optical fiber link under the limitation of dispersion. This length is known as the rise time budget limited link length (L_{RTmax}). That is, beyond this distance the signal distortion produced in the optical link becomes unacceptable at the given BER.

$$L_{RTmax} = \frac{1}{D\sigma_\lambda}\left\{(0.7T_b)^2 - \left(t_{tx}^2 + t_{rx}^2\right)\right\}^{\frac{1}{2}}$$

Once again, the above relation assumes NRZ kind of data format; for RZ kind of data, 0.7 has to be replaced with 0.35. Equations $L_{Pmax} = \frac{\alpha_{fiber}}{Loss / Km}$ and above equation provide us with two different maximum lengths for the same optical link. One is power budget limited and the other is rise time budget limited. The power budget limited length signifies the distance at which the SNR of the signal becomes almost unacceptable. N the other hand the rise time budget limited length signifies the maximum distance upto which the distortion in the signal remains manageable. In both the cases, at the calculated length a suitable signal processing is extremely important so as to achieve the required BER. Hence, the minimum of the two lengths is considered as the position of the repeater of the optical link and these repeaters are installed at the length intervals of the minimum length. Generally, the power budget limited length is smaller than the rise time budget limited and length and so the power budget limited length determines the position of repeaters.

An optical repeater is basically an optical transmitter and an optical receiver put back-to-back with some additional signal processing circuitry.

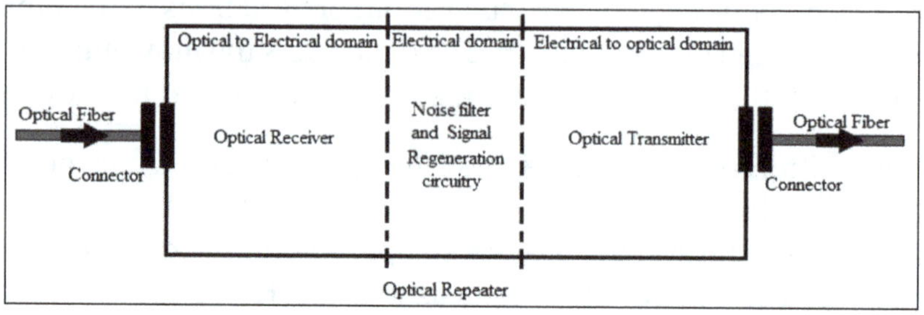

Optical Receiver.

An optical repeater is a costly module as it incorporates both- an Optical Receiver module and an Optical Transmitter module and also some additional signal processing circuitry. That is why installation of optical repeaters at regular intervals of power budget limited length, is rather an expensive option. However, due to non-availability of other alternatives up to a decade ago, older optical links installed optical repeaters at regular length intervals of the optical link to ensure reliable quality of performance of the system. Careful thought shows that an optical repeater is necessary only where the signal distortion of the transmitted signal becomes almost unmanageable. If the signal distortion in the optical fiber is manageable and only the SNR of the transmitted signal deteriorates, there is no need of using a repeater circuit merely to improve the SNR. Transmission distance Vs. Data rate plots are shown in figure below.

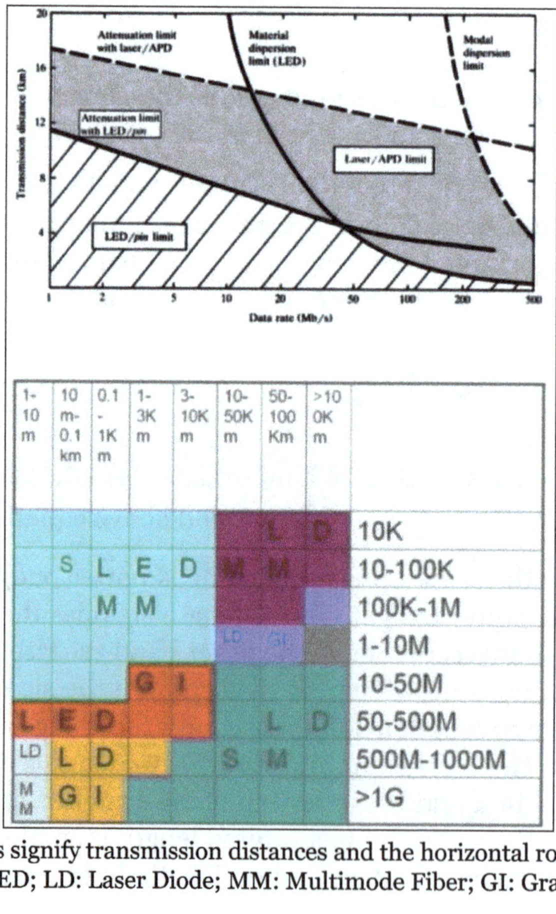

The vertical columns signify transmission distances and the horizontal rows signify data rates [SLED: Surface LED; LD: Laser Diode; MM: Multimode Fiber; GI: Graded Index Fiber].

An amplifier installed at the location would serve the job. Only when the signal distortion becomes almost unmanageable, a repeater may be used to rectify the anomalies. Figure above shows two graphical representations of the preferable and economic optical source-receiver combinations with respect to the data rate and the link length. The second figure serves as a rather more comprehensive figure to determine the best source-receiver combination based on the data rate of transmission and the distance between the transmitter and receiver (link length). These figures enable us to decide the appropriate source receiver combination which best suits the intended application for which the optical link is designed.

Design of Point to Point Optical Link

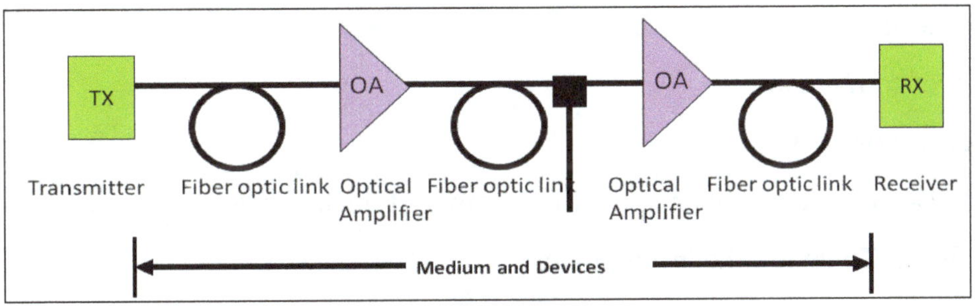

Simple point to point fiber optic link.

The link has primarily three components to design: Optical Transmitter, Optical Fiber and Optical receiver. The typical optical power which LEDs and Lasers can deliver and the photo-detector needs for a BER of 10^{-9}.

- Optical Transmitter: Optical transmitter includes optical sources like LEDs or LASERs. Most Important parameters associated with these sources are as follows:
  - Operating wavelength (λ): This should be suitable for transmission through fiber,
  - Linewidth (Δλ),
  - Rise time,
  - Bit-rate,
  - Line format,
  - Power level.
- Optical Fiber: Communication link is the optical fiber having following parameters:
  - Type of fiber: Single mode optical fiber or mulitimode optical fiber.
  - Fiber type – SMF28, DSF, etc.
  - Cable loss: It is form of propagation loss, bending loss, dispersion etc.
  - Spool length: Length of the spool decides the losses involved in the link. If the spool length sufficient then coupling loss can be reduced.

- Optical Receiver: It includes photo detector device such as photodiode, phototransistor etc. One has to look for most important parameters associated with these devices. These are as follows:
  - Sensitivity: P_{SEN}
  - Saturation: P_{SAT}
  - Rise time
- Link Connection: Optical fiber link is characterized by number of losses specifically splices. Splice is welding or connection of two optical fibers. These losses are dependent on following factors such as:
  - No. of splice
  - Splice loss
  - No. of connectors
  - Connector Loss
- In Line Devices: In the fiber optic link number of devices are required such as:
  - Splitter: To divide the input power.
  - Filter: To remove certain signals.
  - Attenuator: To reduce the strength of signals.
  - Amplifier: To increase the strength of the signals.
  - Insertion loss: Characteristic feature.
  - Gain: Decided by amplifier design.

As the data rate increases the power delivering capacity of the source reduces and at the same time the power requirement of the detector increases. Table gives the combination of the sources and fibers for different link capacity and distance.

Table: Sources and fibers for different link capacity and distance.

1-10m	10m-.1km	.1-1km	1-3km	3-10km	10-50km	50-100km	>100km	
						LD		10k
	SLED					MM		10-100K
		MM						100K-1M
					LD	GI		1-10M
			LED					10-50M
			GI			LD		50-500M
LD		LD				SM		500M-1G
MM		GI						>1G

Considering the cost, speed etc., first choose the laser and the detector. Also the type of fiber is chosen from the above table. Generally a multi-core fiber is laid even if the immediate requirement is only one or two fibers. The link design then reduces to finding locations of the repeater on a long link.

Factors for Evaluation of Fiber Optic Link

To achieve high-quality transmission, careful decisions based on operating parameters apply for each component of a fiber optic transmission system. The main questions, given in table, involve data rates and bit error rates in digital systems, bandwidth, linearity, and signal-to-noise ratios in analog systems, and in all systems, transmission distances. These questions of how far, how good, and how fast define the basic application constraints. Once these are decided, it is time to evaluate the other factors involved.

Table: Factors for Evaluating Fiber Optic System Design.

System Factor	Consideration/Choices
Type of Fiber	Single-mode or Multimode.
Dispersion	Regenerators or Dispersion Compensation.
Fiber Nonlinearities	Fiber Characteristics, Wavelengths, and Transmitter Power.
Operating Wavelength	780, 850, 1310, 1550 nm, and 1625 nm typical.
Transmitter Power	Typically expressed in dBm.
Source Type	LED or Laser.
Receiver Sensitivity/Overload Characteristics	Typically expressed in dBm.
Detector Type	PIN Diode, APD, or IDP.
Modulation Code	AM, FM, PCM, or Digital.
Bit Error Rate (BER) (Digital Systems Only)	10^{-9}, 10^{-12} Typical.
Signal-to-Noise Ratio	Specified in decibels (dB).
Number of Connectors	Loss increases with the number of connectors.
Number of Splices	Loss increases with the number of splices.
Environmental Requirements	Humidity, Temperature, Exposure to Sunlight.
Mechanical Requirements	Flammability, Indoor/Outdoor Application.

Many of these considerations are directly related to other considerations. For example, detector choice will impact the receiver sensitivity which will affect the necessary transmitter output power. Output power impacts the transmitter light emitter type which will affect the usable fiber type and connector type. A logical way to proceed with designing a fiber link involves analyzing the fiber optic link power budget or optical link loss budget.

Calculations for Fiber Optic Link

Power Budgets and Loss Budgets

The power budget refers to the amount of fiber optic cable plant loss that a data link (transmitter to receiver) can tolerate in order to operate properly. Sometimes the power budget has both a minimum and maximum value, which means it, needs at least a minimum value of loss so that it does not overload the receiver and a maximum value of loss to ensure the receiver has sufficient signal to operate properly.

The loss budget is the amount of loss that a cable plant should have if it is installed properly. It is calculated by adding the estimated average losses of all the components used in the cable plant to get the estimated total end-to-end loss. The loss budget has two uses, 1) during the design stage it is used to ensure the cabling being designed will work with the links intended to be used over it and 2) after installation, the loss budget for the cabling is compared to the actual test results to ensure the cable plant is installed properly.

Some standards refer to the loss budget as the "attenuation allowance" but there seems to be very limited use of that term. Obviously, the power budget and loss budget are related. A data link will only operate if the cable plant loss is within the power budget of the link. Remember the calculated loss budget is an estimate that assumes the values of component losses and does not take into account the uncertainty of the measurement. Be aware of this because if measurements are close to the loss budget estimates, some judgment is needed to not fail good fibers and pass bad ones."

Power Budget

All datalinks are limited by the power budget of the link. The power budget is the difference between the output power of the transmitter and the input power requirements of the receiver, both of which are defined as power coupled into or out of optical fiber of a type specified by the link. The power budget is not just a straightforward determinant of the maximum loss in the cable plant that the link can tolerate. As shown below, cable plant loss is only a part of the power budget. Distortion impairments, for example from dispersion (modal and chromatic dispersion in MM fiber, chromatic and polarization mode dispersion in SM fiber), reduce the power budget. In multimode gigabit Ethernet networks, for example, transceivers have a dynamic range (transmitter output to receiver sensitivity) of about 5-6 dB before dispersion is factored in, leaving a power budget of about 2 dB.

Noise in transceivers, mainly in the receiver, affect the power budget also. The receiver has an operating range determined by the signal-to-noise ratio (S/N) in the receiver. The S/N ratio is generally quoted for analog links while the bit-error-rate (BER) is used for digital links. BER is practically an inverse function of S/N. Transceivers may also be affected by the distortion of the transmitted signal as it goes down the fiber, a big problem with multimode links at high speeds or very long OSP single-mode links.

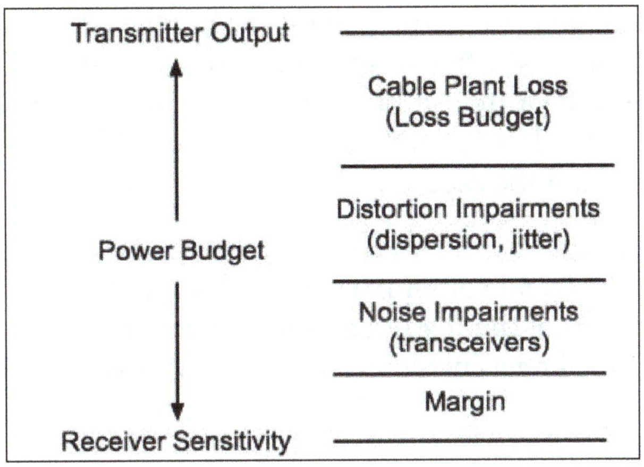

When testing a fiber in a cable plant to determine if the cable plant will allow a specific link to operate over it, the test should be made from transceiver to transceiver, e.g. the cable plant with patchcords installed on either end that would be used to connect the transceivers to the cable plant. When doing a link loss budget (below) for the cabling to be used with a given link to determine if the link will operate over that link, the loss of the patchcords may also be included.

Testing the Power Budget for a Link

How is the power budget determined? You test the link under operating conditions and insert loss while watching the data transmission quality. The test setup is like this:

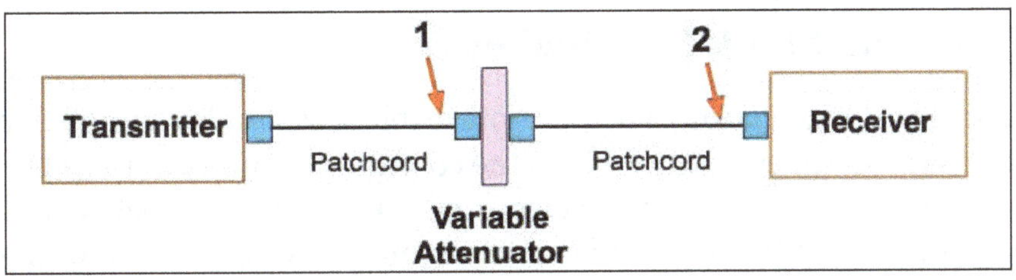

Connect the transmitter and receiver with patchcords to a variable attenuator. Increase attenuation until you see the link has a high bit-error rate (BER for digital links) or poor signal-to-noise ratio (SNR for analog links). By measuring the output of the transmitter patchcord (point 1) and the output of the receiver patchcord (point 2), you can determine the maximum loss of the link and the maximum power the receiver can tolerate. From this test you can generate a graph that looks like this:

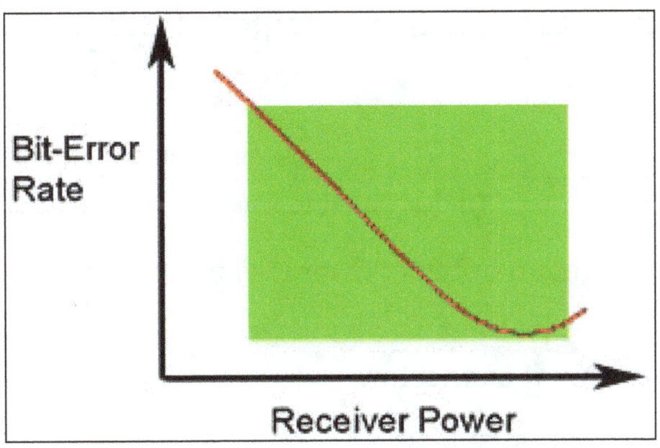

A receiver must have enough power to have a low BER (or high SNR, the inverse of BER) but not so much it overloads and signal distortion affects transmission. We show it as a function of receiver power here but knowing transmitter output, this curve can be translated to loss - you need low enough loss in the cable plant to have good transmission but with low loss the receiver may overload, so you add an attenuator at the receiver to get the loss up to an acceptable level.

You must realize that not all transmitters have the same power output nor do receivers have the same sensitivity, so you test several (often many) to get an idea of the variability of the devices. Depending on the point of view of the manufacturer, you generally error on the conservative side so that your likelihood of providing a customer with a pair of devices that does not work is low. It's easier that way.

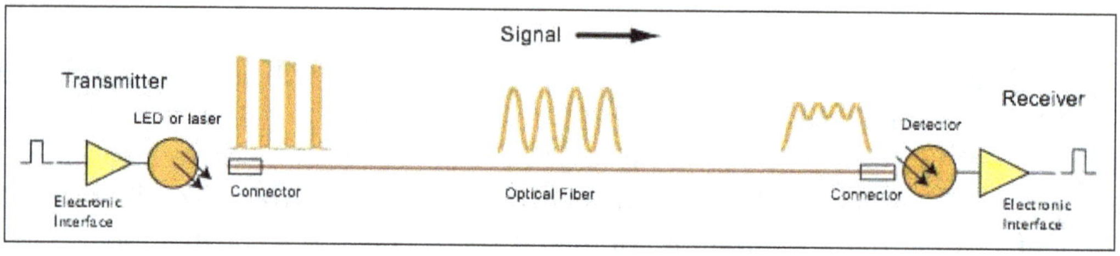

Furthermore, if your link uses multimode fiber at high bit rates (or single-mode on long links at very high bit rates), there will be dispersion. Dispersion spreads out the pulses, causing a power penalty. That's why high speed Ethernet at 10G has a loss budget of 2dB while the power budget calculated from transmitter and receiver specifications is about 6dB.

Calculating Cable Plant Link Loss Budget

Loss budget analysis is the calculation of a fiber optic cabling system's estimated loss performance characteristics. This is sometimes confused with the communication system "power budget" which is a specification of the dynamic range of the electronics, the difference between the output power of the transmitter coupled into the fiber and the minimum received power required at the receiver for proper data transmission. The communications system power budget will set a limit for the loss of the cable plant.

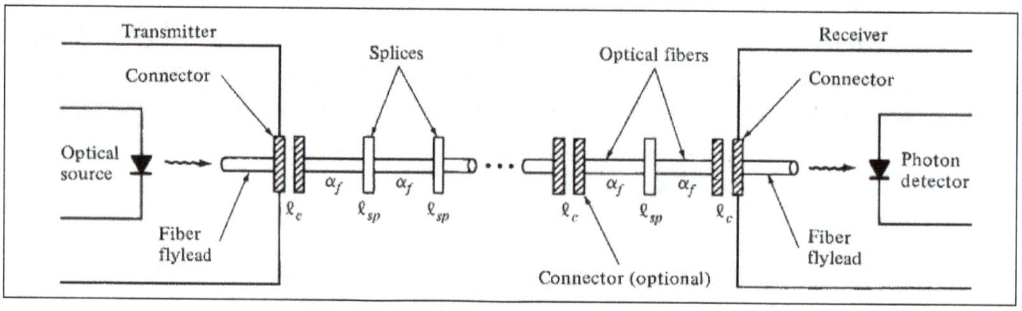

Optical power loss model.

For optimizing link power budget an optical power loss model is to be studied as shown in figure. Let,

- l_c denotes the losses occur at connector.
- L_{sp} denotes the losses occur at splices.

- α_f denotes the losses occur in fiber.
- All the losses from source to detector comprise the total loss (PT) in the system.
- Link power margin considers the losses due to component aging and temperature fluctuations. Usually a link margin of 6-8 dB is considered while estimating link power budget.

  Total optical loss = Connector loss + (Splicing loss + Fiber attenuation) + System margin (Pm) PT = 2lc + αfL + System margin (Pm)

  Where, L is transmission distance.

Example: Design as optical fiber link for transmitting 15 Mb/sec of data for a distance of 4 km with BER of 10^{-9}.

- Solution: Bandwidth x Length = 15 Mb/sec x 4 km = (60 Mb/sec) km.
- Selecting optical source: LED at 820 nm is suitable for short distances. The LED generates – 10 dBm optical powers.
- Selecting optical detector: PIN-FER optical detector is reliable and has – 50 dBm sensitivity.
- Selection optical fiber: Step-index multimode fiber is selected. The fiber has bandwidth length product of 100 (Mb/s) km.
- Links power budget:

  Assuming,
- Splicing loss l_s = 0.5 dB/slice.
- Connector loss l_c = 1.5 dB.
- System link power margin Pm – 8 dB.
- Fiber attenuation αf = 6 dB/km.
- Actual total loss = (2 x lc) + αfL + Pm.
- PT = (2 x 1.5) + (6 x 4) + 8.
- PT = 35 dB.

Maximum allowable system loss:

- P_{max} = Optical source output power- optical receiver sensitivity P_{max} = -10 dBm – (-50 dBm).
- P_{max} = 40 dBm.

Since actual losses in the system are less than the allowable loss, hence the system is functional.

Example: A transmitter has an output power of 0.1 mW. It is used with a fiber having NA= 0.25, attenuation of 6 dB/km and length 0.5 km. The link contains two connectors of 2 dB average loss.

The receiver has a minimum acceptable power (sensitivity) of – 35 dBm. The designer has allowed a 4 dB margin. Calculate the link power budget.

Solution: Source power Ps = 0.1mW:

- Ps = -10dBm.
- Since NA = 0.25.
- Coupling loss = -10log (NA^2) = -10log (0.25^2) = 12 dB.
- Fiber loss = α_f x L.

  l_f = (6dB/km) (0.5km) l_f = 3 dB.
- Connector loss = 2 (2 dB).

  l_c = 4 dB Design margin P_m = 4 dB.
- Actual output power Pout = Source power – (Σ Losses) P_{out} = 10dBm – [12 dB + 3 + 4 + 4].

  P_{out} = -33 dBm.
- Since receiver sensitivity given is – 35 dBm.

  P_{min} = -35 dBm.

As $P_{out} > P_{min}$, the system will perform adequately over the system operating life.

Example: In a fiber link the laser diode output power is 5 dBm, source-fiber coupling loss= 3 dB, connector loss of 2 dB and has 50 splices of 0.1 dB loss. Fiber attenuation loss for 100 km is 25 dB, compute the loss margin for; i) APD receiver with sensitivity – 40 dBm ii) Hybrid PINFET high impedance receiver with sensitivity -32 dBm.

Solution: Power budget calculations.

Source output power	5 dBm
Source fiber coupling loss	3 dB
Connector loss	2 dB
Connector loss	5 dB
Fiber attenuation	25 dB
Total loss	35 dB

- Available power to receiver: (5 dBm – 35 dBm) – 30 dBm.
- APD receiver sensitivity – 40 dBm.
- Loss margin [- 40 – (- 30)] 10dBm.
- H-PIN FET highoimpedance receiver -32 dBm Loss margin [- 32 – (- 30)] 2 dBm.

Rise Time Budget

Rise time gives important information for initial system design. Rise-time budget analysis determines the dispersion limitation of an optical fiber link. Total rise time of a fiber link is the root-sum-square of rise time of each contributor to the pulse rise time degradation.

$$t_{sys} = \sqrt{t_{r1}^2 + t_{r2}^2 + t_{r3}^2 + \cdots}$$

$$t_{sys} = \left(\sum_{i=1}^{N} t_{ri}^2 \right)^{1/2}$$

The link components must be switched fast enough and the fiber dispersion must be low enough to meet the bandwidth requirements of the application adequate bandwidth for a system can be assured by developing a rise time budget.

As the light sources and detectors has a finite response time to inputs. The device does not turn-on or turn-off instantaneously. Rise time and fall time determines the overall response time and hence the resulting bandwidth.

Connectors, couplers and splices do not affect system speed, they need not be accounted in rise time budget but they appear in the link power budget. Four basic elements that contribute to the rise-time are:

1. Transmitter rise-time (ttx).

2. Group Velocity Dispersion (GVD) rise time (tGVD) Modal dispersion rise time of fiber (tmod).

3. Receiver rise time (trx).

$$t_{sys} = \left[t_{tx}^2 + t_{mod}^2 + t_{GVD}^2 + t_{rx}^2 \right]^{1/2}$$

Where,

- B_{rx} is 3 dB – bW of receiver (MHz).
- Rise time due to modal dispersion is given as,

$$t_{mod} = \frac{440}{B_M} = \frac{440\,Lq}{B_o}$$

where,

- B_M is bandwidth (MHz),
- L is length of fiber (km),
- q Is a parameter ranging between 0.5 and 1.
- B_o is bandwidth of 1 km length fiber,

- Rise time due to group velocity dispersion is,

$$t_{GVD} = D^2\sigma_\lambda^2 L^2$$

Where,

- D is dispersion [ns/(nm.km)]
- Σ_λ is half-power spectral width of source L is length of fiber.
- Receiver front end rise-time in nanoseconds is,

$$t_{rx} = \frac{350}{B_{rx}}$$

Equation $t_{sys} = \left[t_{tx}^2 + t_{mod}^2 + t_{GVD}^2 + t_{rx}^2\right]^{1/2}$ can be written as,

$$t_{sys} = \left[t_{tx}^2 + t_{mod}^2 + t_{GVD}^2 + t_{rx}^2\right]^{1/2}$$

$$t_{sys} = \left[t_{tx}^2 + \left(\frac{440\,Lq}{B_o}\right)^2 + D^2\sigma_\lambda^2 L^2 + \left(\frac{350}{B_{rx}}\right)\right]^{1/2}$$

Example: For a multimode fiber following parameters are recorded.

- LED with drive circuit has rise time of 15 ns.
- LED spectral width = 40 nm.
- Material dispersion related rise time degradation = 21 ns over 6 km link.
- Receiver bandwidth = 235 MHz
- Modal dispersion rise time = 3.9 nsec.
- Calculate system rise time.

Solution: $t_{tx} = 15\,n\sec tT_{mat} = 21\,n\sec t_{mod} = 3.9\,n\sec$

Now, $t_{rx} = \left(\frac{350}{B_{rx}}\right)$

$$t_{rx} = \frac{350}{25}$$

$$t_{rx} = 14\,n\sec$$

Since,

$$t_{sys} = \left(\sum_{i=1}^{N} t_{ri}^2\right)^{1/2}$$

$$t_{sys} = \left[15^2 + 21^2 + 3.9^2 + 14^2\right]^{1/2}$$

$$t_{sys} = 29.61\,n\sec$$

Example: A fiber link has following data:

Component	BW	Rise time (tr)
Transmitter	200MHxz	1.75 nsec
LED (850 nm)	100 MHz	3.50 nsec
Fiber cable	90 MHz	3.89 nsec
PIN detector	350 MHz	1.00 nsec
Receiver	180 MHz	1.94 nsec

Compute the system rise time and bandwidth.

Solution: System rise time is given by,

$$_{sys} \left(\sum_{i\ 1} {}_{ri}\right)^{1/2}$$

$$ct_{sys} = \sqrt{\left(1.75^2 + 3.5^2 + 3.89^2 + 1.00^2 + 1.94^2\right)}$$

$$t_{sys} = 5.93\,n\sec$$

System BW is given by,

$$BW = \frac{0.35}{5.93\,n\sec}$$

$$BW = \frac{0.35}{t_{sys}}$$

Fiber Optic Instrumentation

Testing fiber optic components and cable plants requires making several measurements with the most common measurement parameters listed in the Table below. Optical power, required for measuring source power, receiver power and, when used with a test source, loss or attenuation, is the most important parameter and is required for almost every fiber optic test. Backscatter and wavelength measurements are the next most important and bandwidth or dispersion are of lesser importance. Measurement or inspection of geometrical parameters of fiber is essential for fiber manufacturers. And troubleshooting installed cables and networks is required.

Fiber Optic Testing Requirements

Test Parameter	Instrument
Optical Power (Source Output, Receiver Signal Level)	Fiber Optic Power Meter
Attenuation or Loss of Fibers, Cables & Connectors (Insertion Loss)	FO Power Meter & Source or OLTS (optical loss test set)
Source Wavelength, Spectral Width	FO Spectrum Analyzer
Backscatter For Loss, Length and Fault Location)	Optical Time Domain Reflectometer(OTDR)
Fault Location	OTDR Visual Cable Fault Locator
Bandwidth / Dispersion (MM:Modal & Chromatic, SM: Chromatic and Polarization Mode)	Dedicated Bandwidth Testers
Reflectance	OTDR, OCWR (Optical Continuous Wave Reflectometer)
Fiber Geometry (Core and cladding diameter, concentricity, etc.)	Various mechanical and optical inspection tools

Standard Test Procedures

Most test procedures for fiber optic component specifications have been standardized by national and international standards bodies, including TIA in the US and ISO/IEC internationally. Procedures for measuring absolute optical power, cable and connector loss and the effects of many environmental factors (such as temperature, pressure, flexing, etc.) are covered in these procedures. In order to perform these tests, the basic fiber optic instruments are the FO power meter, test source, OTDR, optical spectrum analyzer and an inspection microscope.

Fiber Optic Power Meters

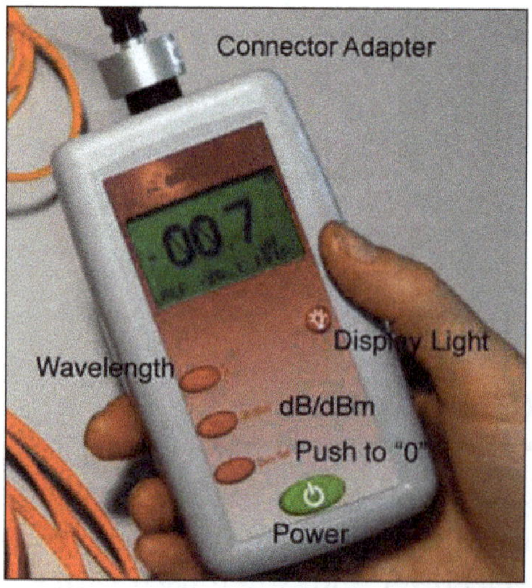

Fiber optic power meters measure the average optical power out of an optical fiber. Power meters typically consist of a solid state detector (silicon for short wavelength systems, germanium or

InGaAs for long wavelength systems), signal conditioning circuitry and a digital display of power. To interface to the large variety of fiber optic connectors in use, some form of removable connector adapter is usually provided.

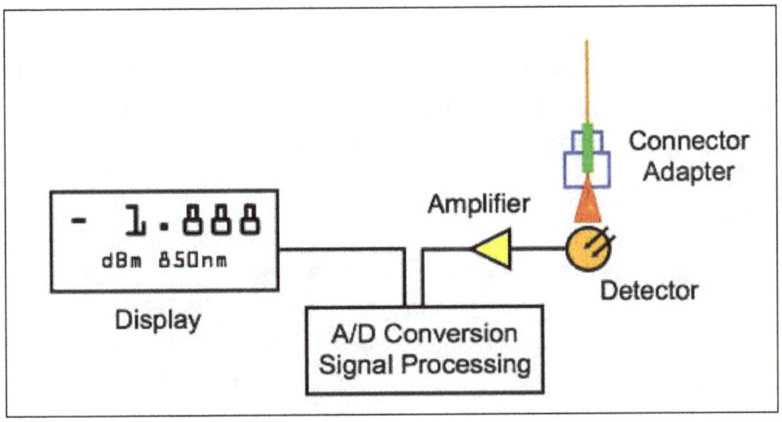

Power meters are calibrated to read in dB referenced to one milliwatt of optical power. Some meters offer a relative dB scale also, useful for loss measurements since the reference value may be set to "0 dB" on the output of the test source. Occasionally, lab meters may also measure in linear units (milliwatts, microwatts and nanowatts). Since all semiconductor detectors have a sensitivity that varies with the wavelength of the light it is measuring, power meters are calibrated at the typical wavelengths used in fiber optics, 850, 1300 and 1550 nm. Meters for POF systems are usually calibrated at 650 and 850 nm, the wavelengths used in POF systems.

Power meters cover a very broad dynamic range, over 1 million to 1 or 60 dB. Although most fiber optic power and loss measurements are made in the range of 0 dBm to -50 dBm, some power meters offer much wider dynamic ranges. For testing analog CATV systems or fiber amplifiers, on needs special meters with extended high power ranges up to +20 dBm (100 mW). Although no fiber optic systems operate at very low power, below about -50 dBm, some lab meters offer ranges to -70 dBm or more, which can be useful in measuring optical return loss or spectral loss characteristics with a monochromator source.

Power meters measure the time average of the optical power, not the peak power, so the meters are sensitive to the duty cycle of an input digital pulse stream. One can calculate peak power if one knows the duty cycle of the input, by dividing the average power by the duty cycle. For most loss measurements, one uses a test source with CW (steady state) or 2 kHz pulsed output. As long as the source modulation doesn't change, no compensation needs to be made. When testing link transmitter power or receiver sensitivity, it is necessary to establish a standard test pattern, generally a 50% duty cycle, called a square wave, to allow accurate measurement of transmitter output or receiver sensitivity.

FO power meters have a typical measurement uncertainty of +/-5% measuring absolute optical power, when calibrated to transfer standards provided by national standards laboratories like the US National Institute of Standards and Technology (NIST). Sources of errors are the variability of coupling efficiency of the detector and connector adapter, reflections off the shiny polished surfaces of connectors, unknown source wavelengths (since the detectors are wavelength sensitive), nonlinearities in the electronic signal conditioning circuitry of the FO power meter and detector noise at very low signal levels. Since most of these factors affect all power meters, regardless of their sophistication, expensive laboratory meters are hardly more accurate that the most inexpensive handheld portable units. Meters should be recalibrated frequently by labs with NIST traceable calibration systems.

Measuring loss, which is a relative measurement over a much smaller range of optical powers, has a much lower uncertainty. Generally the instrument uncertainty is much smaller than the uncertainty caused by the fiber optic components and test setup.

Fiber Optic Test Sources

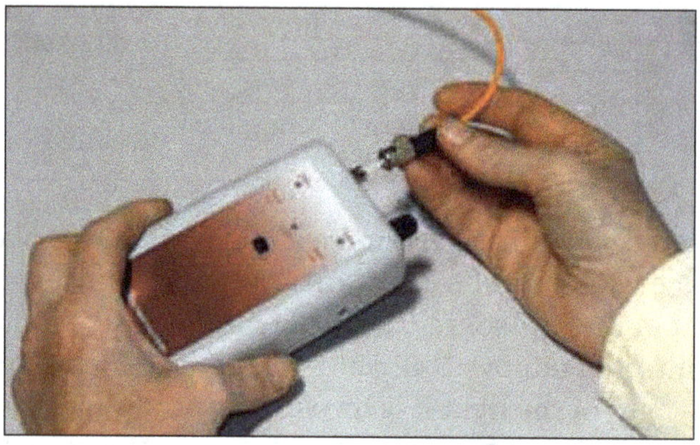

In order to make measurements of optical loss or attenuation in fibers, cables and connectors, one must have a test source as well as a FO power meter. The test source must be chosen for compatibility with the type of fiber in use (single-mode or multimode with the proper core diameter) and the wavelength desired for performing the test. Most sources are either LED's or lasers of the types commonly used as transmitters in actual fiber optic systems, making them representative of actual applications and enhancing the usefulness of the testing. Some tests, such as measuring spectral attenuation of fiber require a variable wavelength source, which is usually a tungsten lamp with a monochromator to vary the output wavelength.

Typical wavelengths of sources are 650 or 665 nm (LEDs for plastic fiber), 850 and 1300 nm (LEDs for multimode fiber) and 1310 nm and 1550 nm (lasers for single-mode fiber). LED's are typically used for testing multimode fiber and lasers are used for single-mode fiber, although there is some crossover, especially in high speed LANs which use multimode fiber with lasers and the testing of short single-mode jumper cables with LED's. The source wavelength can be a critical issue in making accurate loss measurements on long links, since attenuation of the fiber is wavelength sensitive especially at short wavelengths. Thus all test sources should be calibrated for wavelength.

Test sources almost always have fixed connectors. Hybrid test jumpers with connectors compatible with the source on one end and the connector being tested on the other must be used as reference cables. This may affect the type of reference setting mode used for loss testing.

Other source-related factors affecting measurement accuracy are the stability of the output power and the modal distribution launched into multimode fiber. For extremely accurate measurements, the source may need optical feedback stabilization to maintain output power at a precise level for long times required for some measurements. Industry standards have requirements or recommendations on the modal output of test sources for multimode fiber that are aimed at the manufacturers of the test sources. Mode scramblers, filters and strippers may be required to adjust the modal distribution in the fiber to approximate actual operating conditions.

Optical Loss Test Sets/Test Kits

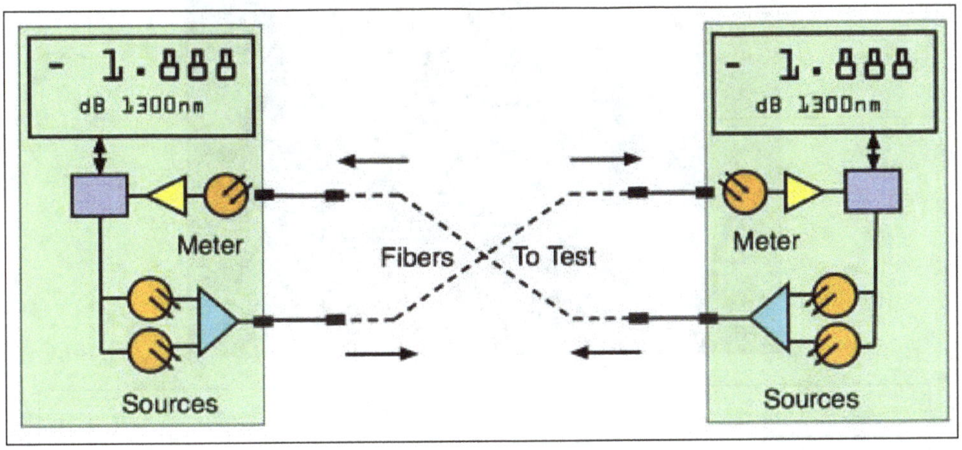

The optical loss test set is an instrument formed by the combination of a fiber optic power meter and source which is used to measure the loss of fiber, connectors and connectorized cables. Early versions of this instrument were called attenuation meters. A test kit has a similar purpose, but is usually comprised of separate instruments and includes accessories to customize it for a specific application, such as testing a FO LAN, telco or CATV.

The OLTS may have several optional features that affect its use. Some have individual source outputs and meter inputs like a separate power meter and test source, but may have two wavelengths from one source output (MM: 850/1300, SM:1310/1550.) Some offer bidirectional testing on a single fiber and some have two bidirectional ports. Some manufactures of premises copper cabling testers offer modules to convert these testers to an OLTS, allowing fiber and copper testing with one instrument.

The combination OLTS instrument which contains both a meter and source may be less convenient than an individual source and power meter, since the ends of the fiber and cable are usually separated by long distances, which would require two OLTSs instead of one source and one meter. An OLTS often has a single port for bidirectional measurements also. This port usually has a fixed connector which may cause problems when testing cable plants with connector styles different than those on the instrument, requiring a 2- or 3-cable reference for loss testing which may not meet industry standards. The bidirectional port may also have problems meeting standards for modal power distribution in multimode fibers. Ask the OLTS manufacturer about these issues before purchasing an instrument.

OLTS sometimes display loss in a different way than if you test with a meter and source. With a meter and source, if you set the reference value for "0dB" at "0" and test cables for loss, loss will be displayed as a negative number, since lower optical powers (after the loss) are more negative numbers. Some but not all OLTS will display the loss as a positive number which can be confusing to those who learn to test using a source and power meter.

Optical Time Domain Reflectometer

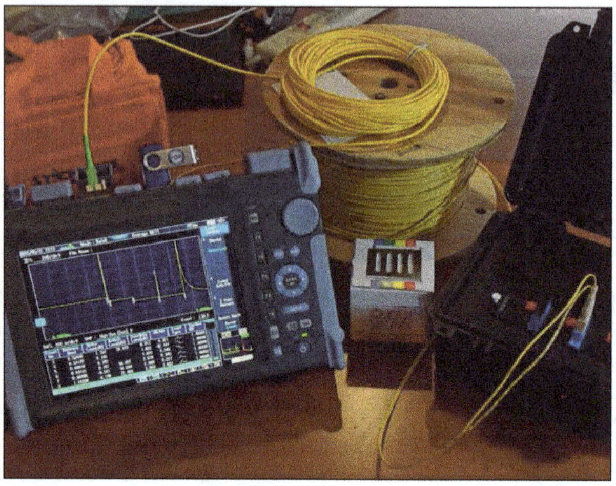

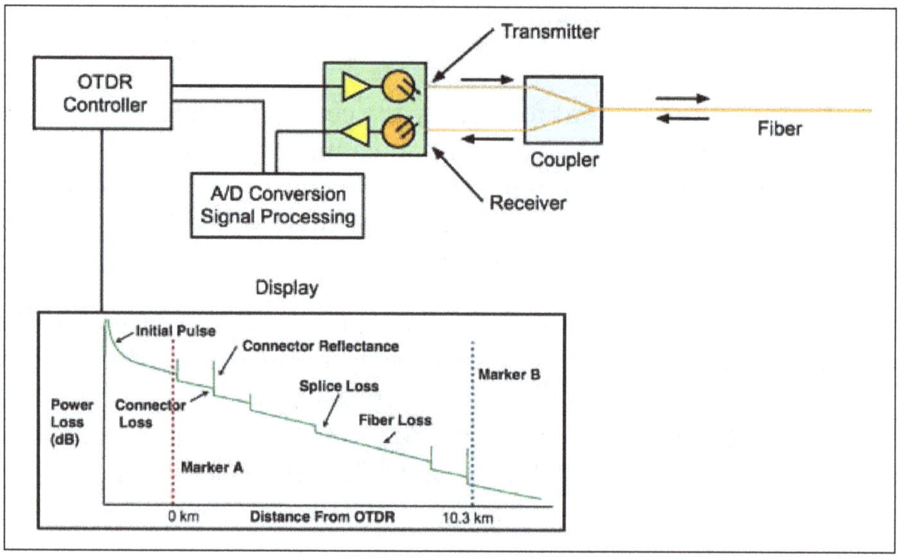

The optical time domain reflectometer (OTDR) uses the phenomena of fiber backscattering to characterize fibers and installed cables, find faults and optimize splices. Since scattering is one of the primary loss factors in fiber (the other being absorption), the OTDR can send out into the fiber a high powered pulse and measure the light scattered back toward the instrument. The pulse is attenuated on the outbound leg and the backscattered light is attenuated on the return leg, so the returned signal is a function of twice the fiber loss and the backscatter coefficient of the fiber.

If one assumes the backscatter coefficient is constant, the OTDR can be used to measure loss as well as locate fiber breaks, splices and connectors. In addition, the OTDR gives a graphic display of the status of the fiber being tested. And it offers another major advantage over the source/FO power meter or OLTS, in that it requires access to only one end of the fiber.

The uncertainty of the OTDR measurement is heavily dependent on the backscatter coefficient, which is a function of intrinsic fiber scattering characteristics, core diameter and numerical aperture. It is the variation in backscatter coefficient that causes many splices to show a "gain" instead of the actual loss. OTDRs must also be matched to the fibers being tested in both wavelength and fiber core diameter to provide accurate measurements. Thus many OTDRs have modular sources to allow substituting a proper source for the application.

While most OTDR applications involve finding faults in installed cables or verifying splices, they are very useful in inspecting fibers for manufacturing faults. Development work on improving the short range resolution of OTDRs for LAN applications and new applications such as evaluating connector return loss promise to enhance the usefulness of the instrument in the future.

OTDRs come in three basic versions. Full size OTDRs offer the highest performance and have a full complement of features like data storage, but are very big and high priced. MiniOTDRs provide the same type of measurements as a full OTDR, but with fewer features to trim the size and cost. Fault finders use the OTDR technique, but greatly simplified to just provide the distance to a fault, to make the instruments more affordable and easier to use.

Visual Cable Tracers and Fault Locators

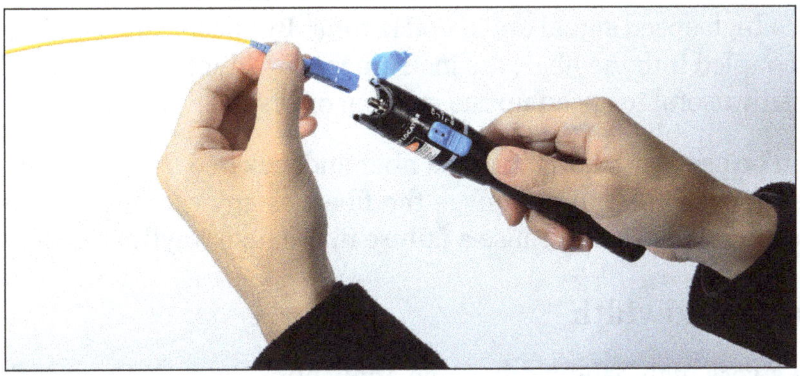

Many of the problems in connection of fiber optic networks are related to making proper connections. Since the light used in systems is invisible, one cannot see the system transmitter light. By injecting the light from a visible source, such as a LED or incandescent bulb, one can visually trace the fiber from transmitter to receiver to insure correct orientation and check continuity besides. The simple instruments that inject visible light are called visual fault locators.

If a powerful enough visible light, such as a HeNe or visible diode laser is injected into the fiber, high loss points can be made visible. Most applications center on short cables such as used in telco central offices to connect to the fiber optic trunk cables. However, since it covers the range where OTDRs are not useful, it is complementary to the OTDR in cable troubleshooting. This method will work on buffered fiber and even jacketed single fiber cable if the jacket is not opaque to the visible light. The yellow jacket of single-mode fiber and orange of multimode fiber will usually pass the visible light. Most other colors, especially black and gray, will not work with this technique, nor will most multifiber cables. However, many cable breaks, macrobending losses caused by kinks in the fiber bad splices etc. can be detected visually. Since the loss in the fiber is quite high at visible wavelengths, on the order of 9-15 dB/km, this instrument has a short range, typically 3-5 km.

Fiber Identifiers

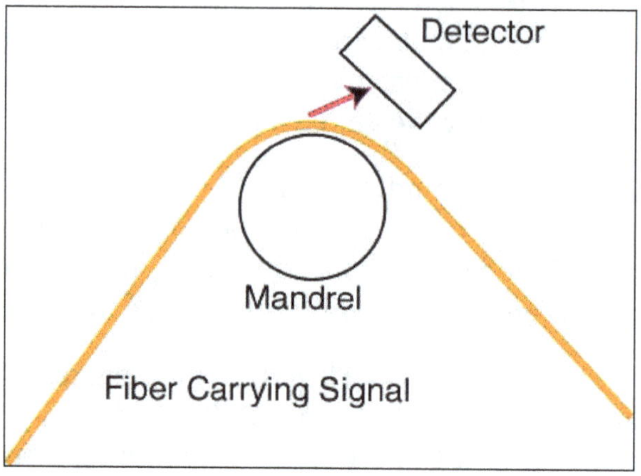

Telco technicians often need to identify a fiber in a splice closure or at a patch panel. If one carefully bends a single-mode fiber enough to cause loss, the light that couples out can also be detected by a large area detector. A fiber identifier uses this technique to detect a signal in the fiber at normal transmission wavelengths. These instruments usually function as receivers, able to discriminate between no signal, a high speed signal and a 2 kHz tone. By specifically looking for a 2 kHz "tone" from a test source coupled into the fiber, the instrument can identify a specific fiber in a large multifiber cable, especially useful to speed up the splicing or restoration process.

Fiber identifiers can be used with both buffered fiber and jacketed single fiber cable. With buffered fiber, one must be very careful to not damage the fiber, as any excess stress here could result in stress cracks in the fiber which could cause a failure in the fiber anytime in the future.

Measuring Fiber Bandwidth

Although fiber has a very high bandwidth, some applications actually approach its limits, requiring performance evaluation. Two factors limit multimode fiber bandwidth: modal dispersion and chromatic dispersion. Long single-mode links require concern over chromatic dispersion or polarization-mode dispersion. Specialized instruments are available for testing each of these specifications but are expensive and rarely used outside the laboratory.

O/E and E/O Converters

Optical to electrical (O/E) and electrical to optical (E/O) converters have other uses besides testing fiber bandwidth. O/E converters can be used with high speed oscilloscopes to analyze pulses in fiber optic links to see if the waveforms are of the proper shape. This means measuring rise and fall times of the pulse and the depth of modulation (the difference between the peak power of the pulse and the lowest power reached between pulses. They can be used for testing lasers and LEDs used in transmitters and link dispersion in long links. E/O converters are used to test receivers for bandwidth and margin, usually in conjunction with a bit error rate tester and attenuator.

Optical Continuous Wave Reflectometers (OCWR)

The OCWR or reflectance tester was originally proposed as a special purpose instrument to measure the reflectance or optical return loss of connectors installed on patch cords or jumpers. Unfortunately, its purpose became muddled between conception and inception. As actual instruments came on the market, they had much higher measurement resolution than appropriate for the measurement uncertainty (0.01 dB resolution vs. 1 dB uncertainty), leading to much confusion on the part of users as to why measurements were not reproducible. In addition, several instruments were touted as a way to measure the optical return loss of an installed cable plant, obviously in ignorance of the fact that they would also be integrating the backscatter of the fiber with any reflections from connectors or splices. Since the measurement of return loss from a connector can be made equally well with any power meter, laser source and calibrated coupler, and an OTDR is the only way to test installed cable plants for return loss, the OCWR has seen little use in fiber optic testing.

Test Equipment for Long Distance Fiber Links

Long distance fiber links may suffer from chromatic dispersion or polarization mode dispersion. Generally they also use DWDM so need testing for spectral attenuation. These tests use very specialized instruments.

Optical Fiber Analyzers

There are many parameters of optical fiber that require testing by the manufacturer. These include attenuation (as a function of source wavelength), bandwidth/dispersion, numerical aperture and all the physical dimensions such as core and cladding diameter, ovality, and concentricity. Automated laboratory instruments are available to measure all these parameters automatically, but many fiber manufacturers prefer to build their own. The most difficult part of fiber measurements is the fact that subtle differences in test setup and instrumentation can cause differences in measured values.

Visual Inspection with Microscopes

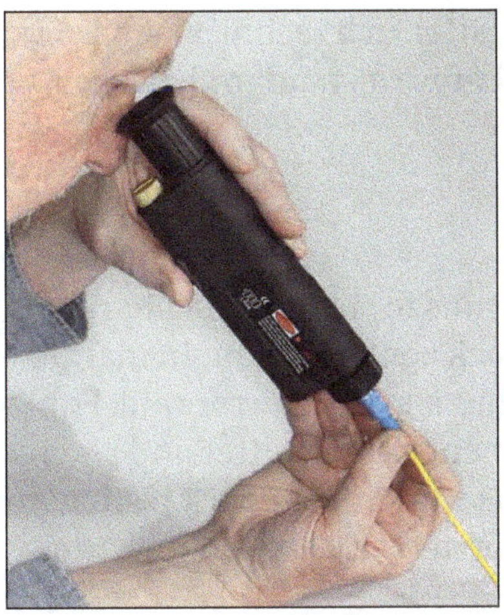

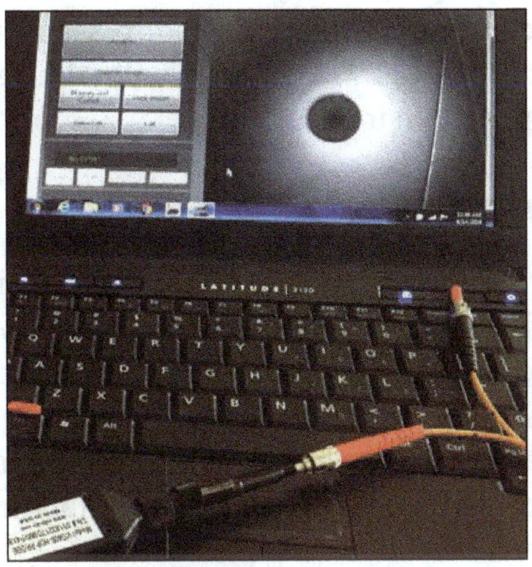

Optical (L) and (R) video microscopes.

Cleaved fiber ends prepared for splicing and polished connector ferrules require visual inspection to find possible defects. This is accomplished using a microscope which has a stage modified to hold the fiber or connector in the field of view. Fiber optic inspection microscopes vary in magnification from 30 to 800 powers with 30-100 power being the most widely used range. Cleaved fibers are usually viewed from the side, to see break over and lip. Connectors are viewed end-on or at a small angle to find polishing defects such as scratches.

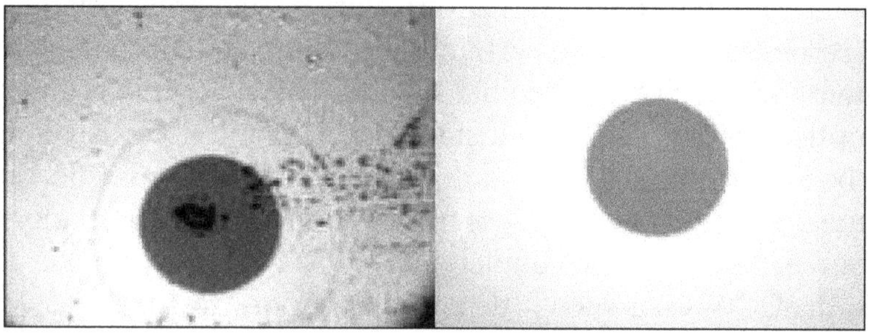

Fiber Optic Talksets

While technically not a measuring instrument, FO talksets are useful for FO installation and testing. They transmit voice over fiber optic cables already installed, allowing technicians splicing or testing the fiber to communicate effectively. Talksets are especially useful when walkie-talkies and telephones are not available, such as in remote locations where splicing is being done, or in buildings where radio waves will not penetrate.

The way to use talksets most effectively is to set up the talksets on one fiber (or pairs appropriate) and leave them there while all testing or splicing work is done. Thus, there will always be a communications link between the working crew, which facilitates deciding which fibers to work with next. The continuous communications capability will greatly speed the process. Recent developments in talksets include talksets for networking multi-party communications, especially helpful in restoration, and system talksets for use as intercoms in installed systems. There are also combination testers and talksets.

There are no standards for the way talksets communicate. Some use simple AM transmission, some FM and some proprietary digital schemes. Thus no two manufacturers' talksets can communicate with each other. Bellcore has addressed this matter in a technical advisory that proposes a FM method at 80 and 120 kHz, but it will take years before a standard has been set and manufacturers offer compatible instruments.

Attenuators

Attenuators are used to simulate the loss of long fiber runs for testing link margin in network simulation in the laboratory or self-testing links in a loopback configuration. In margin testing, variable attenuators are used to increase loss until the system has a high bit error rate. For loopback testing, an attenuator is used between a single piece of equipment's transmitter and receiver to test for operation under maximum specified fiber loss. If systems work in loopback testing, they should work with a proper cable plant. Thus many manufacturers of network equipment specify a loopback test as a diagnostic/troubleshooting procedure.

Attenuators can be made by gap loss, or a physical separation of the ends of the fibers, inducing bending losses or inserting calibrated optical filters. Both variable and fixed attenuators are available, but variable attenuators are usually used for testing. Fixed attenuators may be inserted in the system cables where distances in the fiber optic link are too short and excess power at the receiver causes transmission problems.

Reference Test Jumper Cables and Mating Adapters

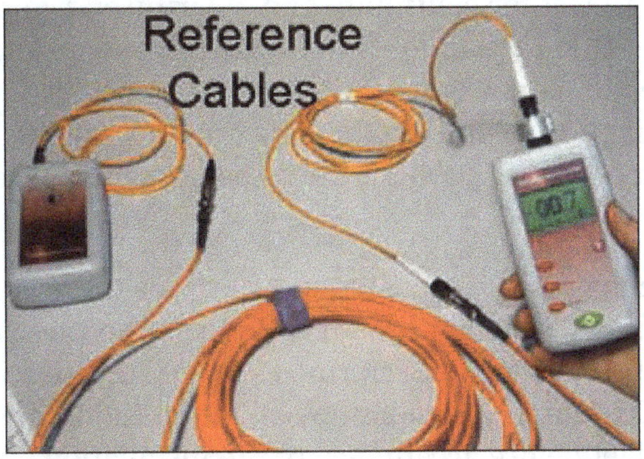

In order to test cables in an insertion loss test, one needs to establish test conditions. This requires reference launch jumper cables to connect the test source to the cable under test and receive cables to connect the fiber optic power meter. For accurate measurements, the launch and receive cables must be made with fiber and connectors matching the cables to be tested and terminated carefully to ensure low loss. To provide reliable measurements, launch and receive cables must be in good condition and kept very clean. They can easily be tested against each other to insure their performance. Connector mating adapters are used to connect the cables under test to the launch and receive cables. Only the highest performance bulkhead splices should be used, and their condition checked regularly, since they are vitally important in obtaining low loss connections.

Fiber Optic Testing

Testing is used to evaluate the performance of fiber optic components, cable plants and systems. As the components like fiber, connectors, splices, LED or laser sources, detectors and receivers are being developed, testing confirms their performance specifications and helps understand how they will work together. Designers of fiber optic cable plants and networks depend on these specifications to determine if networks will work for the planned applications.

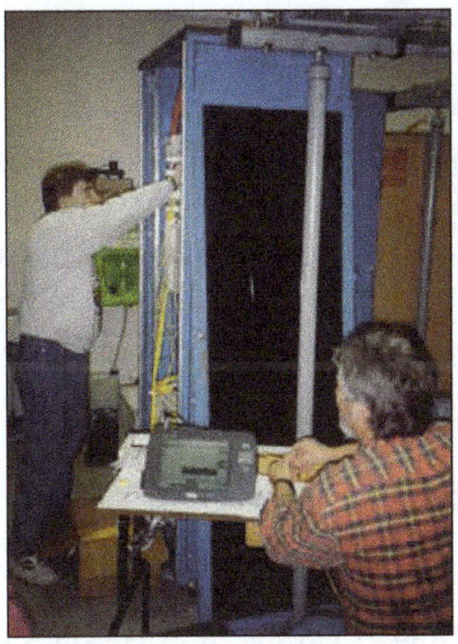

After fiber optic cables are installed, spliced and terminated, they must be tested. For every fiber optic cable plant, you need to test for continuity and polarity, end-to-end insertion loss and then troubleshoot any problems. If it's a long outside plant cable with intermediate splices, you will probably want to verify the individual splices with an OTDR test also, since that's the only way to make sure that each splice is good. If you are the network user, you may also be interested in testing transmitter and receiver power, as power is the measurement that tells you whether the system is operating properly.

Testing is the subject of the majority of industry standards, as there is a need to verify component and system specifications in a consistent manner. Perhaps the most important test is insertion loss of an installed fiber optic cable plant performed with a light source and power meter (LSPM) or optical loss test set (OLTS) which is required by all international standards to ensure the cable plant is within the loss budget before acceptance of the installation. Testing fiber optics requires special tools and instruments which must be chosen to be appropriate for the components or cable plants being tested.

Getting Started

Even if you're an experienced fiber optic tech, make sure you remember these things:

- Optical inspection microscope, 100-200X video scope recommended.
- Source and power meter, optical loss test set (OLTS) or test kit with proper equipment adapters for the cable plant you are testing.
- Reference test cables that match the cables to be tested and mating adapters, including hybrids if needed.
- Fiber Tracer or Visual Fault Locator.
- Cleaning materials: Dry cleaning kits or lint free cleaning wipes and pure alcohol.
- Optional: OTDR with launch and/or receive cables for outside plant jobs and troubleshooting.

Know how to use your Test Equipment

Before you start, get together all your tools and make sure they are all working properly and you and your installers know how to use them. It's hard to get the job done when you have to call the manufacturer from the job site on your cell phone to ask for help. Try all your equipment in the office before you take it into the field. Use it to test every one of your reference test jumper cables in both directions using the single-ended loss test to make sure they are all good. If your power meter has internal memory to record data is sure you know how to use this also. You can often customize these reports to your specific needs - figure all this out before you go it the field - it could save you time and on installations.

Know the Cabling Network you're Testing

Make sure you have cable layouts for every fiber you have to test and have calculated a loss budget so you know what test results to expect. Prepare a spreadsheet of all the cables and fibers before you go in the field and print a copy for recording your test data. You may record all your test data either by hand or if your meter has a memory feature, it will keep test results in on-board memory that can be printed or transferred to a computer when you return to the office.

Fiber optic sources, including test equipment, are generally too low in power to cause any eye damage, but it's still advisable to check connectors with a power meter before looking into it. Besides, most fiber optic sources are at infrared wavelengths that are invisible to the eye, making them more dangerous. Connector inspection microscopes focus all the light into the eye and can increase the danger. Some telco DWDM and CATV systems have very high power and they could be harmful.

Visual Inspection

Visual Tracing

Continuity checking with a visual fiber tracer makes certain the fibers are not broken and to trace a path of a fiber from one end to another through many connections, verifying duplex connector polarity for example. It looks like a flashlight or a pen-like instrument with a light bulb or LED source that mate to a fiber optic connector. Attach the fiber to test to the visual tracer and look at the other end of the fiber to see the light transmitted through the core of the fiber. If there is no light at the end, go back to intermediate connections to find the bad section of the cable.

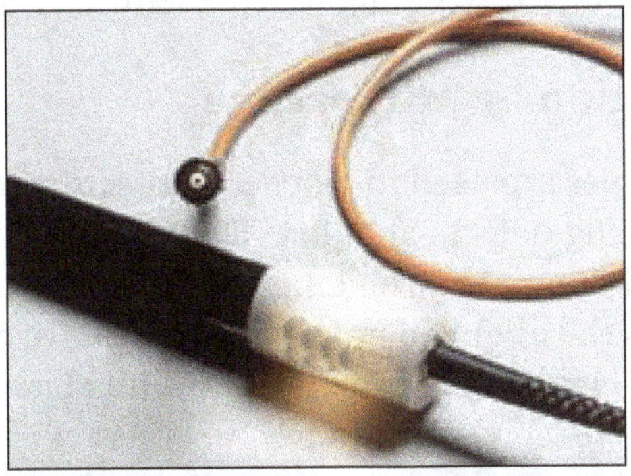

A good example of how it can save time and money is testing fiber on a reel before you pull it to make sure it hasn't been damaged during shipment. Look for visible signs of damage (like cracked or broken reels, kinks in the cable, etc). For testing, visual tracers help also identify the next fiber to be tested for loss with the test kit. When connecting cables at patch panels, use the visual tracer to make sure each connection is the correct two fibers! And to make certain the proper fibers are connected to the transmitter and receiver, use the visual tracer in place of the transmitter and your eye instead of the receiver.

Visual Fault Location

A higher power version of the fiber tracer called a visual fault locator (VFL) uses a visible laser that can also find faults. The red laser light is powerful enough for continuity checking or to trace fibers for several kilometers, identify splices in splice trays and show breaks in fibers or high loss connectors. You can actually see the loss of light at a fiber break by the bright red light from the VFL through the jacket of many yellow or orange simplex cables (excepting black or gray jackets, of course.) It's most important use is finding faults in short cables or near the connector where OTDRs cannot find them.

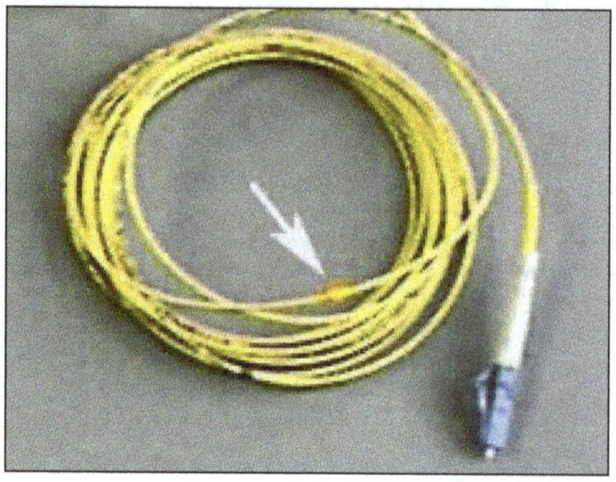

You can also use this gadget to visually verify and optimize mechanical splices or prepolished-splice type fiber optic connectors. By visually minimizing the light lost you can get the lowest loss splice. In fact don't even think of doing one of those prepolished-splice type connectors without one. No other method will assure you of high yield with those connectors.

VFLs use visible light. You will find it uncomfortable to look at the output of a fiber illuminated by a VFL. That's good, because the power level is high and you should not be looking at it. When tracing fibers, look from the side of the fiber to see if light is present.

Visual Connector Inspection by Microscope

Fiber optic inspection microscopes are used to inspect connectors to confirm proper polishing and find faults like scratches, polishing defects and dirt. They can be used both to check the quality of the termination procedure and diagnose problems. A well-made connector will have a smooth , polished, scratch free finish and the fiber will not show any signs of cracks, chips or areas where the fiber is either protruding from the end of the ferrule or pulling back into it (pistoning).

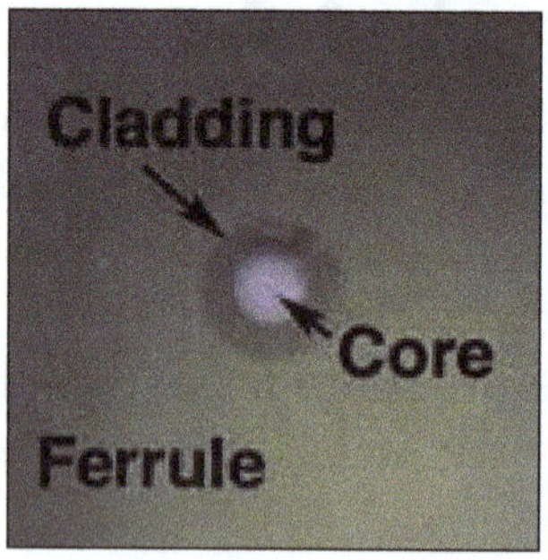

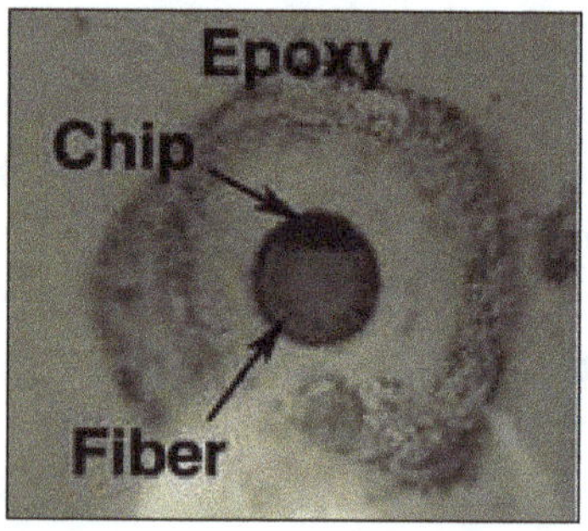

Microscopes are also used to inspect connectors before mating or testing them to ensure the connectors are clean - no dirt or contamination. The process is to inspect - clean - then inspect again to confirm proper cleaning. Repeat as necessary. The magnification for viewing connectors can be 30 to 400 powers but it is best to use a medium magnification. The best microscopes allow you to inspect the connector from several angles, either by tilting the connector or having angle illumination to get the best picture of what's going on. Check to make sure the microscope has an easy-to-use adapter to attach the connectors of interest to the microscope.

Video readout microscopes are now available that allow easier viewing of the end face of the connector and some even have software that analyzes the finish. While they are much more expensive than normal optical microscopes, they will make inspection easier and greatly increase productivity.

And remember to check that no power is present in the cable before you look at it in a microscope to protect your eyes. The microscope will concentrate any power in the fiber and focus it into your eye with potentially hazardous results. Some microscopes have filters which remove infrared light from sources to be safe.

Optical Power - Power or Loss

Practically every measurement in fiber optics refers to optical power measured in dB. The power outputs of a transmitter or the input to receiver are "absolute" optical power measurements, that is, you measure the actual value of the power. Loss is a "relative" power measurement, the difference between the power coupled into a component like a cable, splice or a connector and the power that is transmitted through it. This difference in power level before and after the component is what we call optical loss and defines the performance of a cable, connector, splice, etc. Even an OTDR measures optical power to measure loss.

Measuring Power

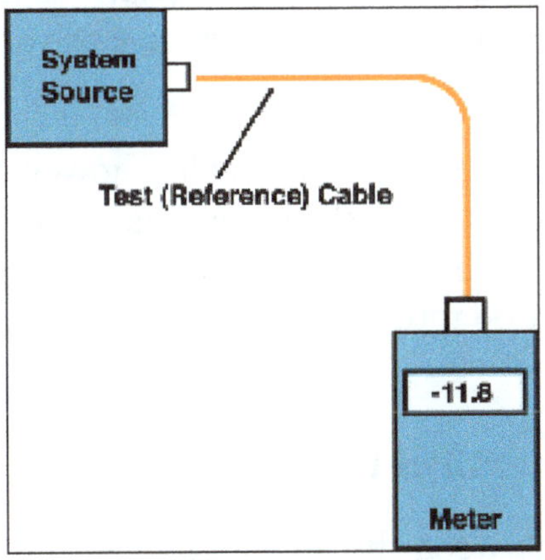

Power in a fiber optic system is like voltage in an electrical circuit - it's what makes things happen! It's important to have enough power, but not too much. Too little power and the receiver may not be able to distinguish the signal from noise; too much power overloads the receiver and causes errors too.

Measuring power requires only a power meter (most come with a screw-on adapter that matches the connector being tested), a known good fiber optic cable (of the right fiber size, as coupled power is a function of the size of the core of the fiber) and a little help from the network electronics to turn on the transmitter. Remember when you measure power, the meter must be set to the proper range (usually dBm, sometimes microwatts, but never "dB" - that's a relative power range used only for testing loss!) and the proper wavelength, matching the source being used in the system (850, 1300, 1550 nm for glass fiber, 650 or 850 nm for POF). Refer to the instructions that come with the test equipment for setup and measurement instructions.

To measure power, attach the meter to the cable attached to the source that has the output you want to measure. That can be at the receiver to measure receiver power, or using a reference test cable (tested and known to be good) that is attached to the transmitter to measure output power. Turn on the transmitter/source and give it a few minutes to stabilize. Set the power meter for the matching wavelength and note the power the meter measures. Compare it to the specified power for the system and make sure it's enough power but not too much.

It is important to remember that dB is for measuring loss, dBm is for measuring power and the more negative a number is, the higher the loss. Set your zero reference before measuring loss and check it occasionally while making measurements.

Calibration of Fiber Optic Power Measurements

Calibrating fiber optic power measurement equipment requires setting up a reference standard traceable to a national standards lab like the National Institute of Standards and Technology in the US for comparison purposes while calibrating every power meter or other instrument. Calibration is done at 850, 1300 and 1550 nm. NIST currently uses a laboratory optical power meter which is sent around to labs as a transfer standard.

Meters calibrated in this manner have an uncertainty of calibration of about +/- 5%, compared to the NIST primary standards. Limitations in the uncertainty are the inherent inconsistencies in optical coupling, about 1% at every transfer, and slight variations in wavelength calibration. NIST is working continuously with instrument manufacturers and private calibration labs to try to reduce the uncertainty of these calibrations.

Recalibration of instruments should be done annually, however experience has shown that the accuracy of meters rarely changes significantly during that period, as long as the electronics of the meter do not fail. The calibration of fiber optic power meters requires considerable investment in capital equipment so meters must be returned to the original manufacturer or private calibration labs for calibration.

Testing Optical Loss or Insertion Loss

Loss of a cable is the difference between the power coupled into the cable at the transmitter end and what comes out at the receiver end. Testing for loss (also called "insertion loss") requires measuring the optical power lost in a cable (including fiber attenuation, connector loss and splice loss) with a fiber optic light source and power meter (LSPM) or optical loss test set (OLTS.) Loss testing is done at wavelengths appropriate for the fiber and its usage. Generally multimode fiber is tested at 850 nm and optionally at 1300 nm with LED sources. Single-mode fiber is tested at 1310 nm and optionally at 1550 nm with laser sources. The measured loss is compared to the estimated loss calculated for the link, called a "loss budget".

The insertion loss measurement is made by mating the cable being tested to known good reference cables with a calibrated launch power that becomes the "0 dB" loss reference. Why do you need reference cables to measure loss? Why can't you just plug the cable to test into a source and power meter and measure the power? There are several reasons:

- You need a cable to measure the output power of the source for calibration of "0 dB" loss.

- In order to measure the loss of the connectors you must mate them to a similar, known good, connector.
- This is an important point often not fully explained. When we say connector loss, we really mean "connection" loss - the loss of a mated pair of connectors. Thus, testing connectors requires mating them to reference connector which must be high quality connectors themselves to not adversely affect the measured loss when mated to an unknown connector.
- Testing with reference cables on each end simulates a cable plant with patchcords connecting to transmission equipment.

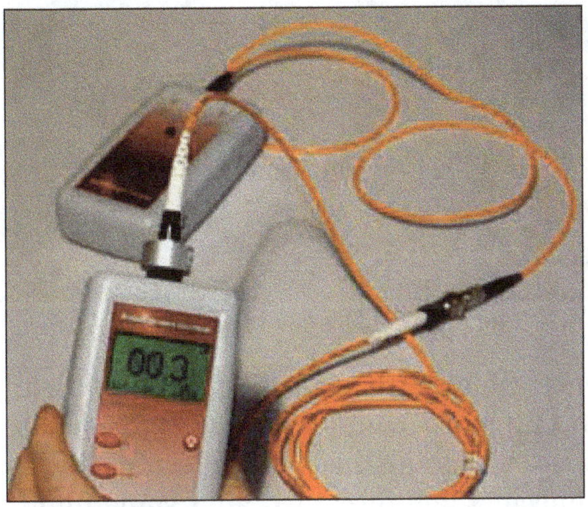

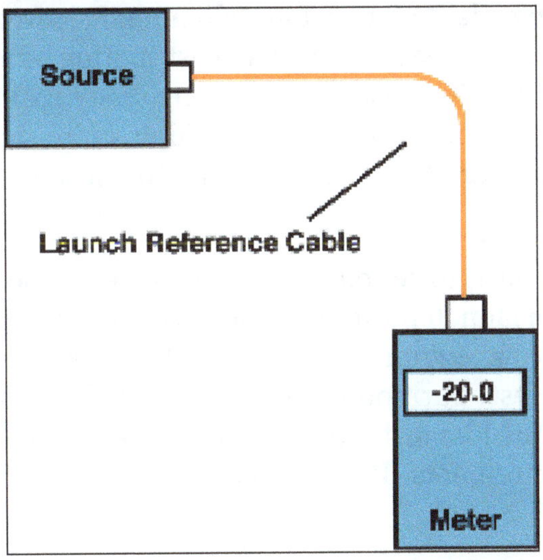

In addition to a power meter, you need a test source. The test source should match the type fiber (generally LED for MM or laser for SM) and wavelength (850, 1300, 1550 nm) that will be used on the fiber optic cable you are testing. If you are testing to some standards, you may need to add some mode conditioning, like a mandrel wrap, to meet the standard launch conditions.

You generally need one or two reference cables, depending on the test we wish to perform. The accuracy of the measurement you make will depend on the quality of your reference cables, since they will be mated to the cable under test. The quality and cleanliness of the connectors on the launch and receive cables is the most important factor in the accuracy of loss measurements. Always test your reference cables by the patchcord or single ended method shown below to make sure they're good before you start testing other cables.

Standards groups have never been able to successfully specify the quality of reference cables in terms of tightly tolerance components like the fiber and connectors. The best recommendation for qualifying reference cables is to choose cables with low loss, tested "single-ended" per FOTP-171 below.

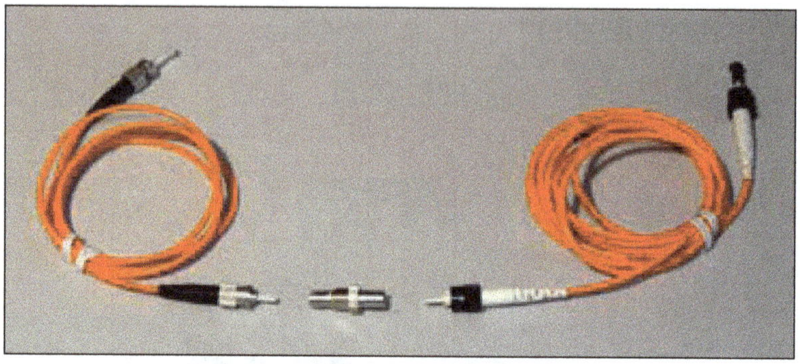

Reference cables and mating adapter.

In order to mate the reference cables to the cables you want to test, you need mating adapters. Mating adapters are as important to low connection loss as the quality of the connectors since the mating adapter is responsible for aligning the two connector ferrules correctly. Mating adapters must be kept clean, like connectors and discarded after some number of uses as they wear out from repeated matings. Mating adapters may have alignment sleeves made from plastic, metal or ceramic. Plastic alignment sleeves used on the cheapest mating adapters should not be used for testing as they wear out in only a few insertions, leaving dusty residue on the connectors. Metal adapters are good for many more insertions and provide a better alignment, so they are acceptable. Ceramic alignment sleeves are the best, providing the best alignment and practically never wearing out.

In order to measure loss, we need to set our reference power for loss our "0 dB" value. Correct setting of the "0 dB' reference launch power is critical to making good loss measurements. Clean your connectors and set up your equipment like this: Turn on the source and select the wavelength you want for the loss test. Turn on the meter, select the "dBm" or "dB" range and select the wavelength you want for the loss test. Measure the power at the meter. This is your reference power level for all loss measurements. If your meter has a "zero" function, set this as your "0" reference.

Some reference books and manuals show setting the reference power for loss using both a launch and receive cable mated with a mating adapter or even three reference cables. Industry standards, in fact, include all three methods of setting a "0dB loss" reference. The two or three cable reference methods are acceptable for some tests and are the only way you can test if the connectors on the cable plant are not the same as your test equipment, but it will reduce the loss you measure by the amount of loss between your reference cables when you set your "0dB loss" reference. Also, if either the launch or receive cable is bad, setting the reference with both cables hides the fact. Then you could begin testing with bad launch cables making all your loss measurements wrong. EIA/TIA 568 and OFSTP-14/7 allows any method as long as the method is reported with the data.

Testing Loss

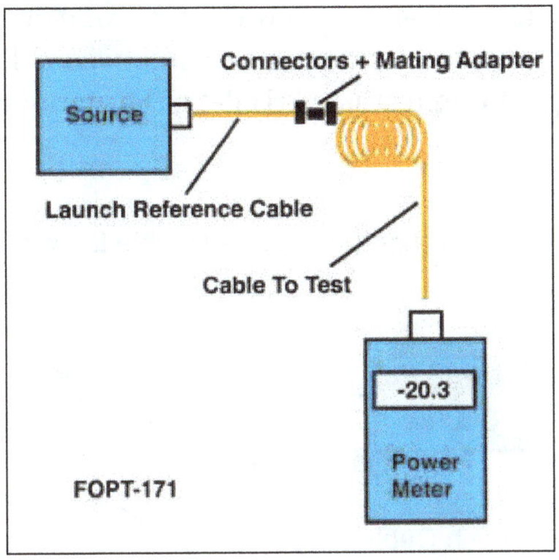

Single-ended testing (e.g. patchcords).

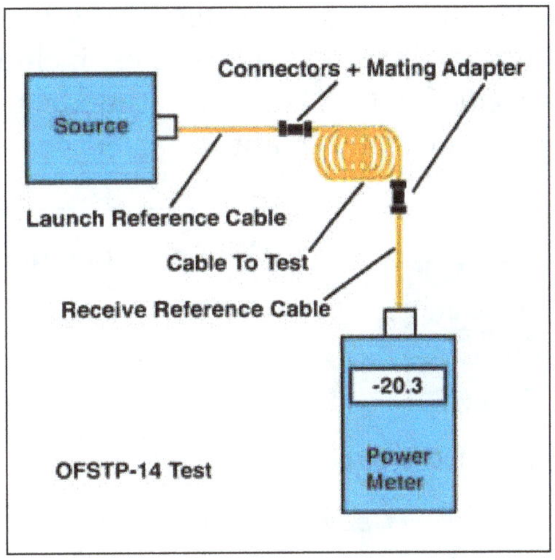

Double-ended testing (installed cable plants).

There are two methods that are used to measure loss, a "patchcord test" which we call "single-ended loss" (TIA FOTP-171) and an "installed cable plant test" we call "double-ended loss" (TIA OFSTP-14 (MM) and OFSTP-7 (SM).) Single-ended loss uses only the launch cable, while double-ended loss uses a receive cable attached to the meter also.

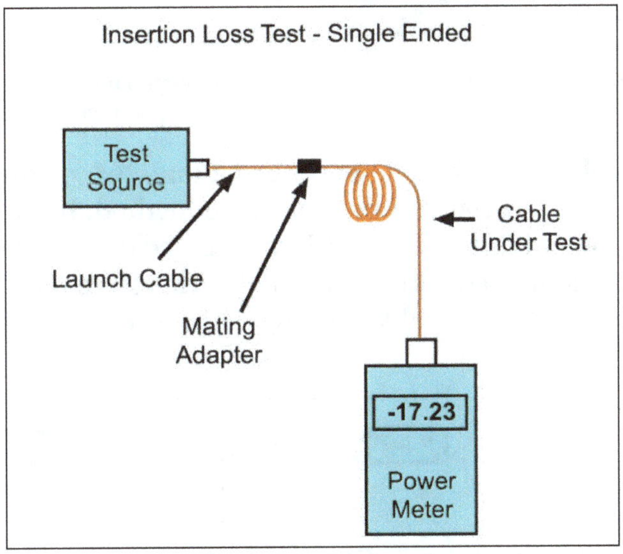

Step-by-step insertion loss test: Single-ended for patchcords or testing one connector on a cable.

Single-ended loss is measured by mating the cable you want to test to the reference launch cable and measuring the power out the far end with the meter. When you do this you measure the loss of the connector mated to the launch cable and the loss of any fiber, splices or other connectors in the cable you are testing. Since you are aiming the connector on the far end of the cable mated to the power meter at a large area detector instead of mating it to another connector, it effectively has no loss so it is not included in the measurement. This method is described in FOTP-171 and is shown in the drawing. An advantage to this test is you can troubleshoot cables to find a bad connector since you can reverse the cable to test the connectors on the each end individually.

In a double-ended loss test, you attach the cable to test between two reference cables, one attached to the source and one to the meter. This way, you measure two connectors' loses, one on each end, plus the loss of all the cable or cables, including connectors and splices, in between. This is the method specified in OFSTP-14 (multimode, the singlemode test is OFSTP-7), the standard test for loss in an installed cable plant.

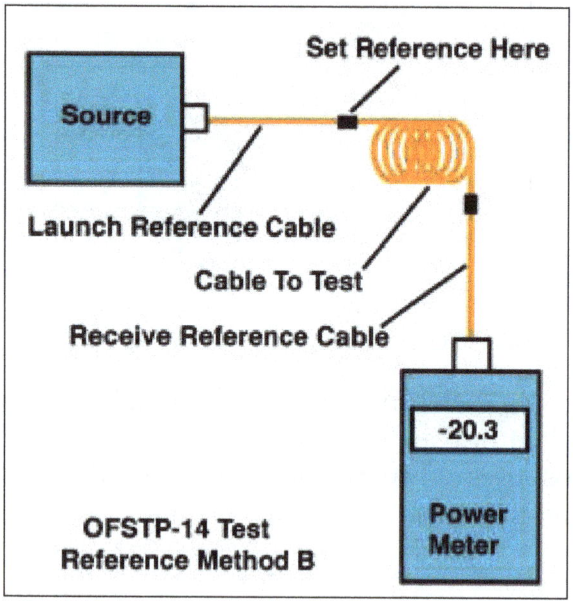

Single cable reference method preferred in some standards.

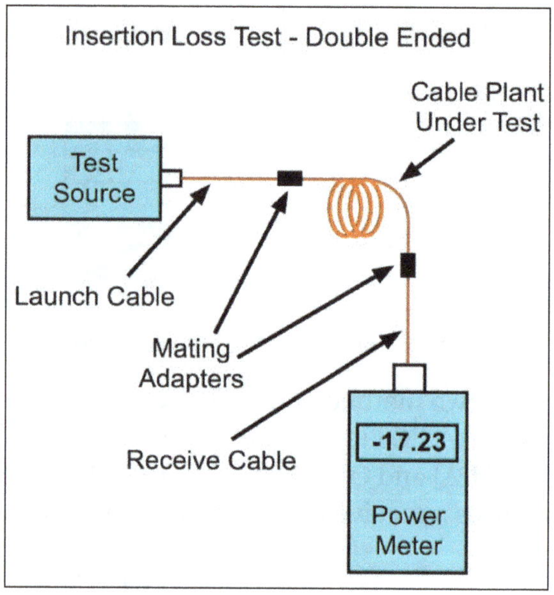

Step-by-step measuring insertion loss of a cable plant with a 1-cable reference.

There are three methods of setting the reference, using one, two or three reference cables. The method originally called for in TIA-568 is the one cable method, but that method doesn't work with every type of connector and test equipment interfaces, so the standards now allow any method as long as it has been documented with the test data. Therefore most standards now allow for using either one, two or three reference cables as long as the test method is documented along with the test data.

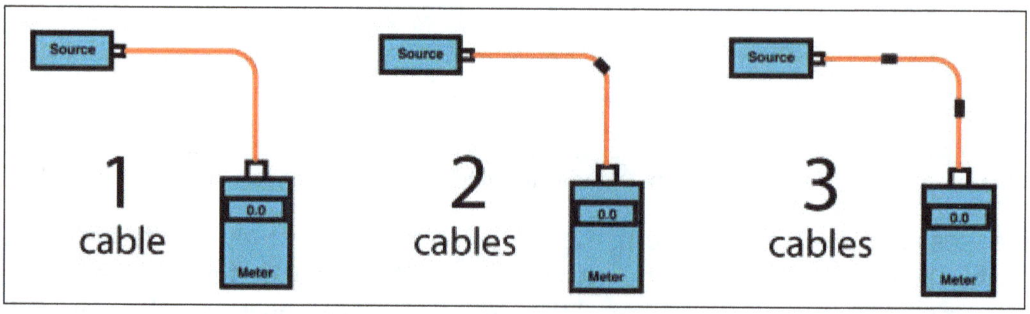

The test results you get when using each of these methods will be different due to the connections (0, 1 or 2) included in setting the "0 dB" reference.

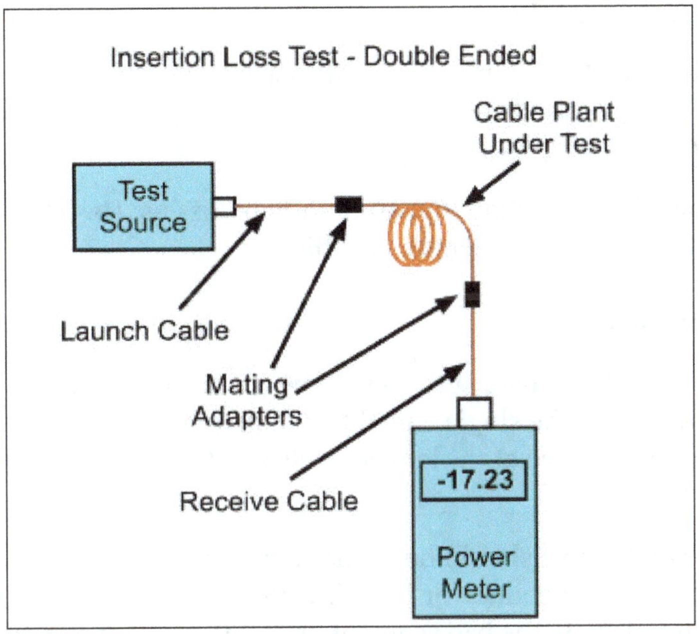

Testing the same cable with different methods of setting the "0dB" reference gives different test results.

Modal Conditioning for Multimode Fibers

Most standards for multimode fiber tests include some modal conditioning to create standardized test conditions to ensure repeatable measurement results. The usual method is to use a source whose output meets standard criteria, coupled to a reference launch cable, on which a mandrel wrap is used to remove higher order modes. Mode conditioners are also available from test equipment manufacturers.

Standards may have different methods, including different requirements for different fibers. The international standards require a mode conditioning metric called encircled flux but TIA-568 only requires it for OM3/4/5 fibers, allowing legacy fibers (OM1, OM2) to use an earlier method. Reference launch cables using bend-insensitive fiber may not respond to the usual methods of mode conditioning and are generally not recommended for launch cables but since most multimode fiber is now BI fiber, this may not be a relevant requirement.

Loss when Testing Cables

Before testing, preferable during the design phase, you should calculate a loss budget for the cable plant to be tested to understand the expected measurement results. Besides providing reference loss values to test against, it will confirm that the network transmission equipment will work properly on this cable. While it is difficult to generalize, here are some guidelines:

- For each connector, figure 0.3-0.5 dB loss for adhesive/polish connectors, 0.75 for prepolished/splice connectors (0.75 max from TIA-568).
- For single-mode fiber, the loss is about 0.5 dB per km for 1300 nm sources, 0.4 dB per km for 1550 nm. This roughly translates into a loss of 0.1 dB per 600 feet for 1300 nm, 0.1 dB per 750 feet for 1300 nm.

- For each splice, figure 0.2 dB.
- For multimode fiber, the loss is about 3 dB per km for 850 nm sources, 1 dB per km for 1300 nm. This roughly translates into a loss of 0.1 dB per 100 feet for 850 nm, 0.1 dB per 300 feet for 1300 nm.

So for the loss of a cable plant, calculate the approximate loss as: (0.5 dB X # connectors) + (0.2 dB x # splices) + fiber loss on the total length of cable.

Troubleshooting Hints

If you have high loss in a cable, make sure to reverse it and test in the opposite direction using the single-ended method. Since the single ended test only tests the connector on one end, you can isolate a bad connector - it's the one at the launch cable end (mated to the launch cable) on the test when you measure high loss.

High loss in the double ended test should be isolated by retesting single-ended and reversing the direction of test to see if the end connector is bad. If the loss is the same, you need to either test each segment separately to isolate the bad segment or, if it is long enough, use an OTDR. If you see no light through the cable (very high loss - only darkness when tested with your visual tracer), it's probably one of the connectors, and you have few options. The best one is to isolate the problem cable, cut the connector of one end (flip a coin to choose) and hope it was the bad one (well, you have a 50-50 chance).

OTDR Testing

OTDRs are powerful test instruments for fiber optic cable plants, if one understands how to properly set the instrument up for the test and interpret the results. When used by a skillful operator, OTDRs can locate faults, measure cable length and verify splice loss. Within limits, they can also measure the loss of a cable plant. About the only fiber optic parameters they don't measure is optical power at the transmitter or receiver.

OTDRs are almost always used on OSP cables to verify the loss of each splice and pinpoint stress areas caused by installation. They are also widely used as OSP troubleshooting tools since they can locate problems in the cables. Most ODTRs lack the distance resolution for use on the shorter cables typical of premises networks.

In order to use an OTDR properly, it's necessary to understand how it works, how to set the instrument up properly and how to analyze traces. Most OTDRs offer an "auto testing" option. Using that option without understanding the OTDR and manually checking its work often leads to problems. Let's look at how an OTDR works and see how it can help testing and troubleshooting.

How OTDRs Work

Unlike sources and power meters which measure the loss of the fiber optic cable plant directly, the OTDR works indirectly. The source and meter duplicate the transmitter and receiver of the fiber optic transmission link, so the measurement correlates well with actual system loss. The OTDR, however, uses backscattered light of the fiber to imply loss. The OTDR works like RADAR, sending a high power laser light pulse down the fiber and looking for return signals from backscattered light in the fiber itself or reflected light from connector or splice interfaces.

At any point in time, the light the OTDR sees is the light scattered from the pulse passing through a region of the fiber. Only a small amount of light is scattered back toward the OTDR, but with wider test pulses, sensitive receivers and signal averaging, it is possible to make measurements over relatively long distances. Since it is possible to calibrate the speed of the pulse as it passes down the fiber, the OTDR can measure time, calculate the pulse position in the fiber and correlate what it sees in backscattered light with an actual location in the fiber. Thus it can create a display of the amount of backscattered light at any point in the fiber.

Since the pulse is attenuated in the fiber as it passes along the fiber and suffers loss in connectors and splices, the amount of power in the test pulse decreases as it passes along the fiber in the cable plant under test. Thus the portion of the light being backscattered will be reduced accordingly, producing a picture of the actual loss occurring in the fiber. Some calculations are necessary to convert this information into a display, since the process occurs twice, once going out from the OTDR and once on the return path from the scattering at the test pulse.

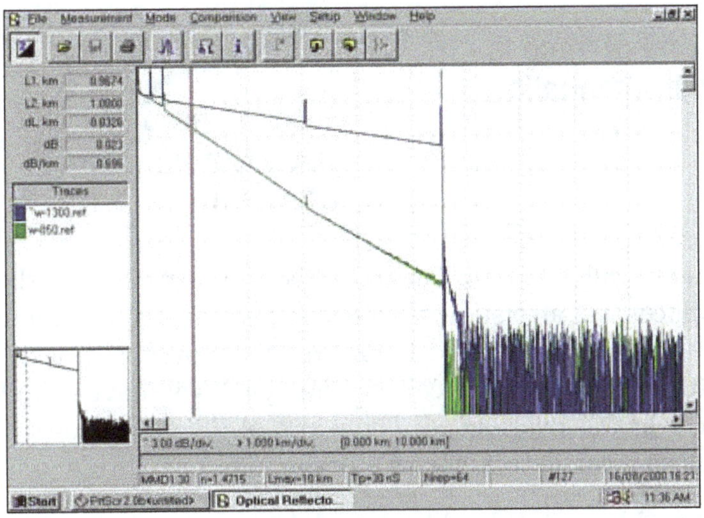

Actual OTDR Trace.

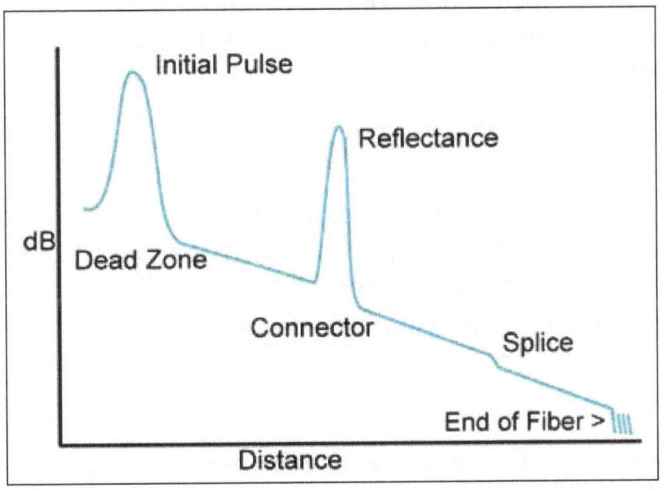

Diagram of OTDR trace with events shown.

There is a lot of information in an OTDR display. The slope of the fiber trace shows the attenuation coefficient of the fiber and is calibrated in dB/km by the OTDR. In order to measure fiber attenuation, you need a fairly long length of fiber with no distortions on either end from the OTDR resolution or overloading due to large reflections. If the fiber looks nonlinear at either end, especially near a reflective event like a connector, avoid that section when measuring loss.

Note the large initial pulse? That is caused by the high-powered test pulse reflecting off the OTDR connector and overloading the OTDR receiver. The recovery of the receiver causes the "dead zone" near the OTDR. To avoid problems caused by the dead zone, always use a launch cable of sufficient length when testing cables.

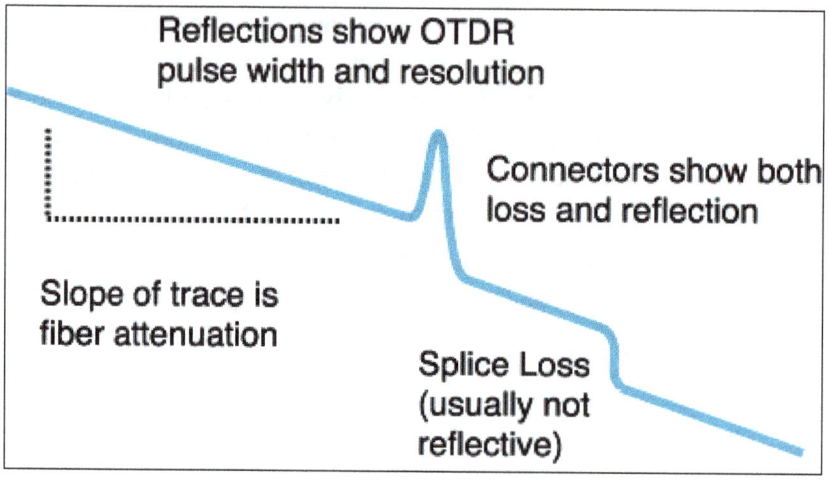

Connectors and splices are called "events" in OTDR jargon. Both should show a loss, but connectors and mechanical splices will also show a reflective peak so you can distinguish them from fusion splices. Also, the height of that peak will indicate the amount of reflection at the event, unless it is so large that it saturates the OTDR receiver. Then peak will have a flat top and tail on the far end, indicating the receiver was overloaded. The width of the peak shows the distance resolution of the OTDR, or how close it can detect events.

OTDRs can also detect problems in the cable caused during installation. If a fiber is broken, it will show up as the end of the fiber much shorter than the cable or a high loss splice at the wrong place. If excessive stress is placed on the cable due to kinking or too tight a bend radius, it will look like a splice at the wrong location.

OTDR Limitations

The limited distance resolution of the OTDR makes it very hard to use in a LAN or building environment where cables are usually only a few hundred meters long. The OTDR has a great deal of difficulty resolving features in the short cables of a LAN and is likely to show "ghosts" from reflections at connectors, more often than not simply confusing the user.

Using the OTDR

When using an OTDR, there are a few cautions that will make testing easier and more understandable. First always use a long launch cable, which allows the OTDR to settle down after the initial pulse and provides a reference cable for testing the first connector on the cable. Always start with the OTDR set for the shortest pulse width for best resolution and a range at least 2 times the length of the cable you are testing. Make an initial trace and see how you need to change the parameters to get better results.

Component Testing

Manufacturers of fiber optic components do extensive testing to qualify their component designs, verify manufacturing procedures and test the products before shipment to customers. Fibers are tested for dimensions (core and cladding size, ovality and concentricity,) performance (attenuation coefficient, bandwidth or dispersion,) physical characteristics (strength, flexibility, etc.) and ability to withstand environmental conditions (temperature, humidity, and many more, including over long times.) Cables add even more stringent environmental tests.

Connectors and splices are tested in large batches to determine average losses expected in normal installations. Environmental testing mirrors that for cables, but may add tests for special applications like vibration for use on vehicles, ships or aircraft. Transceivers, WDMs, fiber amplifiers and other fiber optic components will have testing for both fiber-related performance and electrical performance. Most of these tests have been standardized to allow fair comparison among various manufacturers' products.

Fiber Characterization

As network speeds increase beyond 10 Gb/s and distances become more than 20km, it has become important to test single-mode outside plant fibers for bandwidth. This is especially important for older fiber optic cable plants where network speeds are being upgraded because the older fibers were manufactured to different standards for slower networks. New installations are also usually tested for these same requirements to ensure they will also support speeds higher than 100 Gb/s and dense wavelength division multiplexing. The field tests for cable plants include chromatic dispersion (CD), polarization mode dispersion (PMD) and spectral attenuation (SA).

Restoration

The time may come when you have to troubleshoot and fix the cable plant. If you have a critical application or lots of network cable, you should be ready to do it yourself. Smaller networks can rely on a contractor. If you plan to do it yourself, you need to have equipment ready (extra cables, mechanical splices, quick termination connectors, etc., plus test equipment) and someone who knows how to use it.

We cannot emphasize more strongly the need to have good documentation on the cable plant. If you don't know where the cables go, how long they are or what they tested for loss, you will be spinning you wheels from the get-go. And you need tools to diagnose problems and fix them, and spares including a fusion splicer or some mechanical splices and spare cables. In fact, when you install cable, save the leftovers for restoration. And the first thing you must decide is if the problem is with the cables or the equipment using it. A simple power meter can test sources for output and receivers for input and a visual tracer will check for fiber continuity. If the problem is in the cable plant, the OTDR is the next tool needed to locate the fault.

Factors Affecting the Analysis

- Source Intensity Noise: Refers to noise generated by the LED or Laser.
- Phase Noise: The difference in the phases of two optical wave trains separated by time, cut out of the optical wave.
- Amplitude Noise: Caused by the laser emission process.
- Fiber Noise: Relates to modal partition noise.
- Receiver Noise: Photodiode, conversion resistor.
- Time Jitter and Inter symbol Interference:
    - Time Jitter: Short term variation or instability in the duration of a specified interval.
    - Inter symbol Interference result of other bits interfering with the bit of interest.
    - Inversely proportional to the bandwidth.
    - Eye diagrams: To see the effects of time jitter and inter symbol interference.
    - Bit error rate: Main quality criterion for a digital transmission system.

      $BER = Q\,[\sqrt{I_{MIN}^{2}/(4 \cdot N_o \cdot B)}]$

      where,

        - N_o = Noise power spectral density (A_2/Hz),
        - I_{MIN} = Minimum effective signal amplitude (Amps),
        - B = Bandwidth,
        - Q(x) = Cumulative distribution function (Gaussian distribution).

Wavelength Division Multiplexing

WDM (Wavelength-division Multiplexing) is the technology of combing a number of wavelengths onto the same fiber simultaneously. A powerful aspect of WDM is that each optical channel can carry any transmission format. WDW increases the capacity of a fiber network dramatically. Thus it is recognized as the Layer 1 transport technology in all tiers of the network.

Need of WDM

Due to the rapid growth in telecommunication links, high capacity and faster data transmission rates over farther distances are required. To meet these demands, network managers are relying more and more on fiber optics. Typically, there are three methods for expanding capacity: installing more cables, increasing system bitrate to multiplex more signals and wavelength division multiplexing.

- The first method, installing more cables, will be preferred in many cases, especially in metropolitan areas, since fiber has become incredibly inexpensive and installation methods more efficient. But when conduit space is not available or major construction is necessary, this may not be the most cost-effective.
- Another way for capacity expansion is to increase system bitrate to multiplex more signals. But increasing system bitrate may not prove cost effective either. Since many systems are already running at SONET OC-48 rates (2.5 GB/s) and upgrading to OC-192 (10 GB/s) is expensive, requires changing out all the electronics in a network, and adds 4 times the capacity, may not be necessary.
- Thirdly, the WDM has been proved to be the more cost-effective technology. It does not only support current electronics and fibers but also can share fibers by transmitting channels at different wavelengths (colors) of light. Besides, systems are already using fiber optic amplifiers as repeaters also do not require upgrading for most WDM.

From the above comparison of three methods for expanding capacity, we can easily draw a conclusion that WDM is the best solution to meet the demand for more capacity and faster data transmission rates.

Actually, it is not difficult to understand the operating principle of WDM. Consider the fact that you can see many different colors of light: red, green, yellow, blue, etc. The colors are transmitted through the air together and may mix, but they can be easily separated by using a simple device like a prism. It's like we separate the "white" light from the sun into a spectrum of colors with the prism. WDM is equivalent to the prism in the operating principle. A WDM system uses a multiplexer at the transmitter to join the several signals together. At the same time, it uses a demultiplexer at the receiver to split them apart. With the right type of fiber, it is possible to function as an optical add-drop multiplexer.

This technique was originally demonstrated with optical fiber in the early 80s. The first WDM systems combined only two signals. Modern systems can handle up to 160 signals and can thus expand a basic 10 Gbit/s system over a single fiber pair to over 1.6 Tbit/s. Because WDM systems can expand the capacity of the network and accommodate several generations of technology development in optical infrastructure without having to overhaul the backbone network, they are popular with telecommunications companies.

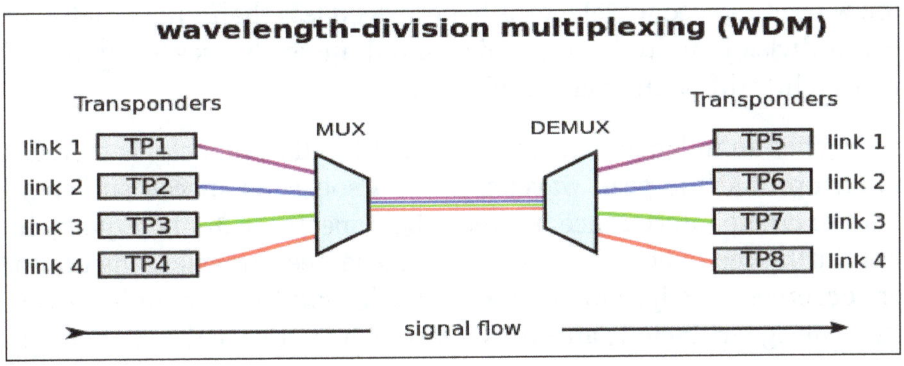

CWDM Vs DWDM

WDM systems are divided into different wavelength patterns: CWDM (Coarse Wavelength Division Multiplexing) and DWDM (Dense Wavelength Division Multiplexing). There are many differences between CWDM and DWDM: spacings, DFB lasers, and transmission distances. The channel spacings between individual wavelengths transmitted through the same fiber serve as the basis for defining CWDM and DWDM. Typically, the spacing in CWDM systems is 20 nm, while most DWDM systems today offer 0.8 nm (100 GHz) wavelength separation according to the ITU standard.

Due to wider CWDM channel spacing, the number of channels (lambdas) available on the same link is significantly reduced, but the optical interface components do not have to be as precise as DWDM components. CWDM equipment is thus significantly cheaper than DWDM equipment. Both CWDM and DWDM architectures utilize the DFB (Distributed Feedback Lasers). However, CWDM systems use DFB lasers that are not cooled. These systems typically operate from 0 to 70 °C with the laser wavelength drifting about 6 nm over this range. Coupled with the laser wavelength of up to ±3 nm, the wavelength drift yields a total wavelength variation of about ±12 nm.

DWDM systems, on the other hand, require the larger cooled DFB lasers, because a semiconductor laser wavelength drifts about 0.08 nm/°C with temperature. DFB lasers are cooled to stabilize the wavelength from outside the passband of the multiplexer and demultiplexer filters as the temperature fluctuates in DWDM systems. Due to the unique attributes of CWDM and DWDM, they are deployed for different transmission distances. Typically, CWDM can travel anywhere up to about 160 km. If we need to transmit the data over a long range, the DWDM system is the best choice. DWDM supports 1550 nm wavelength size, which can be amplified to extend transmission distance to hundreds of kilometers.

Operational Principles of WDM

Since the spectral width of a high-quality source occupies only a narrow slice of optical bandwidth, there are many independent operating regions across the spectrum, ranging from the a-band through the L-band that can be used simultaneously. The original use of WDM was to upgrade the capacity of installed point-to-point transmission links.

This was achieved with wavelengths that were separated from several tens up to 200 nm in order not to impose strict wavelength-tolerance requirements on the different laser sources and the receiving wavelength splitters. Subsequently, the development of lasers that have extremely narrow spectral emission widths allowed wavelengths to be spaced less than a nanometer apart. This is the basis of wavelength-division multiplexing, which simultaneously uses a number of light sources, each emitting at a slightly different peak wavelength.

Each wavelength carries an independent signal, so that the link capacity is increased greatly. The main trick is to ensure that the peak wavelength of a source is spaced sufficiently far from its neighbor so as not to create interference between their spectral extents. Equally important is the requirement that during the operation of a system these peak wavelengths do not drift into the spectral territory occupied by adjacent channels. In addition to maintaining strict control of the wavelength, system designers include an empty guard band between the channels as an operations safety factor. Thereby the fidelities of the independent messages from each source are maintained for subsequent conversion to electrical signals at the receiving end.

WDM Operating Regions

The possibility of having an extremely high-capacity link by means of WDM can be seen by examining the characteristics of a high-quality optical source. As an example, a distributed feedback (DFB) laser has a frequency spectrum on the order of I MHz, which is equivalent to a spectral line width of 10-5 nm. With such spectral widths, simplex systems make use of only a tiny portion of the transmission bandwidth capability of a fiber. This can be seen from Figure, which depicts the attenuation of light in a silica fiber as a function of wavelength. The curve shows that the two low-loss regions of a standard G.652 single-mode fiber extend over the O-band wavelengths ranging from about 1270 to 1350 nm (originally called the second window) and from 1480 to 1600nm (originally called the third window). We can view these regions either in terms of spectral width (the wavelength band occupied by the light signal) or by means of optical bandwidth (the frequency band occupied by the light signal).

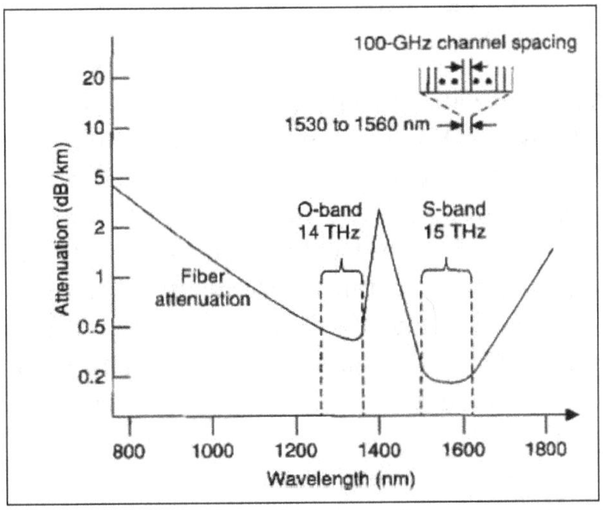

Genetic representation of the attenuation of light in a silica fiver as a function of wavelength.

To find the optical bandwidth corresponding to a particular spectral width in these regions, we use the fundamental relationship c=Lamda*v, which relates the wavelength Lamda. To the carrier frequency v, where c is the speed of light. Differentiating this, we have,

$$\Delta v = \frac{c}{\lambda^2}\Delta\lambda$$

Where the frequency deviation Δv corresponds to the wavelength deviation $\Delta\lambda$ around λ.

Now suppose we have a fiber that has the attenuation characteristic shown in Figure. From above equation the optical bandwidth is .Deltav= 14THz for a usable spectral band .DeltaLamda= 80 nm in the center of the O-band. Similarly, .Deltav= 15 THz for a usable spectral band DeltaLamda= 120 nm in the low-loss region running from near the beginning of the S-band to almost the end of the L-band. This yields a total available fiber bandwidth of about 30THz in the two low-loss windows.

Prior to about 2000, the peak wavelengths of adjacent light sources typically were restricted to be separated by 0.8 to 1.6 nm (100 to 200 GHz) in a WDM system. This was done to take into account possible drifts of the peak wavelength due to aging or temperature effects, and to give both the manufacturer and the user some leeway in specifying and choosing the precise peak emission wavelength. The next generation of WDM systems specified both narrower and much wider channel spacings depending on the application and on the wavelength region being used. The much narrower spacings thus require strict wavelength control of the optical source. On the other hand, the wider wavelength separations offer inexpensive WDM implementations since wavelength control requirements are relaxed significantly.

Generic WDM Link

The implementation of WDM networks requires a variety of passive and/or active devices to combine, distribute, isolate, add, drop, attenuate, and amplify optical power at different wavelengths. Passive devices require no external electric power or control for their operation, so they have a fixed application in WDM networks. These passive components are used to separate and combine wavelength channels, to divide optical power onto a number of fiber lines, or to tap off part of an optical signal for monitoring purposes.

The performance of active devices can be controlled electronically, thereby providing a large degree of network flexibility, Active WDM components include tunable optical filters, tunable light sources, configurable add/drop multiplexers, dynamic gain equalizers, and optical amplifiers. The transmitting side has a series of independently modulated fixed-wavelength light sources, each of which emits signals at a unique wavelength. Here a multiplexer (popularly called a mux) is needed to combine these optical outputs into a continuous spectrum of signals and couple them onto a single fiber. Within a standard telecommunication link there may be various types of optical amplifiers, a variety of specialized active components (not shown), and passive optical power splitters. The operations and maintenance benefits of PONs are that no active devices are used between the transmitting and receiving endpoints.

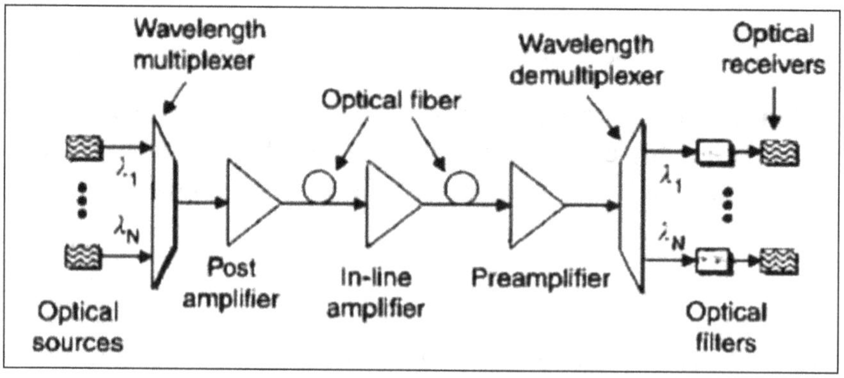

Implementation of a simple WDM link.

At the receiving end a demultiplexer is required to separate the individual wavelengths of the independent optical signals into appropriate detection channels for signal processing. At the transmitter the basic design challenge is to have the multiplexer provide a low-loss path from each optical source to the multiplexer output. A different requirement exists for the demultiplexer, since photodetectors usually are sensitive over a broad range of wavelengths, which could include all the WDM channels.

To prevent spurious signals from entering a receiving channel, that is, to give good channel isolation of the different wavelengths being used, the demultiplexer must exhibit narrow spectral operation or very stable optical filters with sharp wavelength cutoffs must be used. The tolerable crosstalk levels between channels can vary widely depending on the application. In general, a -lOdB level is not sufficient, whereas a level of - 30 dB is acceptable. In principle, any optical demultiplexer can also be used as a multiplexer. For simplicity, the word multiplexer is used as a general term to refer to both combining and separating functions, except when it is necessary to distinguish the two devices or functions.

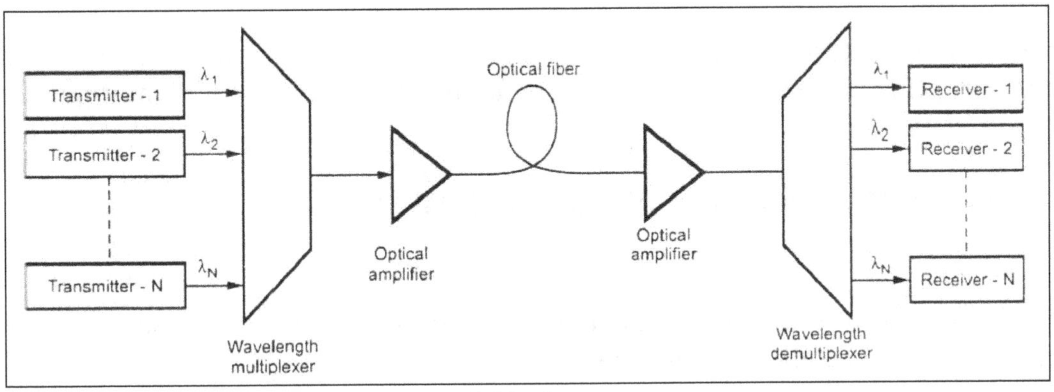

WDM scheme.

Optical signals of different wavelength (1300-1600 nm) can propagate without interfering with each other. The scheme of combining a number of wavelengths over a single fiber is called wavelength division multiplexing (WDM). Each input is generated by a separate optical source with a unique wavelength. An optical multiplexer couples light from individual sources to the transmitting fiber. At the receiving station, an optical demultiplexer is required to separate the different carriers before photo detection of individual signals. To prevent spurious signals to enter into receiving channel, the demultiplexer must have narrow spectral operation with sharp wavelength cut-offs. The acceptable limit of crosstalk is – 30 dB.

Features of WDM

Important advantages or features of WDM are as mentioned below:

- Capacity upgrade: Since each wavelength supports independent data rate in Gbps.
- Transparency: WDM can carry fast asynchronous, slow synchronous, synchronous analog and digital data.
- Wavelength routing: Link capacity and flexibility can be increased by using multiple wavelength. Wavelength switching: WDM can add or drop multiplexers, cross connects and wavelength converters.

Passive Components

For implementing WDM various passive and active components are required to combine, distribute, isolate and to amplify optical power at different wavelength. Passive components are mainly used to split or combine optical signals. These components operate in optical domains. Passive components don't need external control for their operation. Passive components are fabricated by using optical fibers by planar optical waveguides. Commonly required passive components are:

- N × N couplers,
- Power splitters,
- Power taps,
- Star couplers.

Most passive components are derived from basic stat couplers. Stat coupler can person combining and splitting of optical power. Therefore, star coupler is a multiple input and multiple output port device.

Dense Wavelength Division Multiplexing (DWDM)

DWDM

- DWDM (Dense wavelength – division multiplexing) is a data transmission technology having very large capacity and efficiency.
- Multiple data channels of optical signals are assigned different wavelengths, and are multiplexed onto one fiber.
- DWDM system consists of transmitters, multiplexers, optical amplifer and demultiplexer. Figure shows typical application of DWDM system.
- DWDM used single mode fiber to carry multiple light waves of different frequencies.
- DWDM system uses Erbium: Doped Fiber Amplifers (EDFA) for its long haul applications, and to overcome the effects of dispersion and attenuation channel spacing of 100 GHz is used.

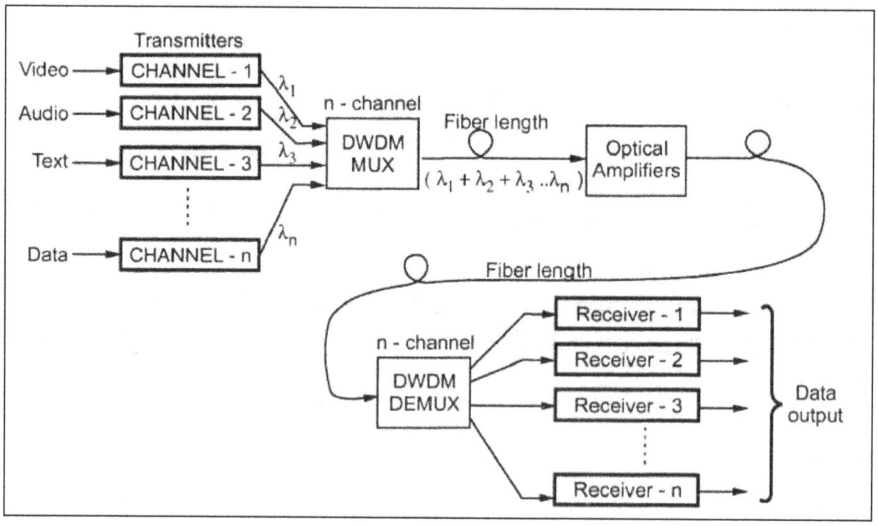

DWDM system.

DWDW is short for dense wavelength division multiplexing. It is an optical multiplexing technology used to increase bandwidth over existing fiber networks. DWDM works by combining and transmitting multiple signals simultaneously at different wavelengths on the same fiber. It has revolutionized the transmission of information over long distances. DWDM can be divided into passive DWDM and active DWDM.

Passive DWDM

Passive DWDM systems have no active components. The line functions only due to the optical budget of transceivers used. No optical signal amplifiers and dispersion compensators are used. Passive DWDM systems have a high channel capacity and potential for expansion, but the transmission distance is limited to the optical budget of transceivers used. The main application of passive DWDM system is metro networks and high speed communication lines with a high channel capacity.

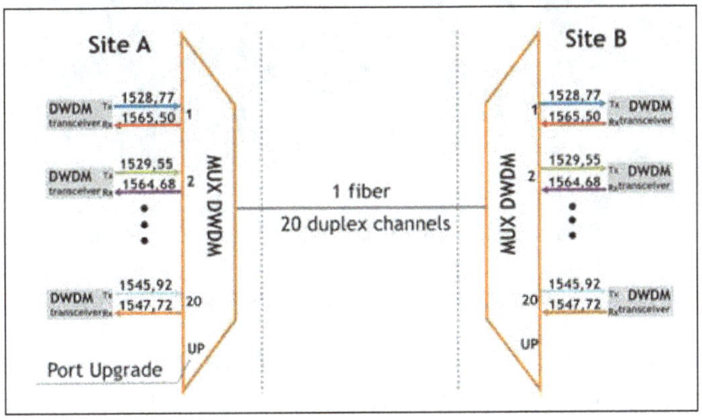

Active DWDM

Active DWDM systems commonly refer to as a transponder-based system. They offer a way to transport large amounts of data between sites in a data center interconnect setting. The transponder takes the outputs of the SAN or IP switch format, usually in a short wave 850nm or long wave 1310nm format, and converts them through an optical-electrical-optical (OEO) DWDM conversion. When creating long-haul DWDM networks, several EDFA amplifiers are installed sequentially in the line. The number of amplifiers in one section is limited and depends on the optical cable type, channel count, data transmission rate of each channel, and permissible OSNR value.

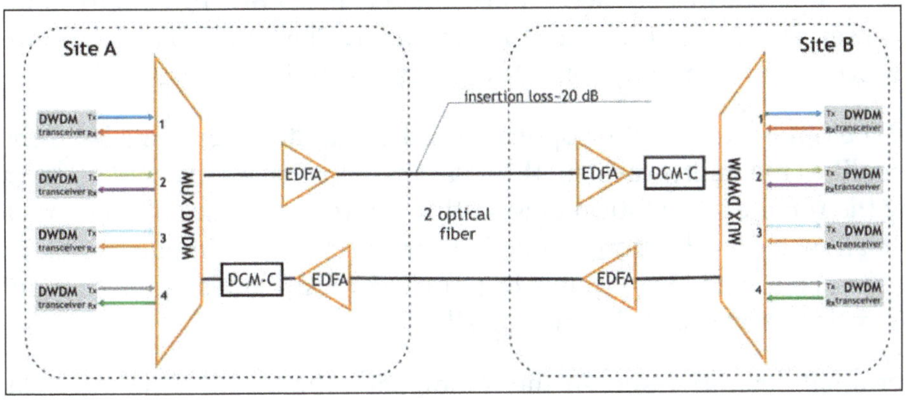

The possible length of lines when using active DWDM system is determined not only with installed optical amplifiers and the OSNR value, but also with the influence of chromatic dispersion—the distortion of transmitted signal impulses, on transmitted signals. At the design stage of the DWDM network project, permissible values of chromatic dispersion for the transceivers are taken into account, and, if necessary, chromatic dispersion compensation modules (DCM) are included in the line. DCM introduces additional attenuation into the line, which leads to a reduction of the amplified section length.

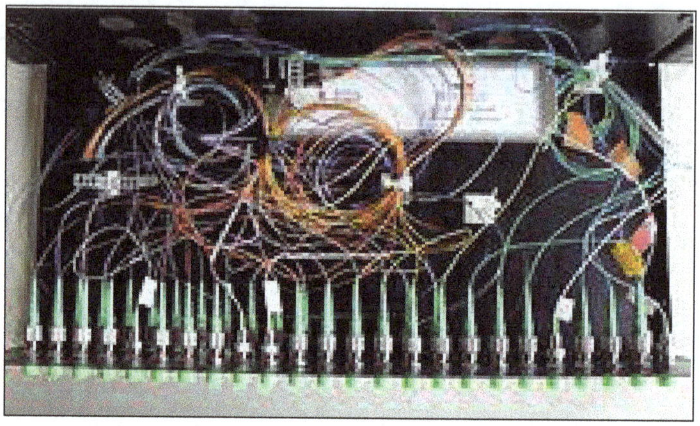

WDM multiplexer for DWDM communications.

At this stage, a basic DWDM system contains several main components:

1. A DWDM terminal multiplexer: The terminal multiplexer contains a wavelength converting transponder for each data signal, an optical multiplexer and where necessary an optical amplifier (EDFA). Each wavelength-converting transponder receives an optical data signal from the client-layer, such as Synchronous optical networking [SONET /SDH] or another type of data signal, converts this signal into the electrical domain and retransmits the signal at a specific wavelength using a 1,550 nm band laser. These data signals are then combined together into a multi-wavelength optical signal using an optical multiplexer, for transmission over a single fiber (e.g., SMF-28 fiber). The terminal multiplexer may or may not also include a local transmit EDFA for power amplification of the multi-wavelength optical signal. In the mid-1990s DWDM systems contained 4 or 8 wavelength-converting transponders; by 2000 or so, commercial systems capable of carrying 128 signals were available.

2. An intermediate line repeater is placed approximately every 80–100 km to compensate for the loss of optical power as the signal travels along the fiber. The 'multi-wavelength optical signal' is amplified by an EDFA, which usually consists of several amplifier stages.

3. An intermediate optical terminal or optical add-drop multiplexer: This is a remote amplification site that amplifies the multi-wavelength signal that may have traversed up to 140 km or more before reaching the remote site. Optical diagnostics and telemetry are often extracted or inserted at such a site, to allow for localization of any fiber breaks or signal impairments. In more sophisticated systems (which are no longer point-to-point), several signals out of the multi-wavelength optical signal may be removed and dropped locally.

4. A DWDM terminal demultiplexer: At the remote site, the terminal de-multiplexer consisting of an optical de-multiplexer and one or more wavelength-converting transponders separates the

multi-wavelength optical signal back into individual data signals and outputs them on separate fibers for client-layer systems (such as SONET/SDH). Originally, this de-multiplexing was performed entirely passively, except for some telemetry, as most SONET systems can receive 1,550 nm signals. However, in order to allow for transmission to remote client-layer systems (and to allow for digital domain signal integrity determination) such de-multiplexed signals are usually sent to O/E/O output transponders prior to being relayed to their client-layer systems. Often, the functionality of output transponder has been integrated into that of input transponder, so that most commercial systems have transponders that support bi-directional interfaces on both their 1,550 nm (i.e., internal) side, and external (i.e., client-facing) side. Transponders in some systems supporting 40 GHz nominal operation may also perform forward error correction (FEC) via digital wrapper technology.

5. Optical Supervisory Channel (OSC): This is data channel which uses an additional wavelength usually outside the EDFA amplification band (at 1,510 nm, 1,620 nm, 1,310 nm or another proprietary wavelength). The OSC carries information about the multi-wavelength optical signal as well as remote conditions at the optical terminal or EDFA site. It is also normally used for remote software upgrades and user (i.e., network operator) Network Management information. It is the multi-wavelength analogue to SONET's DCC (or supervisory channel). ITU standards suggest that the OSC should utilize an OC-3 signal structure, though some vendors have opted to use 100 megabit Ethernet or another signal format. Unlike the 1550 nm multi-wavelength signal containing client data, the OSC is always terminated at intermediate amplifier sites, where it receives local information before re-transmission.

References

- Fiber-optical-communications: mrcet.com, Retrieved 19, May 2020
- Power-Budgets-And-Loss-Budgets: thefoa.org, Retrieved 23, January 2020

4

Optical Couplers

An electronic component that is used to transfer electric signals between two circuits by using light is termed as an optical coupler. There are various types of optical couplers such as directional couplers, star couplers and T couplers. The topics elaborated in this chapter will help in gaining a better perspective about these types of optical couplers.

Optical coupler is a semiconductor device, which is designed to transfer electrical signals by using light waves in order to provide coupling with electrical isolation between circuits or systems. The main purpose of an optical coupler is to prevent rapidly changing voltages or high voltages on one side of a circuit from distorting transmissions or damaging components on the other side of the circuit, and to separate the high-voltage side from the low-voltage side.

Directional Couplers

A directional coupler is an electronic component having four-port circuits with one port being isolated from the input port and another being considered as a through port. The device is normally used to split the input signal and distributed power. The device couples part of the transmission power by a specific factor through one port. Directional couplers are used in a wide range of applications which involve measurement, power monitoring and other utilities.

Directional couplers are categorized as passive reciprocal networks. A directional coupler is used for isolating, eliminating or combining signals in microwave signal routing and radio frequency. The ports in the directional coupler are:

- Coupled
- Input
- Transmitted
- Isolated

A special design is put into use by which the input power is split between the coupled and output ports in a specific ratio known as the coupling ratio. Depending on the application for which it is used, the key specifications of the directional coupler varies. The parameters/specifications which are mostly varied are the coupling factor, transmission loss, low variation of the coupling attenuation, high directivity and input power. For most directional couplers, the features which are desired are high directivity, good impedance and wide operational bandwidth. But the performance of a directional coupler is computed using the directivity factor. There are different types of directional couplers like single, dual directional, coaxial, waveguide and even combination types.

Basic Principle of a Directional Coupler

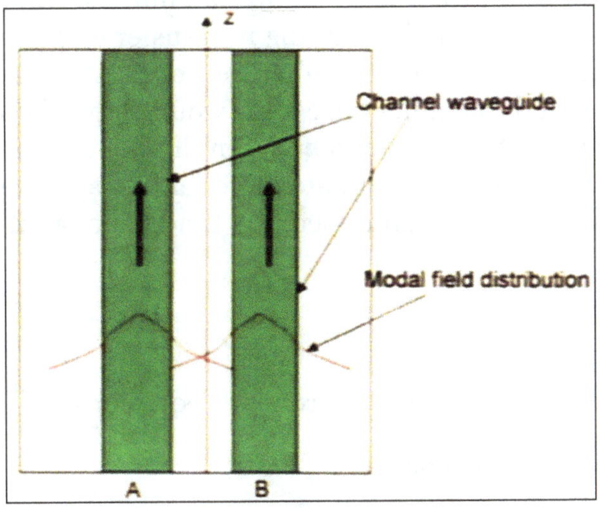

Integrated Directional Coupler.

Let us assume the channel A and B to be exactly identical in all respects and we excite only channel A at z = 0. According to the assumed geometry, a wave would now propagate in the positive z direction as a result of excitation by a source. However, the power of the wave will not remain confined only to channel A as the wave propagates. Due to the phenomenon of Evanescent coupling, there would be coupling of power into the channel B as well and if the length of the channels are sufficiently large, then at some z, the entire power launched into A may get coupled to B. Since we assume the channels A and B to be identical, the field distribution of the travelling wave in both the channels would also be exactly identical.

If 'b' is the isolated phase constant of each of the two channels in the absence of the other channel, then a forward propagating wave in channel A and channel B can be represented by the following two equations:

$$a = a_o e^{-j\beta z}$$

$$b = b_o e^{-j\beta z}$$

Here, a and b are the signals in the two channels. Since we assume only one channel to be present, there would be no coupling of power and the wave propagation in the two channels is governed by the following differential equations:

$$\frac{da}{dz} = -j\beta a_o e^{-j\beta z} = -j\beta a$$

$$\frac{db}{dz} = -j\beta b_o e^{-j\beta z} = -j\beta b$$

Now, in presence of both the channels, the fields of the waves in the two channels would get coupled. Let us define a quantity κ known as the coupling coefficient which gives a measure of the amount of coupling between the fields in the two channels. In other words, κ is a measure of the amount of overlap between the modal fields in the two waveguides. More the overlap more is the coupling. Thus the quantity κ is governed by an overlap integral which indicates the behaviour of the coupling between the modal fields resulting in transfer of energy from one waveguide to the other. From our earlier discussion on filed distribution in an optical fiber, we can infer that the modal field distribution in the channel waveguide would depend on various parameters such as the width of the channel (d), the separation between the two channels (S), the refractive indices of the two waveguides (n_1) and the substrate material (n_2) and the wavelength of operation (λ). So, the coupling coefficient can be expressed as a function of all these parameters as shown below:

$$\kappa = f(d, S, n_1, n_2, \lambda)$$

In the presence of coupling, the differential equations $\frac{da}{dz} = -j\beta a_o e^{-j\beta z} = -j\beta a$ and $\frac{db}{dz} = -j\beta b_o e^{-j\beta z} = -j\beta b$ would now change to,

$$\frac{da}{dz} = -j\beta a - j\kappa b$$

$$\frac{db}{dz} = -j\beta b - j\kappa a$$

The above two equations are the coupled equations between a and b which govern the propagation of energy in the two channels in the presence of coupling. The analysis of these equations is, hence, known as coupled mode analysis. Analytical solutions of the two differential equations above w.r.t. appropriate boundary conditions give the field distribution in the two waveguides in the presence of coupling between the two channels. To solve the above equations, we need to de-couple them which can be done by taking a second derivative followed by a substitution. Following the same procedure for both the equations, we obtain the following decoupled equations:

$$\frac{d^2a}{dz^2} + j2\beta\frac{da}{dz} + \left(\kappa^2 + \beta^2\right)a = 0$$

$$\frac{d^2b}{dz^2} + j2\beta\frac{db}{dz} + \left(\kappa^2 - \beta^2\right)b = 0$$

The decoupled equations are second order differential equations with constant coefficients and can be easily solved to obtain the following solutions:

$$a(z) = \left(A_1 \cos\kappa z + A_2 \sin\kappa z\right)e^{-j\beta z}$$

$$b(z) = \left(B_1 \cos\kappa z + B_2 \sin\kappa z\right)e^{-j\beta z}$$

The quantities A_1, A_2, B_1 and B_2 are arbitrary constants that need to be determined with the help of appropriate boundary conditions. The analysis, so far, has been addressed to in a general manner, but for generating appropriate boundary conditions we need to consider some specific conditions.

The initial condition is that at z = 0, only the channel A has been excited with a unity source and there is no excitation applied to channel B. Therefore:

$$\text{Initial Condition: at } z = 0,\ a(0) = 1;\ b(0) = 0$$

Applying the initial conditions to the equations $a(z) = (A_1 \cos \kappa z + A_2 \sin \kappa z)e^{-j\beta z}$ and $b(z) = (B_1 \cos \kappa z + B_2 \sin \kappa z)e^{-j\beta z}$ and solving for the arbitrary constants, we obtain the following solutions:

$$a(z) = \cos \kappa z\ e^{-j\beta z}$$

$$b(z) = -j \sin \kappa z\ e^{-j\beta z}$$

The above equations suggest that the field propagation in the two channels occur in a sinusoidal fashion where the amplitudes of the waves undergo sinusoidal variations and the phase variation is indicated by the exponential term. If we plot these variations, we obtain the following plot:

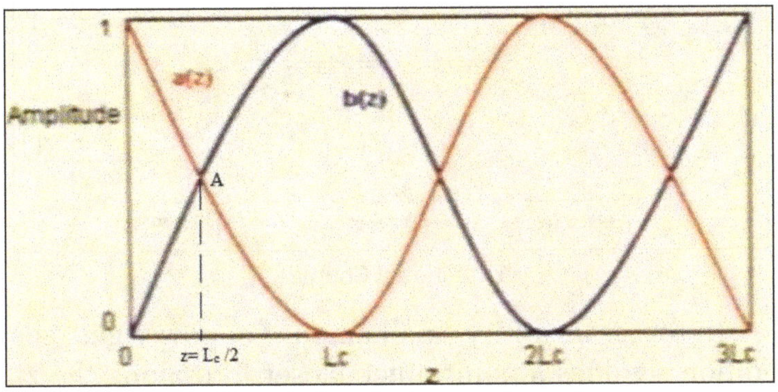

Field Variations in the two channels.

As seen from the plot of the field variation, at z = 0, the entire power lies in the channel A and the amplitude of the field is 1(maximum) at z = 0 in Channel A whereas in B it is zero indicating no power in channel B. However, as the fields propagate (as Z increases), the coupling begins and the power in the channel A starts coupling into channel B which is indicated by decreasing amplitude of the field in A and a corresponding increase field amplitude in channel B. As the wave moves further ahead, at a distance $z = L_c$, then entire power that was launched into A gets coupled into B which is indicated by the field amplitude in B becoming maximum whereas in A it goes to zero. As the wave moves beyond L_c, the power in B gets re-coupled to the channel A and at a distance $2L_c$ the entire power is available in A again. This cycle of exchange of power between the two channels goes on as the wave moves further ahead and this fact is indicated by the sinusoidal field amplitude variations shown in the above figure. The distance L_c is, in fact, the point where $\kappa z = \pi / 2$. This length L_c is called the coupling length of the channels. We can, now, relate the coupling coefficient κ and the coupling length L_c as shown below:

$$\kappa = \frac{\pi}{2L_c}$$

This relation shows that the coupling coefficient is inversely proportional to the coupling length of the two channels. Stronger the coupling more is the value of the coupling coefficient and shorter is the coupling length. The important thing to note is the fact that however small the coupling coefficient may be, as long as the fields of the two channels interact, there would always be a length at which the entire power in one channel would get coupled to the other channel. If we take a channel length equal to L_c and install a detector at the other end, we would observe that even if energy was launched into channel A initially, yet the detector at A shows no output detected energy whereas the detector at B would detect the entire energy initially launched channel A which has been coupled into B over the coupling length of the channel.

The directional coupler is another basic integrated device in WDM networks and a variety of other complex devices are fundamentally based on the directional coupler. For example, with appropriate modifications to the basic design, a directional coupler can be made to act as an optical switch or even a power divider or a router. Let us now look into some of the applications that a directional coupler can be put into.

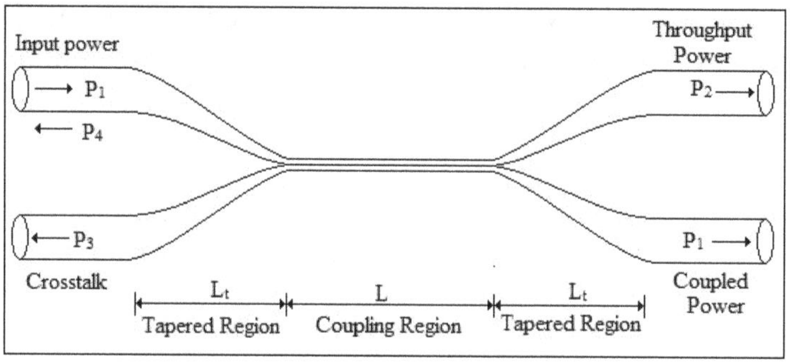

Directional Coupler.

- Case: If L<< L_c - In this case, a very small amount of power gets coupled to channel B (in figure this condition signifies a point much closer to 0 along the z axis) and most of the power will remain confined to channel A. That is if coupled power P_1 = a, then P_2 = 1-a for a launched power P = 1. This observation is made use in tapping of optical signals.
- Case: If L=L_c/2 - This condition signifies a operating condition corresponding to point A In figure. In this case, output power at both the waveguides is equal to half of the total power launched into A (3dB power). Thus the directional coupler, now, acts as a 3dB power divider for optical signals.
- Case: If L=L_c - In this case Optical power launched into channel A would be available at the output of channel B and power launched into channel B would be available at the output of channel A due to evanescent coupling between the two channels. Thus making the length of the channels equal to the coupling length, the directional coupler may be made to function as a signal cross-over module.

If a directional coupler is fabricated on an electro-optic substrate such as $LiNbO_3$, then the electro-optic effect of the substrate material can be combined with the above directional coupler principle and a new range of complex devices can be constructed. Applying an electric field to such kind of device changes the refractive index of the channel and as a result changes the coupling coefficient of the device and by equation $\kappa = \frac{\pi}{2L_c}$ the coupling length of the two channels change too.

Using this observation, a very basic directional coupler fabricated on an electro-optic substrate can now be dynamically converted onto a different module altogether. For example, application of appropriate electric field can result in directional couple being switched to operate in any one of the three cases mentioned above.

Directional Coupler as an Amplitude Modulator

A directional coupler fabricated on an electro-optic modulator can be made to function as an amplitude modulator. For our discussion we shall consider the following device:

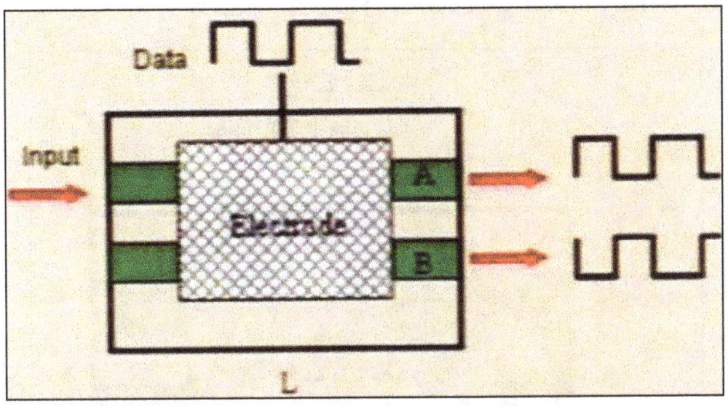

Amplitude Modulator using a Directional Coupler.

Two channel waveguides A and B, each of length L, are fabricated on a substrate of $LiNbO_3$ and a provision of application of electric field to these channels have been made as shown in the above figure. Continuous wave (CW) optical signal is provided as an input to channel A. The data signal which is a stream of rectangular pulses, are fed to the electrode. This signal dynamically alters the nature if the channel and the optical outputs of the two channels are detected on the other end as shown. Adopting proper design methodology, the dimensions of the channels are such that when the data signal goes HIGH, the length L becomes equal to $2L_c$ and when the data signal is low $L=L_c$. With this configuration, the outputs at A and B corresponding to the data signal input are shown in figure.

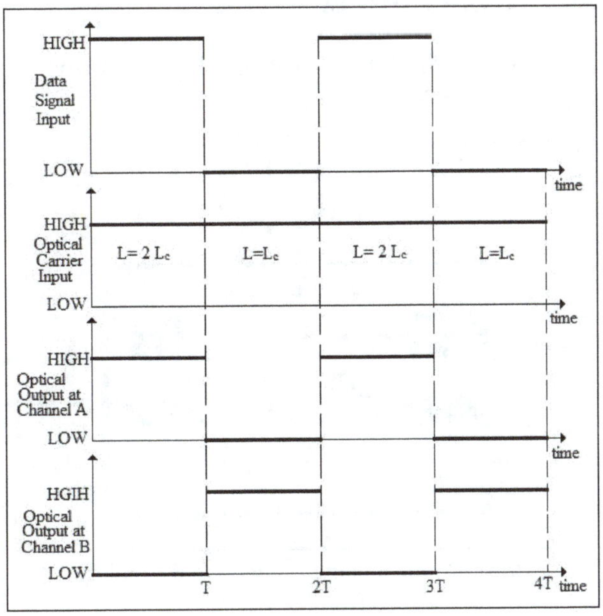

Output of Amplitude Modulator.

As seen from the above figure, when the data signal goes HIGH, the length $L = 2L_c$ and the output at A corresponds to the input at A and hence it goes HIGH and the output at B goes low due to no power being coupled to B at that L. When the data signal goes LOW, the length $L = L_c$ and due to evanescent coupling entire input to A gets coupled to B. Consequently, B goes HIGH and A goes low. Thus, the output at B may be assumed to be the complement of the output at A.

Directional Coupler as an Optical Multiplexer

The next device that a directional coupler on an electro-optic material can be made into is an optical multiplexer which is shown in the figure below:

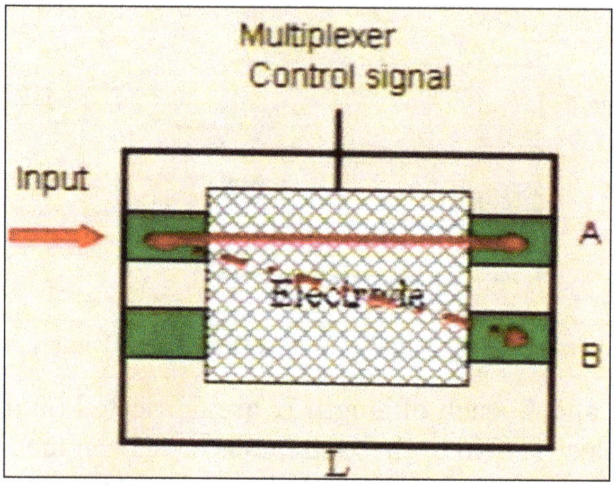

Optical Multiplexer.

The presence or absence of the input voltage changes the coupling length of the two channels so as to route the optical input signal either to the output of channel A or channel B thus creating a multiplexing action in the optical domain.

Directional Coupler as Cross-connect

The same principles may be used to construct another device known as the cross connect which, basically, is a 2x2 optical switch which is shown below:

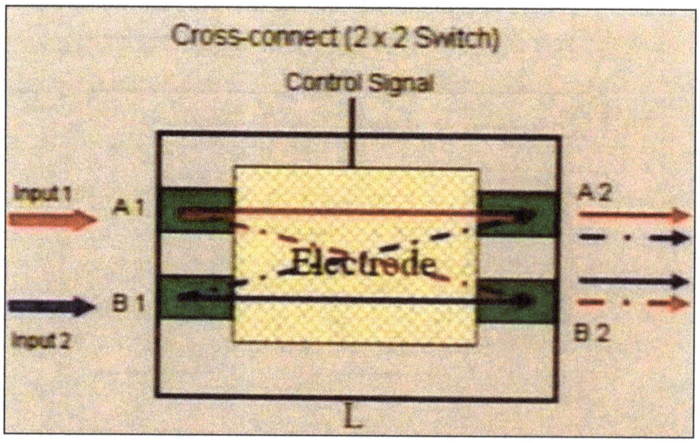

Cross-Connect using a directional coupler.

The cross-connect has a similar operating principle as an optical multiplexer discussed above, the only difference being the presence of simultaneous inputs on both channels A and B. When the control signal goes HIGH, the length L of the device becomes equal to L_c and the signal power in the input of A gets coupled to B and the signal power in the input of B gets coupled to A. Thus the device is in a cross-connected mode and this state of the device is said to be a cross-state. In the absence (or LOW) of the control signal, the length $L = 2L_c$ and the input of A is detected at the output of A and the input to B is detected at the output of B. This state of the device is said to be Bar-state.

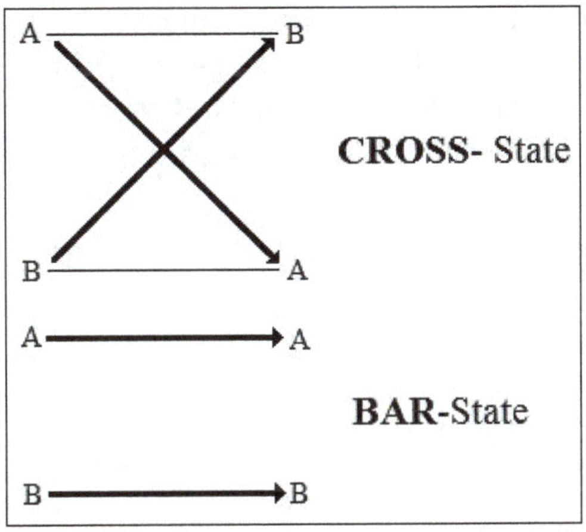

CROSS and BAR states of a cross-connect: 2x2 cross-connects may be intelligently connected in layered fashion to construct higher order (nxn) cross-connects. For example, a 3x3 cross-connect can be constructed using 2x2 cross-connects.

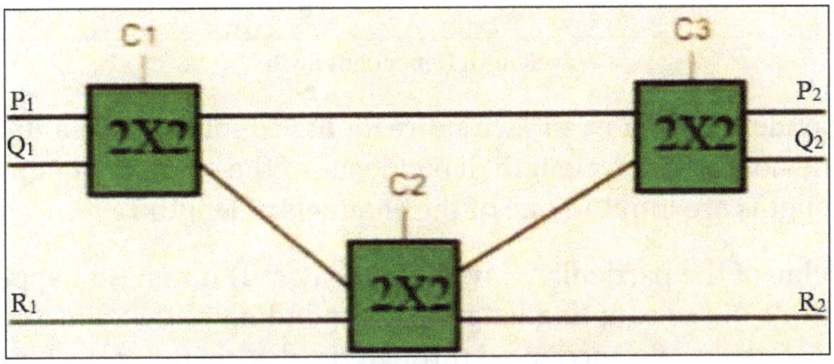

3x3 cross-connect using 2x2 cross-connects.

Depending on the control signals C_1, C_2 and C_3 the 2x2 cross-connects operate in the CROSS (x) and BAR (=) states and the corresponding outputs are produced. The output table corresponding to the different configurations of the control signals is given below:

Table: Output table of 3x3.

Control Signal Configuration			Outputs		
C_1	C_2	C_3	P_2	Q_2	R_2
BAR	BAR	BAR	P_1	Q_1	R_1
BAR	BAR	CROSS	Q_1	P_1	R_1
BAR	CROSS	BAR	P_1	R_1	Q_1
BAR	CROSS	CROSS	R_1	P_1	Q_1
CROSS	BAR	BAR	Q_1	P_1	R_1

CROSS	BAR	CROSS	P_1	Q_1	R_1
CROSS	CROSS	BAR	Q_1	R_1	P_1
CROSS	CROSS	CROSS	R_1	Q_1	P_1

Directional Couplers as Wavelength Dependent Switches

Another device which may be constructed based on evanescent coupling, is the wavelength dependent switch. The fundamental principle behind the design of this device is the dependence of the coupling coefficient κ on the wavelength of operation. Consequently, the coupling length is also dependent on the wavelength of operation. The structure of a wavelength dependent switch has been shown in the following figure:

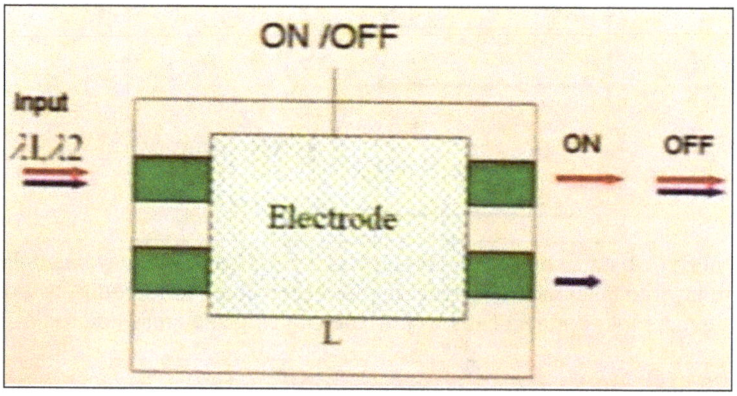

Wavelength Dependent Switch.

The wavelength dependent switch is, in fact a directional coupler put to a different fundamental use on the basis of the notion of wavelength dependence of the coupling length. Signals of two different optical wavelengths are input to one of the channels of length L.

Depending on the value of the particular wavelength (say red) it may so happen that the length L may be an even multiple of the coupling length corresponding to that wavelength and as a result that particular wavelength travels uncoupled through the device. But if the length L is an odd multiple of the coupling length corresponding to that wavelength, the entire energy in that wavelength gets coupled to the other channel and is available at its output. Thus, from number of wavelengths supplied as an input to the channel, the one whose coupling length is such that L is its odd multiple, gets separated out from the pool and is detected separately at the output of the other channel. This operation is similar to the operation of a wavelength selective filter or a wavelength selective dropping switch because it drops one particular wavelength out of many others. This is particularly useful in networks where multiple wavelengths travel together and at a user node such a switch can be connected so that the wavelength corresponding to that user may be dropped at his node and data is thus received by the intended user.

The above integrated devices enable light energy to be switched and routed from the source to the destination in the optical domain itself without creating the need for a conversion into the electrical domain. This has resulted in reduction of costs of optical networks and also increased the speed of communication in the optical domain by not having to convert the signal into equivalent electrical signal for routing, switching and other such signal processing purposes.

A Geometrical Approach of Wavelength Flattened Directional Couplers

The need for wide band optical communication demands optical devices able to cover the whole band of interest. A very basic element of a typical device is, of course, the power splitter. Power exchange between optical waveguides is usually obtained via directional couplers. The simplest splitting device is the synchronous coupler, obtained keeping two identical waveguides close together over a given length. This coupling mechanism is based on the difference in propagation constants of the coupler super-modes. On the other hand this means that it can strongly depend on the wavelength.

Many proposals have been presented to overcome this problem. For example coupling different waveguides (asynchronous coupler), tapering the coupled waveguides (tapered coupler), or implementing interferometric structures (Mach-Zehnder coupler). Takagi et al. fabricated an asynchronous coupler with a coupling ratio of (50±5)% over a wavelength range of 400nm, as well as a tapered coupler achieving (50±5)% over 500nm. Jinguji et al. built a Mach Zehnder coupler that achieves a splitting ratio of (50±1.9)% over 400nm, and with a similar structure Gonthier et al. obtained (50±2.5)% over 300nm. Here, we will focus on wavelength sensitivity of 50/50 splitters, but a similar approach applies to all splitting ratios as well as to changes of parameters other than wavelength.

Unitary Transformations

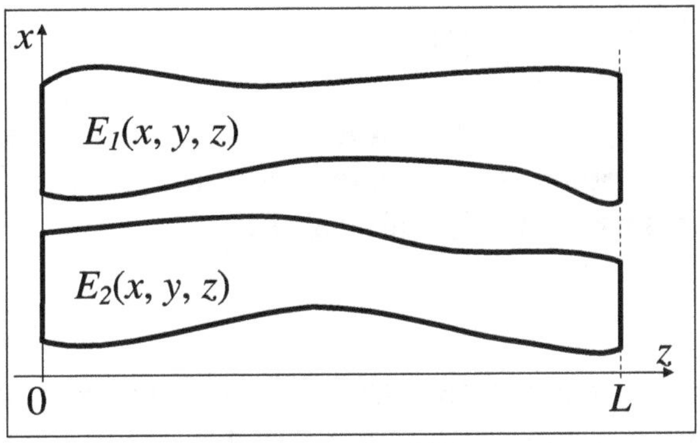

Schematic of a splitter.

We will assume a power splitter as a lossless device made of two waveguides, one beside the other over a length L along the propagation axis z. In general their cross-sections, their distance and their refractive index profiles may vary along z. We will also assume that, for any fixed z_0, the two eigenmodes of the single waveguides cross-section, $E_j[z_0] \equiv E_j(x, y, z_0)$, $(j = 1,2)$ can be considered as a basis for the field at that point, so that it is always possible to write:

$$E[z] \equiv a_1[z]E_1[z] + a_2[z]E_2[z]$$

where $a_j[z]$ are complex numbers so that $|a_1[z]|^2 + |a_2[z]|^2 = 1$ and $E_j[z]$ are normalized so that $|a_j|^2$ represent the power fractions in each of the waveguides. We will regard E[z] as a scalar quantity, since we will consider only singly copolarized modes of the structure. With these assumptions we can always represent the generic state of the system through the complex vector,

$$\mathbf{u}[z] \equiv \begin{pmatrix} a_1[z] \\ a_2[z] \end{pmatrix}.$$

Its evolution can be described by an unitary 2x2 matrix or, up to an overall phase, by an element **U** of SU(2) (i.e. an unitary matrix so that det **U** = 1). From its definition, the most general form for **U** is:

$$\begin{pmatrix} \exp(-i\theta/2)\cos\phi & \exp(i\xi)\exp(-i\theta/2)\sin\phi \\ -\exp(-i\xi)\exp(i\theta/2)\sin\phi & \exp(i\theta/2)\cos\phi \end{pmatrix}.$$

If optical power is launched in just one of the two waveguides, the splitting ratio (defined as the power fraction transferred to the other waveguide) does not depend on which waveguide is chosen and it is clearly given by $\sin^2\phi$.

Global View

Equation above describes a power splitter as a black box which propagates the input state from z=0 to z=L, i.e. so that **u**[L] = **U**[L, 0] **u**[0]. It can be cast in the form **U** = **V**(θ) **W**(φ, ξ) where:

$$\mathbf{V}(\theta) \equiv \begin{pmatrix} \exp(-i\theta/2) & 0 \\ 0 & \exp(i\theta/2) \end{pmatrix}$$

$$\mathbf{W}(\phi,\xi) \equiv \begin{pmatrix} \cos\phi & \exp(i\xi)\sin\phi \\ -\exp(-i\xi)\sin\phi & \cos\phi \end{pmatrix}$$

which physically means that any unitary transformation can be decomposed in the product of a θ-phase shift between the two branches and a coupling in which the field transferred from the upper (or lower) branch acquires a ξ (or π −ξ)-phase shift.

Local View

If we look locally at the whole transformation **U**[L, 0], we can decompose it in an infinite number of infinitesimal unitary transformations $\mathbf{U}_n \equiv \mathbf{U}\left[z_{n+1}, z_n\right]$:

$$\mathbf{U}[L,0] = \lim_{N\to\infty} \overleftarrow{\prod_{n=0}^{N-1}} \mathbf{U}_n$$

where $\overleftarrow{\Pi}$ means matrix product on the left, and we have defined $\Delta z \equiv L/N$, $z_n \equiv n\Delta z$ (from now on $f_n \equiv f(z_n) \forall f(z)$).

From the physical interpretation of equations $\mathbf{V}(\theta) \equiv \begin{pmatrix} \exp(-i\theta/2) & 0 \\ 0 & \exp(i\theta/2) \end{pmatrix}$ and $\mathbf{W}(\phi,\xi) \equiv \begin{pmatrix} \cos\phi & \exp(i\xi)\sin\phi \\ -\exp(-i\xi)\sin\phi & \cos\phi \end{pmatrix}$ we can think of every small section of the splitter as a composition of a local phase shifter $\mathbf{V}(\theta_n)$ with $\theta_n \equiv \Delta\beta_n \Delta z$ (where $\Delta\beta_n \equiv \beta_2[z_n] - \beta_1[z_n]$ is the local difference in propagation constants of the two branches) and a local coupler $\mathbf{W}(\phi_n, \xi_n)$ with $\phi_n = \kappa_n \Delta z$ (where κ_n is the local coupling coefficient between the two waveguides and ξ_n is the phase acquired locally, passing from the first to the second branch).

To 1st order in Δz can rewrite,

$$\mathbf{U}_n \equiv \mathbf{I} + \tau_n \Delta z$$

where **I** is the identity matrix and,

$$\tau_n \equiv \begin{pmatrix} -{}^{i}\!/_{2}\,\Delta\beta_n & \kappa_n \exp(i\xi_n) \\ -\kappa_n \exp(-i\xi_n) & {}^{i}\!/_{2}\,\Delta\beta_n \end{pmatrix}$$

$\mathbf{U}_n$ is the infinitesimal propagator from z_n to z_{n+1}, i.e. the matrix so that,

$$\mathbf{u}_{n+1} = \mathbf{U}_n \mathbf{u}_n.$$

Notice that above equation is nothing but a compact form for writing the standard system of coupled differential equations in the coefficients $a_1[z], a_2[z]$ of equation $\mathbf{u}[z] \equiv \begin{pmatrix} a_1[z] \\ a_2[z] \end{pmatrix}$.

Special Cases

In some special cases equation $\mathbf{U}[L,0] = \lim_{N\to\infty} \overleftarrow{\prod_{n=0}^{N-1}} \mathbf{U}_n$ can be cast in closed form:

1) Null coupling: Setting $\kappa = 0$ gives $\mathbf{U}[L,0] = \mathbf{V}(\Theta)$ with $\Theta \equiv \int_0^L \Delta\beta\,dz$. This result describes a 'pure' phase shifter and it is easily shown by noticing that $\mathbf{V}(\theta_1)\mathbf{V}(\theta_2) = \mathbf{V}(\theta_1 + \theta_2)$.

If we define $\langle\beta[z]\rangle \equiv (\beta_1[z] + \beta_2[z])/2$, the neglected overall phase is $\exp(i\eta)$, with $\eta \equiv \int_0^L \langle\beta[z]\rangle dz$.

2) Null phase shift: Imposing $\beta_1 = \beta_2 = \beta$ and $d\xi/dz = 0$ gives $\mathbf{U}[L,0] = \mathbf{W}(\Phi,\xi)$ with $\Phi \equiv \int_0^L \kappa\,dz$. This describes a 'pure' coupler and, as before, it is a consequence of the property $\mathbf{W}(\phi_1,\xi)\mathbf{W}(\phi_2,\xi) = \mathbf{W}(\phi_1, + \phi_2, \xi)$.

In this case the neglected overall phase amounts to $\exp(i\eta)$, with $\eta \equiv \int_0^L \beta\,dz$.

Actually the only coupling mechanism known to the author is reciprocal, i.e. time reversal. It can be shown that the time reversal combined to the unitarity condition implies $\mathbf{U}^{\dagger} = \mathbf{U}^{*}$, that in equation $\begin{pmatrix} \exp(-i\theta/2)\cos\phi & \exp(i\xi)\exp(-i\theta/2)\sin\phi \\ -\exp(-i\xi)\exp(i\theta/2)\sin\phi & \exp(i\theta/2)\cos\phi \end{pmatrix}$ requires $\xi = \pi/2$. Physically this means that the phase acquired coupling from the first to the second branch must be equal to the phase acquired in the reverse process. Nevertheless we will allow for generic ξ values, so that our formalism holds for any unitary system (e.g. for polarization states, where the Faraday rotation is non-reciprocal).

3) Constant coupling and phase shift (asynchronous coupler): Suppose $d\Delta\beta/dz = 0$, $d\kappa/dz = 0$ and $d\xi/dz = 0$. Equation $\mathbf{U}[L,0]\ \lim_{\to\infty} \overleftarrow{\prod_{n\ 0}^{N\ 1}} \mathbf{U}$ becomes:

$$\mathbf{U}[L,0] \equiv \bar{\mathbf{U}}[L,0] = \lim_{N\to\infty}\left(1 + \frac{L}{N}\tau\right)^N = \lim_{N\to\infty}\sum_{k=0}^{N}\binom{N}{k}\left(\frac{L}{N}\tau\right)^k = \lim_{N\to\infty}\sum_{k=0}^{N}\frac{(L\tau)^k}{k!} = \exp(L\tau).$$

where, from now on, the overbar will denote the solution with constant coupling and phase shift.

Let's now introduce the hermitian matrix $\boldsymbol{\upsilon} \equiv \boldsymbol{\tau}(i\mu)$, where $\mu^2 \equiv \det \boldsymbol{\tau} = (\Delta\beta/2)^2 + \kappa^2$. It is easily verified that $\boldsymbol{\upsilon}^{2n} \equiv (-\mathbf{I})^n$. So the sum splits in even and odd terms giving:

$$\overline{\mathbf{U}}[L] = \cos(\mu L)\mathbf{I} + i\sin(\mu L)\upsilon.$$

The matrix $\boldsymbol{\upsilon}$ is the so called infinitesimal generator of $\overline{\mathbf{U}}$.

In this case the neglected overall phase is clearly $\exp(i\langle\beta\rangle L)$, where,

$$\langle\beta\rangle \equiv (\beta_1 + \beta_2)/2.$$

If we define $\cos\gamma \equiv -\Delta\beta/2(\mu)$ (which implies $\sin\gamma = \kappa/\mu$), Equation $\overline{\mathbf{U}}[L] = \cos(\mu L)\mathbf{I} + i\sin(\mu L)\upsilon$ becomes,

$$\overline{\mathbf{U}}[L] \equiv \overline{\mathbf{U}}[\mu L, \xi, \gamma] = \begin{pmatrix} \cos(\mu L) + i\sin(\mu L)\cos\gamma & \exp(i\xi)\sin(\mu L)\sin\gamma \\ -\exp(-i\xi)\sin(\mu L)\sin\gamma & \cos(\mu L) - i\sin(\mu L)\cos\gamma \end{pmatrix}$$

Since we are neglecting overall phases,

$$\overline{\mathbf{U}}(\mu L + \pi, \xi, \gamma) = -\overline{\mathbf{U}}(\mu L, \xi, \gamma) \cong \overline{\mathbf{U}}(\mu L, \xi, \gamma)$$

i.e. $\overline{\mathbf{U}}(\mu L, \xi, \gamma)$ is π-periodic in μL.

Notice that in solving for the propagator matrix $\mathbf{U}[L, 0]$ (Equation $\mathbf{U}[L,0] = \lim_{N\to\infty} \overleftarrow{\prod_{n=0}^{N-1}} \mathbf{U}_n$), we don't need to impose any boundary conditions, as for usual differential equations, because they are implicit in the input state $u[0]$ on which $\mathbf{U}[L, 0]$ operates.

Eigenstates

In order to interpret the last result, it is useful to determine the eigenstates of $\overline{\mathbf{U}}$. From the determinantal equation the eigenvalues are found to be:

$$\lambda_{\pm} = \exp(\pm i\mu L)$$

and, correspondingly, the eigenvectors can be cast in the form,

$$\mathbf{u}_{+} = \begin{pmatrix} \exp\left(-\tfrac{i}{2}(\pi/2 - \xi)\right)\cos(\gamma/2) \\ \exp\left(\tfrac{i}{2}(\pi/2 - \xi)\right)\sin(\gamma/2) \end{pmatrix}$$

$$\mathbf{u}_{-} = \begin{pmatrix} -\exp\left(-\tfrac{i}{2}(\pi/2 - \xi)\right)\sin(\gamma/2) \\ \exp\left(\tfrac{i}{2}(\pi/2 - \xi)\right)\cos(\gamma/2) \end{pmatrix}.$$

From equation $\mathbf{V}(\theta)\equiv\begin{pmatrix}\exp(-i\theta/2) & 0\\ 0 & \exp(i\theta/2)\end{pmatrix}$, $\mathbf{W}(\phi,\xi)\equiv\begin{pmatrix}\cos\phi & \exp(i\xi)\sin\phi\\ -\exp(-i\xi)\sin\phi & \cos\phi\end{pmatrix}$ defining $\mathbf{R}(\gamma,\xi)\equiv\mathbf{V}(\pi/2-\xi)\mathbf{W}(\gamma/2,0)$, we can also rewrite,

$$\mathbf{u}_{+,-}=\mathbf{R}(\gamma,\xi)\mathbf{u}_{1,2}$$

$$\overline{\mathbf{U}}(\mu L,\xi,\gamma)=\mathbf{R}(\gamma,\xi)\mathbf{V}(-2\mu L)\mathbf{R}^{-1}(\gamma,\xi).$$

Physically equation $\lambda_{\pm}=\exp(\pm i\mu L)$ means that the difference between the propagation constants of the two eigenstates is $\Delta\beta_{\pm}\equiv\beta_{-}-\beta_{+}=-2\mu$. On the other hand above equation means that $\overline{\mathbf{U}}(\mu L,\xi,\gamma)$ is nothing but a $\mathbf{V}(\Delta\beta_{\mp}L)$ diagonal transformation (i.e. a transformation represented in the basis of its eigenstates) represented in the rotated basis $\{\mathbf{R}^{-1}(\gamma,\xi)\mathbf{u}_{+},\mathbf{R}^{-1}(\gamma,\xi)\mathbf{u}_{-}\}$.

Looking at $\overline{\mathbf{U}}$ globally, we can also rewrite the neglected overall phase as $\exp(i\langle\beta_{\mp}\rangle L)$, where $\langle\beta_{\mp}\rangle\equiv(\beta_{+}+\beta_{-})/2$. From equation $\langle\beta\rangle\equiv(\beta_1+\beta_2)/2$ (which was obtained looking at the local $\mathbf{U}_n$) we get $\langle\beta_{\mp}\rangle=\langle\beta\rangle$, which implies $\beta_{\mp}=\langle\beta\rangle\mp\mu$.

For $\kappa = 0$, the eigenstates of $\overline{\mathbf{U}}$ do not depend on $\Delta\beta$, so they must also be the eigenstates of $\mathbf{V}(\Theta)$. Similarly, for $\Delta\beta = 0$ they do not depend on κ, so they must also be the eigenstates of $\mathbf{W}(\Phi,\xi)$.

Generalized Poincaré Sphere

All the results obtained so far are better understood introducing a convenient geometrical representation. It is well known that SU(2) transformations may be mapped into SO(3) transformations through a homomorphism. This means that all the transformation we have analyzed before can be represented as rotations on a spherical surface, analogous of the Poincaré sphere for polarization states.

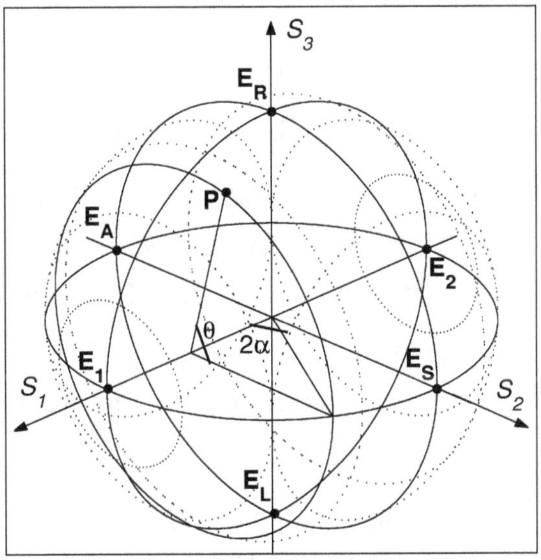

The Generalized Poincaré sphere.

In figure are displayed all the intersection of the S_1, S_2, S_3 axes with the sphere. They represent the single waveguides modes E_1, E_2 and their linear combinations $E_{S,A} \equiv \frac{1}{\sqrt{2}}(E_1 \pm E_2)$ and $E_{R,L} \equiv \frac{1}{\sqrt{2}}(E_1 \pm iE_2)$. Also is plotted the generic normalized mode $P \equiv a_1E + a_2E_2 \equiv \cos\alpha\, E_1 + \exp(i\theta)\sin\alpha\, E_2$, with Stokes parameters,

$$\begin{aligned} S_0 &\equiv |a_1|^2 + |a_2|^2 = 1 \\ S_1 &\equiv |a_1|^2 - |a_2|^2 = \cos 2\alpha \\ S_2 &\equiv a_1a_2^* + a_1^*a_2 = \sin 2\alpha \cos\theta \\ S_3 &\equiv i\left(a_1a_2^* - a_1a_2^*\right) = \sin 2\alpha \sin\theta \end{aligned}$$

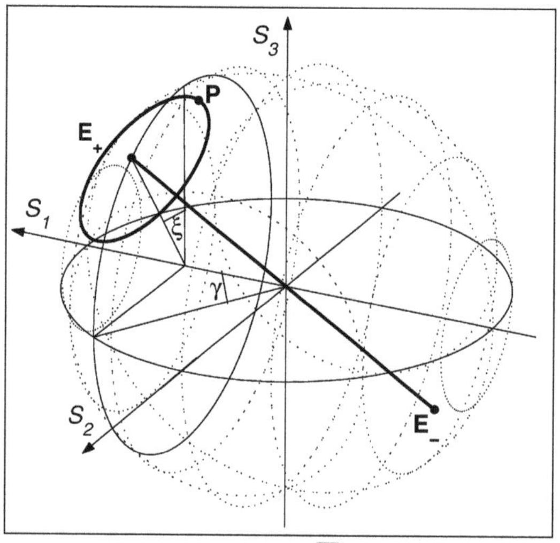

The eigenstates of $\overline{\mathbf{U}}(\mu L, \xi, \gamma)$.

In figure, we have plotted the eigenmodes $E_{+,-}$ corresponding to the eigenstates $\mathbf{u}_{+,-}$ of equation $\mathbf{u}_+ = \begin{pmatrix} \exp\left(-\frac{i}{2}(\pi/2-\xi)\right)\cos(\gamma/2) \\ \exp\left(\frac{i}{2}(\pi/2-\xi)\right)\sin(\gamma/2) \end{pmatrix}$ $\mathbf{u}_- = \begin{pmatrix} -\exp\left(-\frac{i}{2}(\pi/2-\xi)\right)\sin(\gamma/2) \\ \exp\left(\frac{i}{2}(\pi/2-\xi)\right)\cos(\gamma/2) \end{pmatrix}$. It is clear that the generic transformation $\overline{\mathbf{U}}$ (Equation $\overline{\mathbf{U}}[L] = \cos(\mu L)\mathbf{I} + i\sin(\mu L)\upsilon$), corresponds, on the sphere, to a rotation about the axis of its eigenstates $\overline{E_+E_-} \equiv S(\gamma, \pi/2-\xi)$ (which is the rotated of the axis S_1 by an angle γ about S_3 and then by an angle $\pi/2 - \xi$ about S_1). This means that a generic input state P will be rotated by an angle $-2\mu L$ (remember equation $\mathbf{u}_{+,-} = \mathbf{R}(\gamma,\xi)\mathbf{u}_{1,2}$ $\overline{\mathbf{U}}(\mu L,\xi,\gamma) = \mathbf{R}(\gamma,\xi)\mathbf{V}(-2\mu L)\mathbf{R}^{-1}(\gamma,\xi)$) on the circle of revolution about $S(\gamma, \pi/2-\xi)$ passing through P. The three special cases discussed earlier may be represented as rotations on the sphere:

1. $\kappa = 0$ implies $E_{+,-} = E_{1,2}$ and $\mathbf{U} = \mathbf{V}(\Theta)$.

Physically it is clear that $E_{1,2}$ are the eigenmodes of the phase shifter. On the sphere a generic input state P will be rotated by an angle Θ on the circle of revolution about S_1 passing through P. In the special case of constant phase shift will be $\Theta = \Delta\beta L$. The circles of revolution about S_1 represent the loci of constant power splitting, which we will call isodias (ισο = equal, δια = split in two parts).

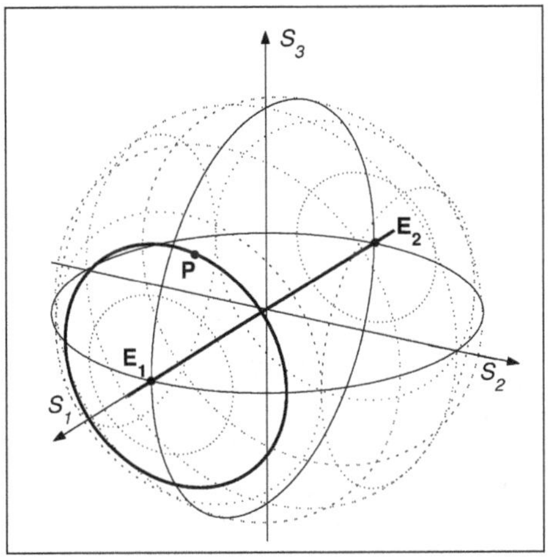

Action of a phase shifter.

2. $\Delta\beta = 0$ and $\xi = \pi/2$ means $E_{+,-} = E_{S,A}$ and $\mathbf{U} = \mathbf{W}(\Phi, \pi/2)$.

Physically it is clear that the symmetric and antisymmetric superpositions of the single waveguide modes are the eigenmodes of the synchronous coupler. On the sphere a generic input state P will be rotated by an angle 2Φ on the circle of revolution about S_2 passing through P. In the special case of constant coupling will be $-2\Phi = -2\kappa L = \Delta\beta_{AS} L$, where we have defined $\Delta\beta_{AS} \equiv \beta_A - \beta_S$.

If one allows for generic ξ ($\mathbf{U} = \mathbf{W}(\Phi, \xi)$), it is easily seen that the loci of constant $(\pi/2 - \xi)$ phase shift are the semicircles with diameter on S_1, which we will call isophases (notice that isodias and isophases can be regarded as a parallels and meridians for the sphere, with poles on E_1 and E_2).

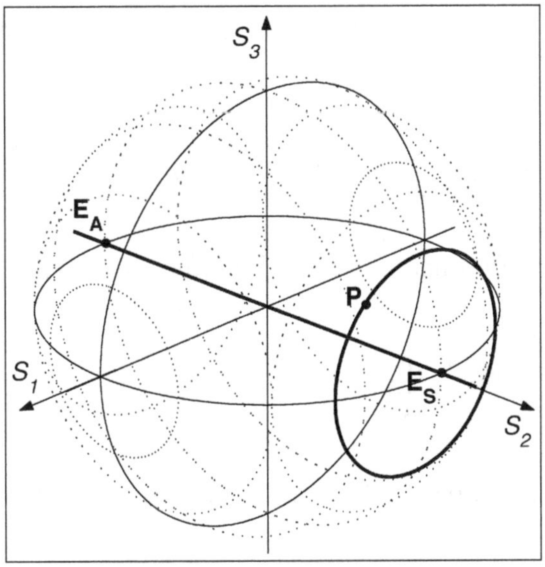

Action of a synchronous coupler.

3. $\Delta\beta \neq 0$ and $\xi = \pi/2$ means $\mathbf{u}_{+,-} = \mathbf{W}(\gamma/2, 0)\mathbf{u}_{1,2}$ and $\mathbf{U} \equiv \overline{\mathbf{U}}(\mu L, \pi/2, \gamma)$.

In the limit of validity of approximation, the eigenmodes of an asynchronous coupler are similar to $E_{S,A}$, but with unbalanced power in the waveguides. On the sphere a generic input state P will be rotated by an angle $-2\mu L = \Delta\beta_{\mp} L$ on the circle of revolution about S(γ, 0) passing through P.

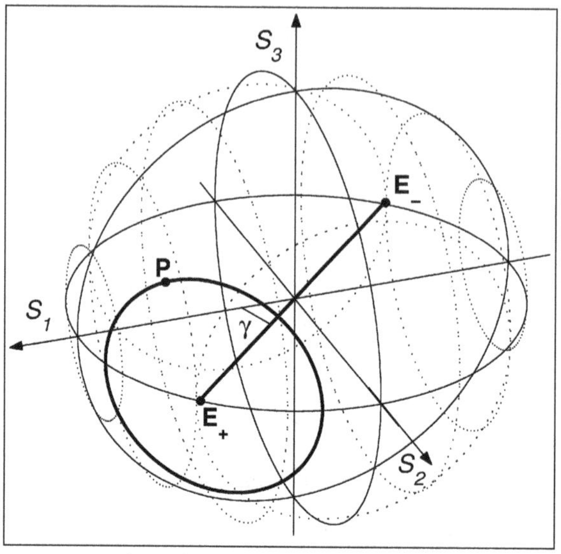

Action of an asynchronous coupler.

50/50 Splitters

We will now analyze the wavelength dependence of some 50/50 splitters. Under a change dλ in the wavelength λ, to any f(λ) will correspond a relative change df/f (relative changes are easier to calculate when working with products and divisions resembling the rules of relative error propagation). On the sphere a 50/50 splitter is any transformation which, starting from E_1 (or, equivalently, from E_2), reaches any point on the isodia Γ with diameter on S_2.

1. The straightest way to do so is by a synchronous coupler, i.e. by a **W**(π/4, π/2) transformation. Power transfer is given by $P_2 = \frac{1}{2}\left(1-\cos\left(\Delta\beta_{AS} L\right)\right)$, and varies as:

$$dP_2 = \frac{1}{2}\sin\left(\Delta\beta_{AS} L\right) L d\Delta\beta_{AS} = \frac{1}{2}\Delta\beta_{AS} L \sin\left(\Delta\beta_{AS} L\right)\left(\frac{d\Delta n_{AS}}{\Delta n_{AS}} - \frac{d\lambda}{\lambda}\right)$$

where $\Delta n_{AS} \equiv n_A - n_S$ is the difference between the effective indexes of the symmetric and antisymmetric modes.

$$dP_2 = \frac{1}{2} L d\Delta\beta_{AS} = \frac{\pi}{4}\left(\frac{d\Delta n_{AS}}{\Delta n_{AS}} - \frac{d\lambda}{\lambda}\right).$$

So the only way to minimize the wavelength sensitivity is by a convenient choice of the waveguides and of their distance.

On the sphere we observe that power splitting is measured by the S_1 parameter (that, by definition, is the difference between optical power in the two waveguides). Since a synchronous coupler approaches the isodia Γ parallel to S_1, a change in the angle $L\Delta\beta_{AS}$ translates immediately into a power change.

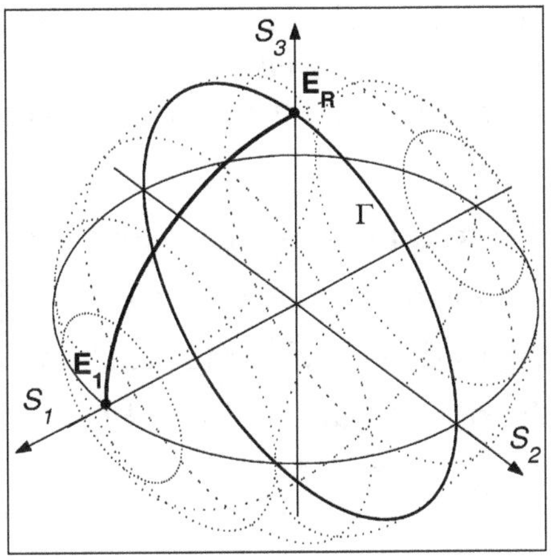

A 50/50 synchronous coupler.

2) An alternative approach could be using an asynchronous coupler. Power transfer in this case is given by $P_2 = P_oF$, where $P_o \equiv \sin^2\gamma = 1-(\Delta\beta/\Delta\beta_\mp)^2$ is the maximum transferable power and $F \equiv {}^1\!/_2(1-\cos(\Delta\beta_\mp L))$, is the power oscillation along propagation. A λ change will now give $dP_2/P_2 = dP_o/P_o + dF/F$ where,

$$dP_o = 2(P_o - 1)\left(\frac{d\Delta n}{\Delta n} - \frac{d\Delta n_\mp}{\Delta n_\mp}\right) \equiv 2(P_o - 1)\delta$$

$$dF = \frac{1}{2}\sin(\Delta\beta_\mp L) L d\Delta\beta_\mp = \frac{1}{2}\Delta\beta_\mp L \sin(\Delta\beta_\mp L)\left(\frac{d\Delta n_\mp}{\Delta n_\mp} - \frac{d\lambda}{\lambda}\right)$$

having defined $\Delta n \equiv n_2 - n_1$ the difference in effective index of the single waveguides modes.

In particular when $\Delta\beta_\mp L = k\pi$, the oscillating contribution vanishes and we can get a 50/50 splitter if $\gamma = \pi/4$. Regarding the dP_o contribution we notice that it is null for $\gamma = m\pi/2$ and it doesn't feature any $d\lambda$ term. So, to get an insensitive coupler, a convenient choice of the coupler parameters must made δ become negligible.

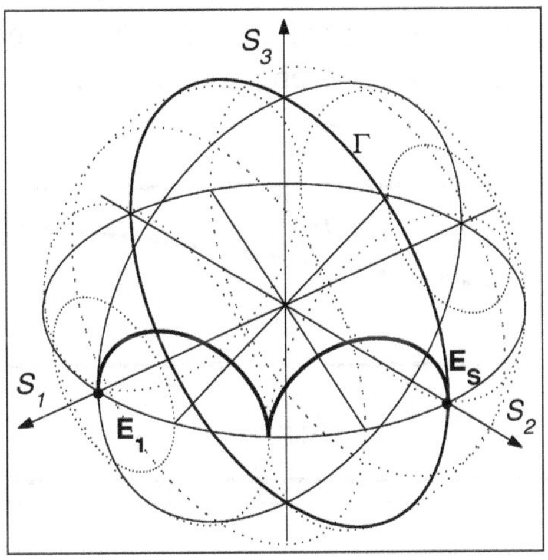

A 50/50 double asynchronous coupler.

On the sphere dP_0 is due to a γ change, i.e. a change of the rotation axis, while dF is related to a $\Delta\beta_\mp$ change, i.e. a change of the rotation angle. It is clear that this asynchronous coupler approaches the isodia Γ perpendicular to S_1 direction, so that, if the rotation axis doesn't change, power splitting it is invariant under $\Delta\beta_\mp$ changes.

This approach can be generalized cascading N asynchronous couplers so that $\Delta\beta_\mp^{(i)}L^{(i)} = k\pi \ \forall i$ and $\sum_{i=1}^{N}(-)^{i+1}\gamma_i = \pi/4$. In figure it is shown the case $\gamma_1 = \pi/8$ and $\gamma_2 = 3\pi/8$. Under the hypothesis that $d\gamma_1 = d\gamma_2$, we expect this configuration to solve the problem of rotation axis changes as shown in figure below, in general it will be true for any choice of $\{\gamma_i\}$ so that $\sum_{i=1}^{N}(-)^{i+N}d\gamma_i = 0$.

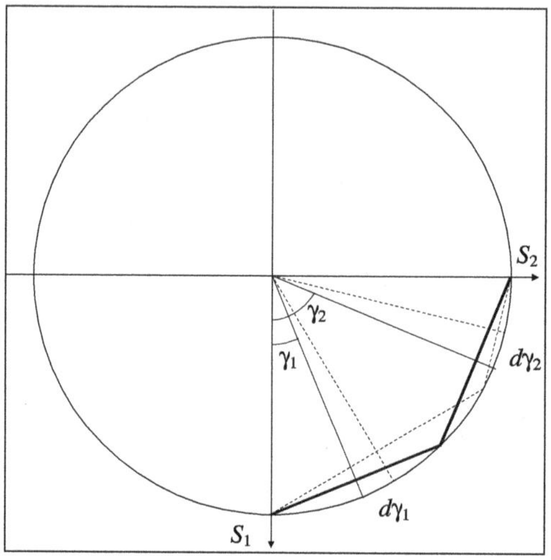

Insensitivity of the configuration of above figure.

Another approach may be, instead of setting to zero each single contribution, to find a combination of γ and $\Delta\beta_\mp L$ reaching a point on Γ that make dP_0/P_0 and dF/F cancel each other out. These simple examples show how using a pictorial view can help not only to interpret well known results, but also to find better solutions.

3. Having understood the working principle of the asynchronous coupler, it seems that even better insensitivity could be obtained if we approached Γ descending from points closer to E_R. This could be done with a structure that starts as a pure synchronous coupler and, along propagation, becomes an almost pure phase shifter. This is the tapered coupler. The trajectory on the sphere can be seen as a composition of small rotations about different axes $S(\gamma(z),0)$, with $\gamma(0) = \pi/2$ and $\gamma(L) \approx 0$.

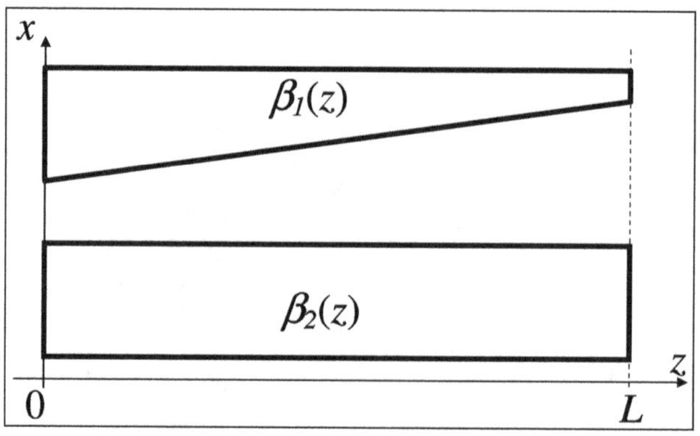

Schematic of a tapered coupler.

This structure is tolerant at the beginning and at the end, but a dilatation (contraction) in the middle part of the trajectory, could shift the ending part of the trajectory on an isodia different by Γ. So, from a pictorial point of view, it is not clear whether a tapered coupler may be better than an asynchronous coupler. The answer can come only from a numerical study, and will depend on the parameters of the specific structure under investigation.

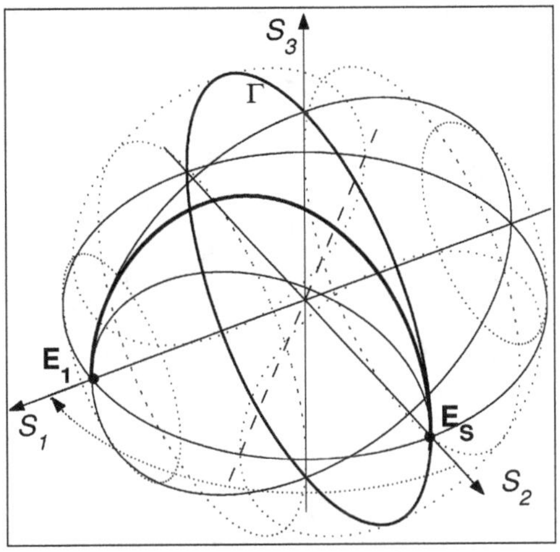

A 50/50 tapered coupler.

4. A completely different approach is based on an interferometric scheme. Consider a synchronous directional coupler so that (at a certain working wavelength λ_0) $\phi_o = \kappa_o L = \pi/2$ (100% power transfer), cascaded with a θ–phase shifter and another directional coupler, identical to the first one, but half the length (50% power transfer). At a generic λ will be $\phi = \phi_o + \Delta\phi$ and the system will be described by the matrix,

$$\mathbf{M} = \mathbf{W}\left({}^{1}\!/_{2}\left(\pi/2+\Delta\phi\right),\pi/2\right)\mathbf{V}(\theta)\mathbf{W}\left(\pi/2+\Delta\phi,\pi/2\right).$$

Of course when $\Delta\phi = 0$ this is perfectly equivalent to a $3\pi/4$ 50% splitter disregarding the θ –value.

Assigned a (non-infinitesimal) fixed value to Δϕ, here we determine a corresponding θ–value that still gives 50/50 power splitting. This means requiring $|M_{11}|^2 = 1/2$, which gives:

$$\cos\theta = \frac{t(t-1)-1/2}{1-t^2}$$

where $t \equiv \sin\Delta\phi$ and the condition $|\cos\theta| \le 1$ implies $t \le 1/2$. For small Δϕ we have $\cos\theta \approx -1/2 - t \rightarrow -1/2$ or $\theta \rightarrow \pm 2\pi/3 + 2k\pi$.

This result is easily understood on the sphere. Since we have a Δϕ angular shift in the first coupler and a Δϕ/2 angular shift in the second coupler, a 2π/3 rotation about S_1, which goes from $\phi_o + \Delta\phi$ to $\phi_o - \Delta\phi/2$, will compensate, at once, the shifts of both couplers (being the change of the circle representing the second coupler a 2nd order effect).

In general, also the phase shifter will be wavelength dependent. In the case of a 'concentrated' phase shifter, made of two identical waveguides of different length and effective index n, it will be

$d\theta/\theta = dn/n - d\lambda/\lambda$. In the case of a 'distributed' phase shifter, made of two different waveguides of the same length and a difference Δn in their effective indexes, it will be $d\theta/\theta = d\Delta n/\Delta n - d\lambda/\lambda$. So it may be convenient to set the working point of the couplers and the working point of the phase shifter at different wavelengths in the desired band. This example shows the power of the geometrical representation that becomes apparent especially when dealing with interferometric schemes.

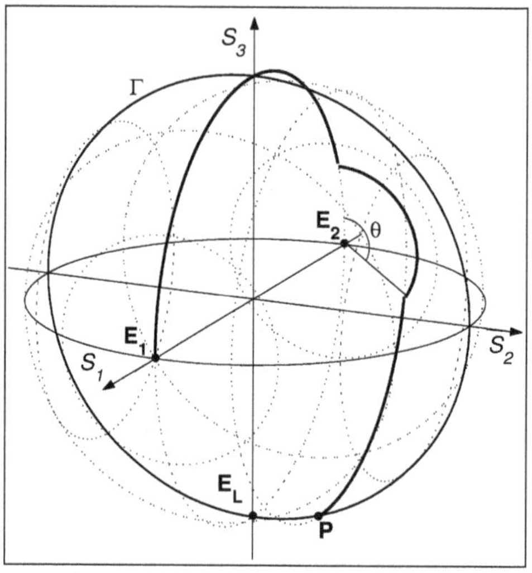

The 50/50 Mach-Zehnder coupler.

Star Couplers

A star coupler is an optical device having multiple input and output ports. It is used for uniformly distributing each input signal to the output ports. A multimode 6x6 star coupler is shown in figure. This coupler was made of photopolymerized polymer. The lightwave from an input port propagates in a planar mixing guide. The lightwave is repeatedly reflected by both sidewalls of the mixing guide. The lightwave is spread by the reflections and is almost uniformly distributed to output ports. The fabricated star coupler with input and output fiber arrays showed 2.6 dB of excess insertion losses and 1.2 dB of loss deviation among the six ports at the 0.84-μm wavelength. The total waveguide length was 54 mm and the mixing guide was 30 mm long. A multimode 16×16 star coupler was also fabricated using ion-exchanged waveguides.

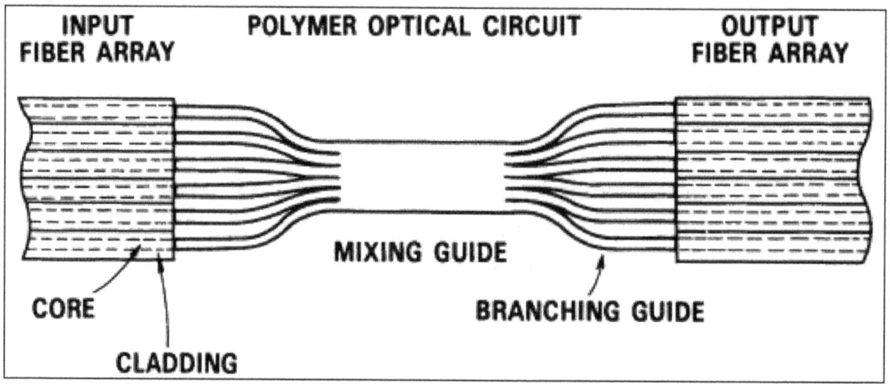

A 6 × 6 star coupler made of polymer multimode waveguides.

Figure shows a reflective multimode star coupler, in which a reflective mirror is placed at one end of the mixing region. In this coupler, no distinction is made between input and output ports. Therefore, three-port couplers such as a conventional Y-branch coupler are added to split inputs and outputs. An eight-port reflective star coupler was demonstrated using Ag-ion-exchanged glass waveguides.

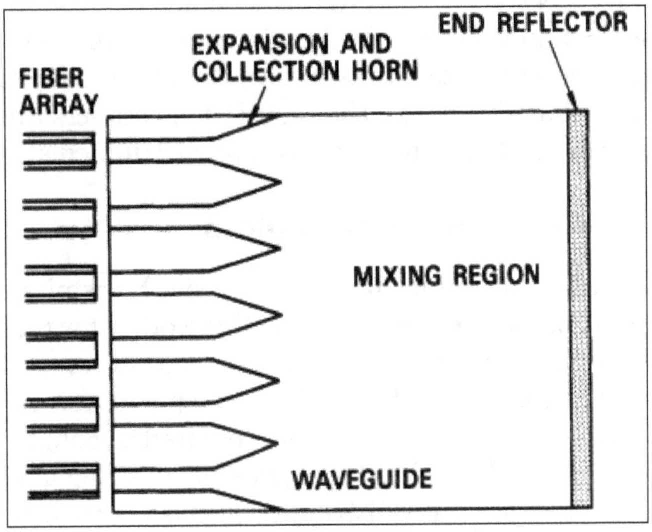

Reflective multimode star coupler.

A single-mode star coupler is shown in figure. This coupler consists of two arrays of channel waveguides separated by a slab waveguide region. The lightwave in each waveguide of the input array is radiated in the slab waveguide (planar waveguide) region and is received efficiently by the output array. Unlike the previously mentioned multimode star coupler, the slab waveguide region has been designed so that the input lightwave is not reflected on the sidewalls. A 19×19 star coupler using P-silica waveguides was demonstrated first, and this showed an excess loss of 3.5 dB at the 1.3-μm wavelength. An 8x8 star coupler based on FHD high-silica waveguides had an excess loss of 1.6 dB at the 1.55-μm wavelength. The 8×8 coupler chip size was 5 × 26 mm.

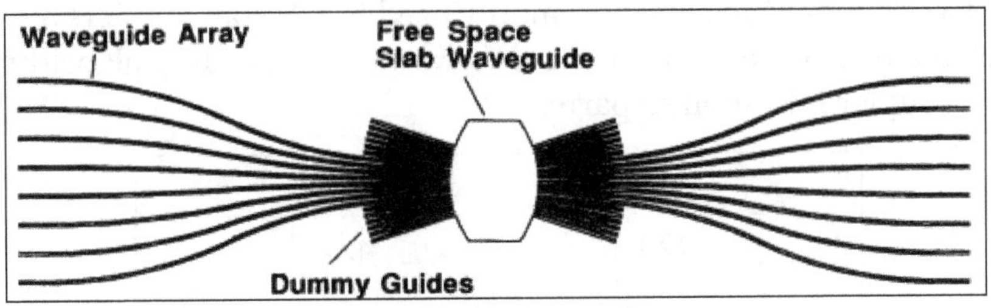

Single-mode star coupler.

Distributed Optical Passive Star Couplers

Optical star networks using wavelength division multiplexing (WDM) have the potential to provide very large throughput and they are expected to find widespread application in upgrading existing networks as well as future photonic networks. The star topology is preferred in these networks because the link attenuation between users (in decibels) grows logarithmically with the number of users, while that for bus and ring topologies is proportional to the number of users. However, the main disadvantage of the star topology is that the number of fibers required to implement the network is very large compared to that of other topologies.

A key component in most of these network architectures is the optical passive star coupler. Two types of star couplers are used in optical star networks; the transmissive N × N star coupler and the reflective N-star coupler. The number of fibers and also the number of components needed to implement a network can be reduced to a half by using a reflective N-star coupler instead of the conventional transmissive N × N star coupler. In this scheme, each user is connected to the reflective N-star coupler by one fiber only. Here, a further reduction in the number of fibers is achieved, using a technique by which the centralized transmissive N × N star coupler or the reflective N-star coupler is replaced by a distributed version. The distributed couplers are nothing more than the original single unit couplers divided into smaller parts and distributed at the nodes of the network.

Distributed Transmissive N × N Star Couplers

In a basic star network, the users are connected to the passive transmissive N × N star coupler by two fibers, one fiber for transmission to the coupler and the other for reception from the coupler. An ideal N × N star coupler uniformly distributes the signal power transmitted by each user across its N output fibers resulting in a loss of 10 Log N dB in the power received by the other users. In this centralized wiring topology, the number of fibers required to connect the N users to the star coupler is 2N fibers, large compared to other topologies such as ring and bus. However, in many practical networks, users are geographically distributed in clusters such as several floors in a building, several buildings on a campus, etc. In this case, it is possible to partition the N × N star coupler into smaller parts and distribute them at the network nodes. It is found that the decentralization of the star coupler results in a tremendous saving in the number of fibers required building the network. Except for the physical separation between its components, the distributed star is identical to the centralized one and therefore will exhibit the same performance.

Let us start with the following example: 16 users are located in two buildings on a campus with eight users in each building and they are separated by a distance L · km. It is required to interconnect these users, using an optical passive star network. In a centralized star network, each user is connected to the 16 × 16 star coupler by two fibers, thus 32 fibers of length 16 L · km are needed to construct the network. Since the most commonly used N × N star couplers are constructed from basic couplers such as the 2 × 2 coupler using a hierarchical cascading algorithm, it is possible to partition the N × N coupler into smaller parts.

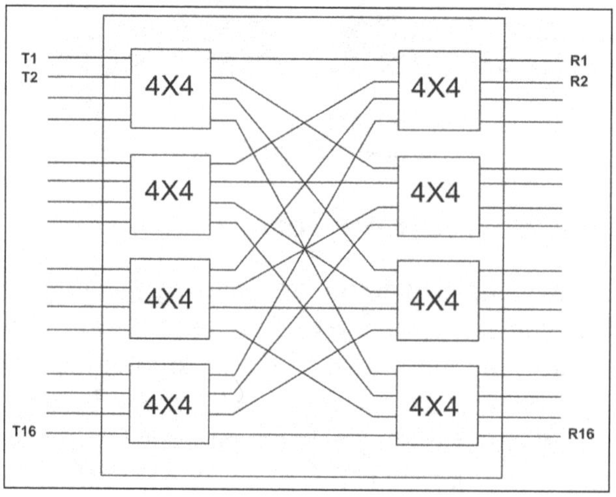

A transmissive 16 × 16 star coupler.

In example, the 16 × 16 star coupler consists of eight 4 × 4 couplers connected. This 16 × 16 star coupler can be divided into two halves and can be located at the two nodes of the network. The number of fibers needed to construct the network is eight with a total length of 8 L · Km compared to 16 L · Km for that of the centralized coupler. The loss in the received power by a user in one location on a signal transmitted from another user location is 10 Log N + aL · dB and it is 10 Log N · dB from a co-located user where U is the fiber loss in dB/km. In a centralized network, the loss is 10 Log N + aL · dB for all users, assuming that the N × N star coupler is located half-way between the two locations.

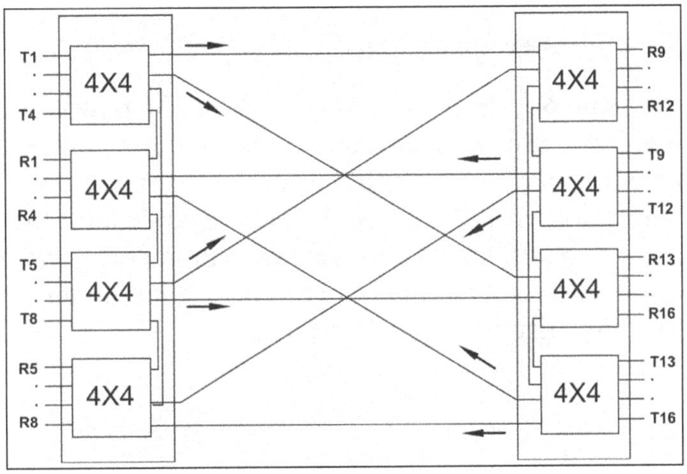

A transmissive 16 × 16 star coupler distributed over two locations.

If the 16 users in example are clustered in four locations, the transmissive 16 × 16 coupler can be distributed over these locations. The number of fibers required to interconnect the star network is 12 as opposed to 32 for the centralized case. To compare the total length of the fibers required to construct each of the two networks, the four locations are assumed to be sited at the comers of a square with a diagonal length of L · km. It is found that the total length of the fibers required to construct the distributed network is 9.66 L · km as opposed to 16 L · km for the centralized one. The procedure can be generalized easily when the number of users is a perfect square; N = m². The transmissive N × N star coupler can be distributed over m locations with m users in each location.

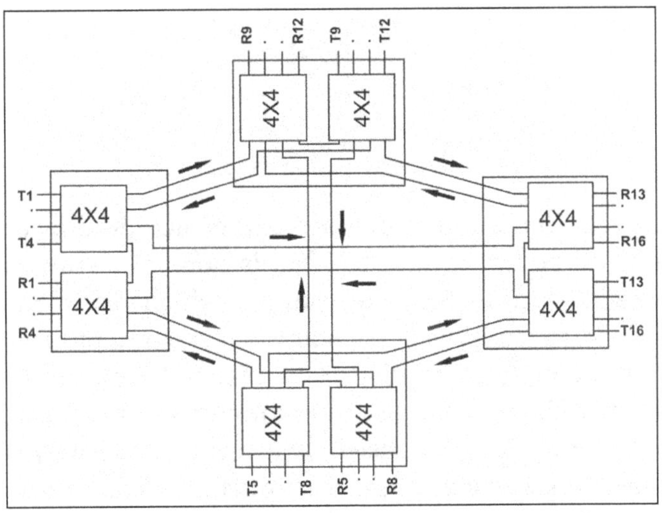

A transmissive 16 × 16 star coupler distributed over four locations.

Each location contains two transmissive m × m star couplers. Two fibers are needed to interconnect each two locations in the network, one for transmission and one for reception. Therefore, the number of fibers needed to interconnect the m locations is m(m - 1) as opposed to $2m^2$ for the centralized network case. Our prime concern was the saving in the number of fibers; however, there are other important advantages to having a distributed coupler. First, the difficulties associated with constructing a large single star coupler for networks with a large number of users is alleviated. Second, any failure occurring at a large single N × N star coupler may result in a complete failure of the network, while for a distributed coupler this is not the case.

Distributed Reflective N-star Couplers

A reflective N-star coupler is an N-port passive device where the power entering any one of the N ports gets divided equally and is reflected back out of all the ports. When this is used as a central node in a passive star network, each user would be connected to the coupler by a single fiber for both transmission and reception of information. Various users would then separate their transmitted and received signals using diplexers at their own remote locations. The main advantage in using a reflective star coupler is that the number of fibers and components required to implement the network is half of that for transmissive star couplers.

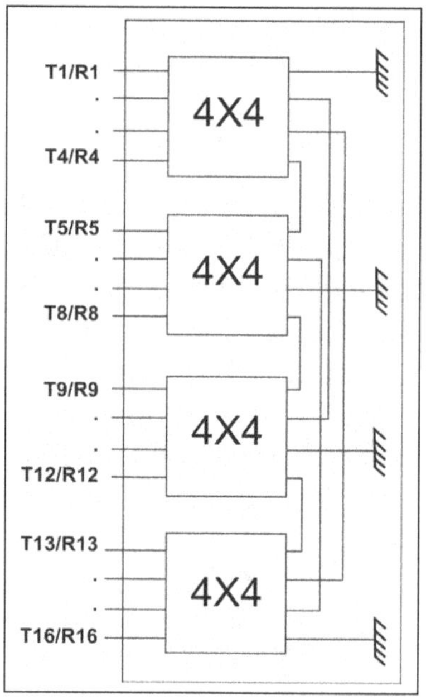

A reflective 16-star coupler.

A further reduction in the required number of fibers can be achieved by distributing the reflective star coupler over the network nodes, similar to what is done to a transmissive star coupler. For comparison purposes, we will continue with the example, which is on a star network with 16 users. The coupler is built from four transmissive 4 × 4 star couplers and four mirrors, i.e., half of that for the transmissive 16 × 16 star coupler. The 16-star coupler of figure can be easily distributed over two locations. The number of fibers needed to interconnect the two locations is four with a total length of 4 L · km where L again is the distance between the two locations. This number is half of that for a centralized reflective star coupler and one fourth that of a centralized transmissive star coupler.

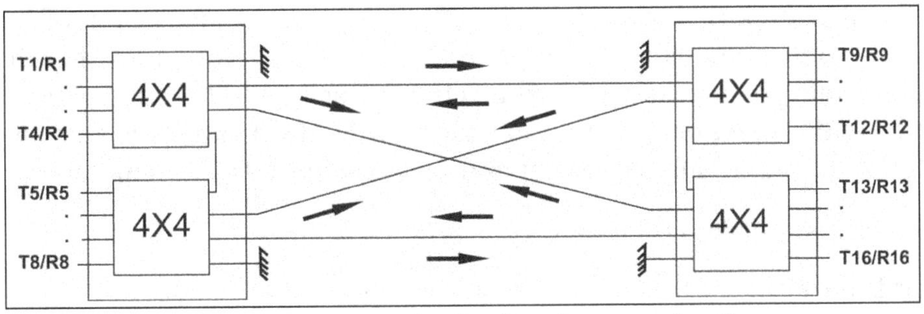

A reflective 16-star coupler distributed over two locations.

The reflective star coupler can be distributed over four locations. The number of fibers needed to interconnect the four locations is six compared to 16 for the case of centralized reflective star coupler. This procedure can be generalized to any reflective N-star coupler, given N as a perfect square number $N = m^2$. The maximum number of locations among which the coupler can be distributed is m with m users at each location. The distributed reflective N-star coupler consists of m transmissive $m \times m$ star couplers and m mirrors. One of the outputs of the transmissive $m \times m$ star coupler is terminated by a mirror and the remaining (m - 1) outputs of each of the transmissive $m \times m$ couplers are connected to the outputs of the couplers in the other (m - 1) locations by (m - 1) fibers. Therefore, the number of fibers needed to interconnect the m locations is $m(m - 1)/2$ compared to m^2 in centralized reflective star coupler case.

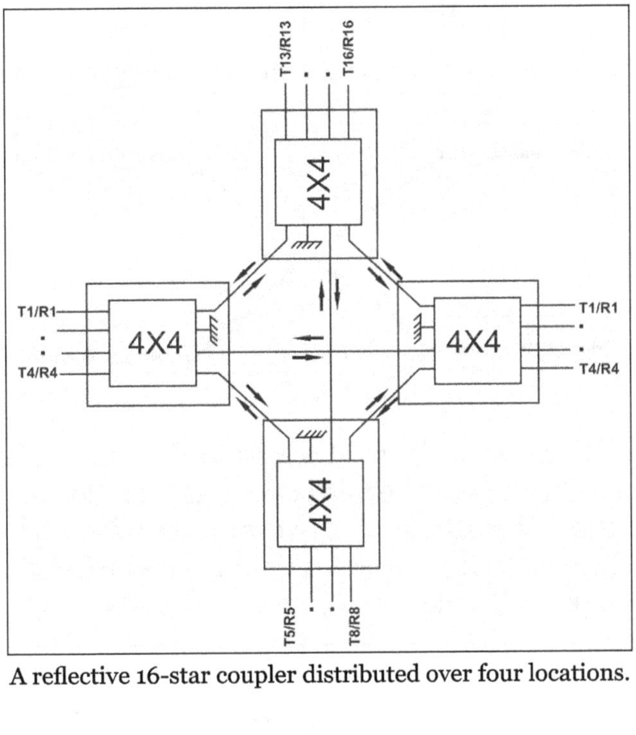

A reflective 16-star coupler distributed over four locations.

T Couplers

A passive optical coupler having three ports (three fibers). Two isolated inputs may be combined into one output; or one input, into two isolated outputs. The amount of coupling loss, usually expressed in dB, between ports is determined by the design and construction of the coupler. The T-Coupler doubles connection capacity by providing two connectors and simplifies installation with fixed coupling nuts. By increasing the connection capacity, the T-Coupler dramatically reduces the number of fieldbus I/O or distribution boxes needed by as much as 50%. It lowers equipment expenditures, makes installation easier and saves manpower and maintenance time.

Power fed into port 1 is split into two parts:

- P2 towards port 2.
- P3 towards port 3.

P2 and P3 need not be equal. Splitting is determined by the detailed coupler structure. The splitting is expressed either in decibel or in percentage. The most important term associated with T coupler is its directivity. The directivity of a T coupler is defined as the power ratio between one port and another which is zero for ideal directivity. It is expected that the power is to be transferred between any two ports on the same side of the coupler. The excess loss is having the same meaning as that of the star coupler and defined as the ratio of total output power input power. There are three types of T couplers:

- Fused T coupler.
- Expanded beam T coupler.
- Expanded beam T coupler s with GRIN lenses.

Fused T couplers are as shown in figure. These are similar to transmissive type star coupler.

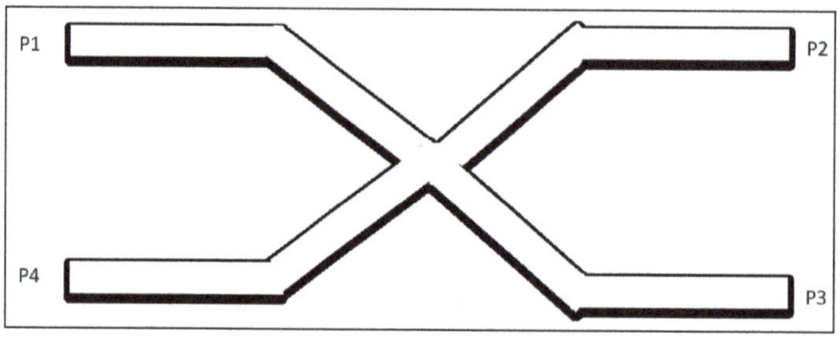

Fused T coupler.

Expanded beam T coupler is as shown in figure. It is expanded beam connector. A partially reflecting mirror is placed between the fiber lenses. The mirror permits a fraction of the incident light to travel from port 1 to port 2 and reflects a part to port 3. In expanded beam with GRIN lenses as shown in figure, GRIN lens and partially reflective surface is used to accomplish the coupling action. In this type of coupler a beam from port 3 is reflected and power coupled to port 4 after reflection.

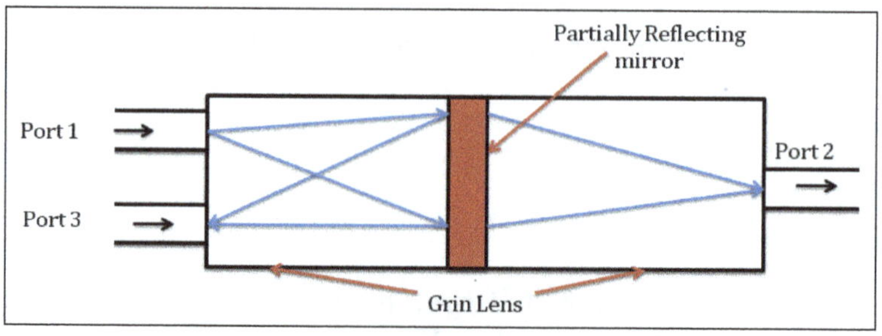

Expanded T coupler with GRIN lenses.

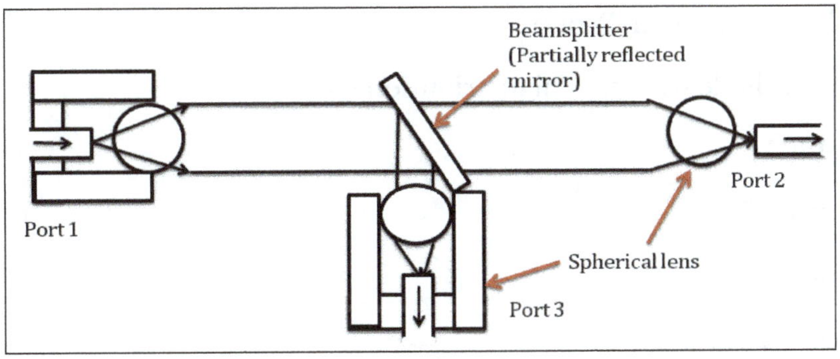

Expanded T coupler.

Calculations on Couplers

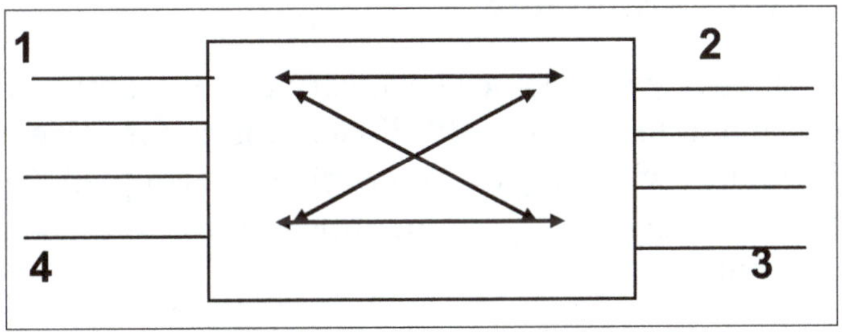

Four port directional coupler.

Consider figure showing directions of allowed power flow. Suppose that power P1 is incident on port 1 of the coupler. This power will split between port 2 and port 3 according to desired splitting ratio. Port 4 is called isolated port as no power will reach this port. The power emerging from port 2 is P2 and that at the port 3 is P3. Let us define some parameters of the coupler.

1. Tap loss: Expressed in Db,

$$Pltap=10 \log(P3/P1)$$

2. Throughput loss: Represents the amount of transmission loss between the input port and the favoured port and it given as,

$$Pltsp=-10\log(P2/P1)$$

3. Directionality: It is defined as the loss between the input port and port we wish to isolate,

$$Pld=-10 \log(P4/P1)$$

4. Excess loss: Power lost within the coupler,

$$Plc=-10 \log(P3+P2)/P1$$

5. Coupling efficiency: It is defined as the ratio of power coupled to the power output of the device. It is the function of the structure of the fiber, light intensity distribution and size and NA of the fiber. It is given by,

$$Pcoup = \left(\frac{x+1}{2}\right)\frac{af}{ac}NA^2 \, Pemi$$

Coupling loss in db =10 log(Pemi/Pcoup)

Here x is LED or laser diode with spatial distribution factor,

- X=1.5 for SLED.
- X= 2 to 3 for ELED.
- X= 6 to 7 for lasers.
- Af is fiber core area.
- Ac is emitting area.

Source to Fiber Coupling Efficiency

To improve the coupling efficiency it becomes necessary to align optical fiber with another optical fiber or an optical source like LED or LASER diode. This can involve either carefully aligning the fiber and placing it in contact with the device to which it is coupled. One can use lens to allow coupling in air gap. The lens gathers the rays from optical source and focuses all rays into the acceptance angle of the optical fiber.

Optocouplers

An optocoupler (or an optoelectronic coupler) is basically an interface between two circuits which operate at (usually) different voltage levels. The key advantage of an optocoupler is the electrical isolation between the input and output circuits. With an optocoupler, the only contact between the input and the output is a beam of light. Because of this it is possible to have an insulation resistance between the two circuits in the thousands of megohms. Isolation like this is useful in high voltage applications where the potentials of two circuits may differ by several thousand volts.

The most common industrial use of the optocouplers (or optically-coupled isolators) is as a signal converter between high-voltage pitot devices (limit switches etc.) and low voltage solid-state logic circuits. Optical isolators can be employed in any situation where a signal must be passed between two circuits which are isolated from each other. Complete electrical isolation between two circuits (i.e. the two circuits have no conductors in common) is often necessary to prevent noise generated in one circuit from being passed to the other circuit. This is especially necessary for the coupling between high-voltage information-gathering circuits and low-voltage digital logic circuits. The information circuits are almost badly exposed to noise sources and the logic circuits cannot tolerate noise signals.

In many applications SCR and triac power circuits are under the control of sensitive electronic systems. For example, it is not unusual to have a microprocessor system programmed to turn motors, lights, and heaters on and off. To reduce the possibility of power-line noise being induced into the control electronics, and to protect it in the event of an SCR or triac failure, it is highly desirable to provide isolation. The ideal isolation scheme should only allow signal flow in one direction, should respond to dc levels, and should offer an extremely large resistance between the input and output circuits. These features are available in a class of optoelectronic devices called optocouplers or optoisolators.

The optical coupling method eliminates the need for a relay-controlled contact or an isolating transformer, which are traditional methods of providing electrical isolation between circuits. The optical coupling method is superior in many applications, because it gets rid of some of the less desirable features of relays and transformers.

The optocouplers works well on either ac or dc high-voltage signals. For this reason, signal converters employing optical coupling are sometimes referred to the universal signal converters. The optocoupler is a device that contains an infra-red LED and a photodetector (such as a photodiode, phototransistor, Darlington pair, SCR or triac) combined in one package.

An autocoupler combining a LED and a photodiode in a single package is shown in figure. It has a LED on the input side and a photodiode on the output side. The left source voltage and the series resistor set up a current through the LED. Then the light from the LED impinges on the photo-diode, and this sets up a reverse current in the output circuit. This reverse current develops a voltage across the output resistor R. The output voltage then equals the output supply voltage V_2 minus the voltage drop across the load resistor R. When the input voltage is varied, the amount of light fluctuates.

Types of Optocouplers

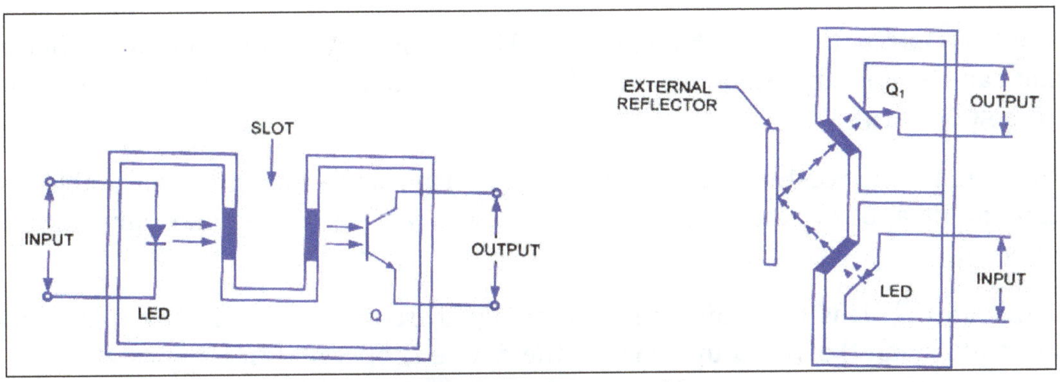

Reflective-slotted-optocoupler.

- Slotted Optocoupler: A slotted optocoupler has a slot moulded into the package between the LED light source and the phototransistor light sensor; the slot houses transparent windows, so that the LED light can normally freely reach the face of transistor, but can be interrupted or blocked via opaque object placed within the slot. The slotted optocoupler can thus be employed in a variety of presence detecting applications, including end-of-tape detection, limit switching, and liquid level detection.
- Reflective Optocoupler: Here the LED and phototransistor are optically screened from each other within the package, and both face outwards (in same direction) from the package. The construction is such that an optocoupled link can be set up by a reflective object (such as metallic paint or tape, or even smoke particles) placed a short distance outside the

package, in line with both the LED. The reflective coupler can thus be employed in applications such as tape-position detection, engine-shaft revolution counting or speed measurement, or smoke or fog detection etc.

Characteristics of an Optocoupler

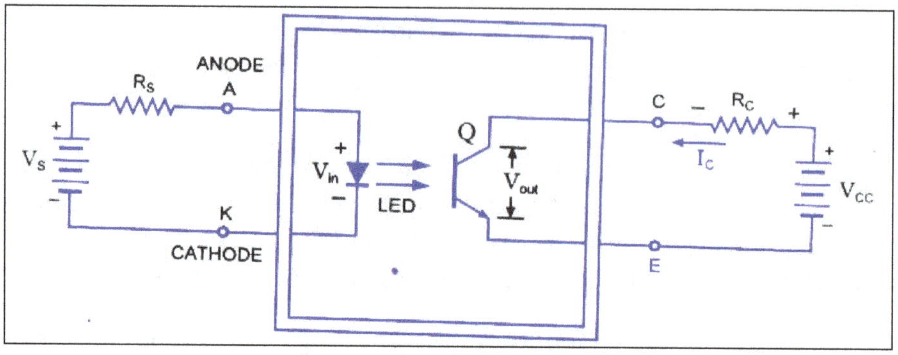

Optocoupler-characteristics.

- Current Transfer Ratio (CTR): One of the most important parameters of an optocoupler device is its optocoupling efficiency. This parameter is maximized by closely matching spectrally the LED and the phototransistor (which usually operate in the infra-red range). The optocoupling efficiency of an optocoupler may be conveniently specified by the output-to-input current transfer ratio (CTR) i.e., the ratio of the output current I_c (measured at the collector terminal of the phototransistor), to the input current I_F flowing into the LED.
- Input-to-Output Isolation Voltage (V_{iso}): This is the maximum potential difference (dc) that can be allowed to exist between the input and output terminals. Typical values range from 500 V to 4 kV.
- Maximum Collector-Emitter Voltage, $V_{CE\ (max)}$: This is the maximum allowable dc voltage that can be applied across the output transistor. Typical values may vary from 20 to 80 volts.
- Bandwidth: This is the typical maximum signal frequency (in kHz) that can be usefully passed through the optocoupler when the device is operated in its normal mode. Typical values vary from 20 to 500 kHz, depending on the type of device construction.
- Response Time: Divided into rise time t_r and fall time t*. For a phototransistor output stages, t_r and t_r are usually around 2 to 5 us.
- A simple isolating optocoupler uses a single phototransistor output stage and is usually housed in a six-pin package, with the base terminal of the phototransistor externally available. In normal use the base is left open circuit, and under such a condition the optocoupler has a minimum CTR value of 20 % and a useful bandwidth of 300 kHz.

Opto Coupled Devices

Optocouplers or opto isolators consisting of a combination of an infrared LED (also IRED or ILED) and an infrared sensitive device such as a photodiode or a phototransistor are widely used to pass information between two parts of a circuit that operate at very different voltage levels. Their main purpose is to provide electrical isolation between two parts of a circuit, increasing safety for users by reducing the risk of electric shocks, and preventing damage to equipment by potential short circuits between high-energy output and low-energy input circuits. They are also used in a number of sensor applications to sense the presence of physical objects.

Transistor Optocouplers

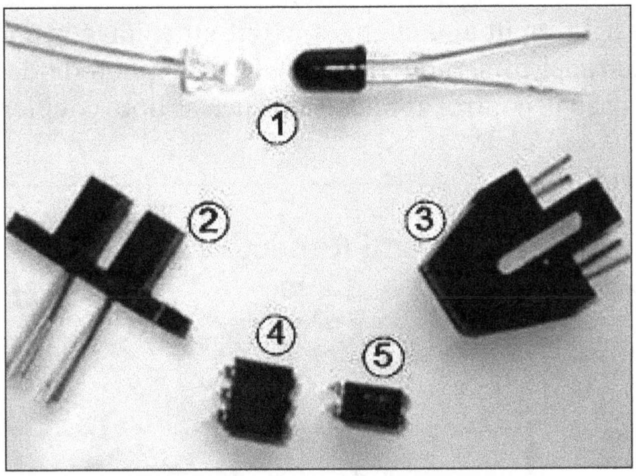

Transistor Optocouplers and Opto Sensors.

The devices shown in figure use phototransistors as their sensing elements as they are many times more sensitive than photodiodes and can therefore produce higher values of current at their outputs. Example 1 in figure illustrates the simplest form of opto coupling consisting of an infrared LED (with a clear plastic case) and an infrared phototransistor with a black plastic case that shields the phototransistor from light in the visible spectrum whilst allowing infrared light to pass through. Notice that the phototransistor has only two connections, collector and emitter, the input to the base being infrared light. Examples 2 and 3 in figure are typical optocoupled devices widely used as position and proximity sensors; these are used as optically activated switches.

Example 4 in figure is a 4N25 optocoupler in a 6 pin DIL integrated circuit from Vishay. It uses an output phototransistor with a base connection that is also connected to an external pin for applying an external circuit if required. This allows the optocoupler to have a DC bias applied to prevent the transistor from producing current at very low light levels. Biasing the phototransistor can also enable it to be used with signals such as analogue audio. In this case the emitter connection can be left unconnected and the base connection used as an output, then the output phototransistor collector/base junction operates as a photodiode, greatly increasing the frequency range of the optocoupler, but at the expense of greatly reducing the available signal amplitude at the output. The 4N25 can also operate as a digital optocoupler with logic 1 and logic 0 inputs. The isolation between input and output on the 4N25 is a minimum of 5.3 kV. Example 5 in figure is a PC817 4 pin single channel opto isolator chip from Sharp, which uses an integral infrared LED and a phototransistor to produce an output of up 50mA and provides electrical isolation up to 5kV. It is also available in 2, 3 and 4 channel versions.

Phototransistors

Figure shows the basic structure of a phototransistor. Its operation is similar to the photodiodes. However because the conversion from light to current takes place in a transistor, the tiny current produced by a particular level of photon input to the base can be amplified to produce a collector current 200 times greater or more, depending on the h_{fe} of the transistor, making the phototransistor much more efficient than a photodiode. However, because of the large junction area (and therefore much higher junction capacitance) of a phototransistor, its response at high frequencies is poor, and the switching time is much slower, compared to a photodiode. Also the relationship between changes in light input and changes in output voltage is not as linear as in photodiodes. Consequently phototransistors, though less useful than photodiodes for high frequency data transmission, are widely used in control applications such as opto couplers/isolators, and position sensors.

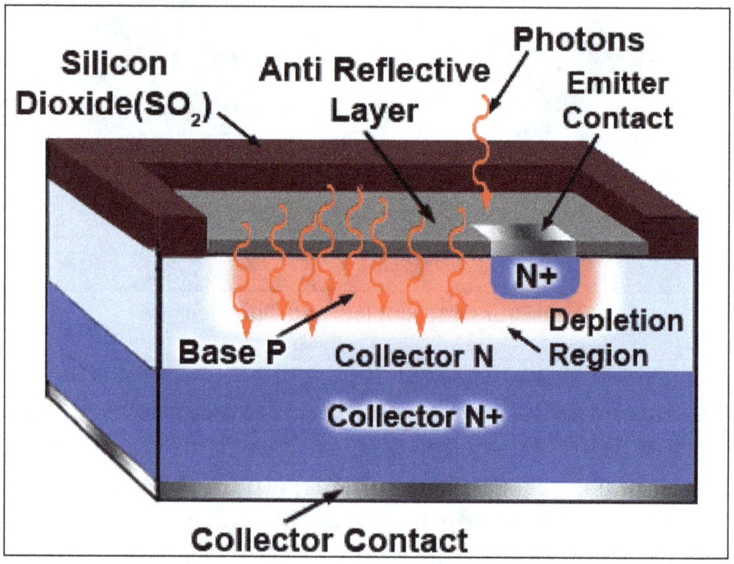

Basic Phototransistor Structure.

Phototransistor Operation

In a phototransistor, light, in the form of photons is collected in the base layer, which occupies the major part of the visible window on the top surface of the device. The emitter area is therefore reduced in size to maximise light absorption in the base. The conversion between photons and current takes place largely in the depletion region around the base/collector PN junction where photons absorbed via the anti-reflective layer into the base layer dislodge electrons to create electron/hole pairs in a similar manner to that in photodiodes, but now the free electrons created by this process are the source of base current in the transistor, and are now amplified by an amount equal to the h_{fe} of the transistor.

The N type collector immediately beneath the depletion layer has a higher resistance than the N+ layer next to the collector terminal. Because of this higher resistance close to the PN junction, there is a large voltage gradient in the collector close to the base/collector junction. This provides a higher positive voltage close to the depletion layer to attract and accelerate the negatively charged electrons in the depletion layer towards the collector terminal.

Compared to photodiodes however, phototransistors do have some drawbacks; their response to varying levels of light is not quite so linear, making phototransistors less suitable than photodiodes for accurate light measurement. Although phototransistors can be used to detect light sources across the visible light spectrum, they are most sensitive to wavelengths in the near infrared range around 800 to 900nm, and are most often used with infrared emitting sources such as infrared emitting LEDs (also called IREDs or ILEDs) as their light source.

Phototransistors are generally not as fast as photodiodes at reacting to abrupt changes in light levels. For example, the time taken for the phototransistor output to change between 10% and 90% in response to a sudden change in light level at the input can be between 30 and 250μs whereas high speed photodiodes can have risen and fall times as low as 20ps (pico seconds) or less. Manufacturers normally quote these figures for rise time (t_r) and fall time (t_f) under particular conditions of temperature and collector current. The main reason for the much slower response in phototransistors is due to the much larger area of the base/collector junction, and the fact that the capacitance that exists across this junction is further magnified by the 'Miller Effect', which causes the junction capacitance to be magnified by the current gain (h_{fe}) of the transistor. In practice this means that the more sensitive the transistor (i.e. the larger the base area) and/or the higher the current gain of the transistor, the longer the rise and fall times will be. For these reasons phototransistors are mostly used for switching DC or low frequency AC applications.

Phototransistor Connections

Phototransistors are available in several forms such as NPN or PNP. Many phototransistors only have connections for the emitter and collector, as the base input is provided by light; however a base connection is provided on some types. Darlington phototransistors are also available; using a Darlington pair transistor configuration gives even greater current gain.

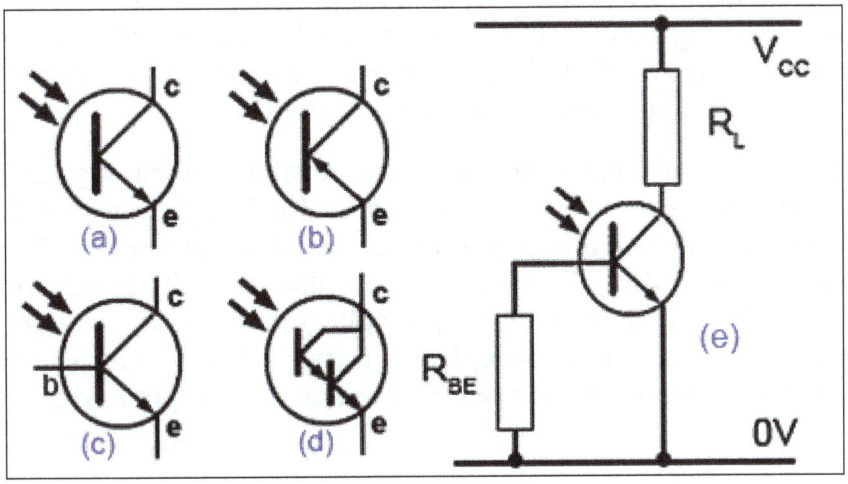

At low or even no light levels, phototransistors can still produce a small amount of current due to random collisions in the depletion layer. Applying base bias can have the effect of preventing this 'Dark Current', so reducing the effect of random noise and giving a better defined on/off level to the output current. Optocouplers have many uses and are available in many varied types; a few examples are illustrated in figure below. Use the type numbers to search for datasheets and use them to identify the purpose of each design.

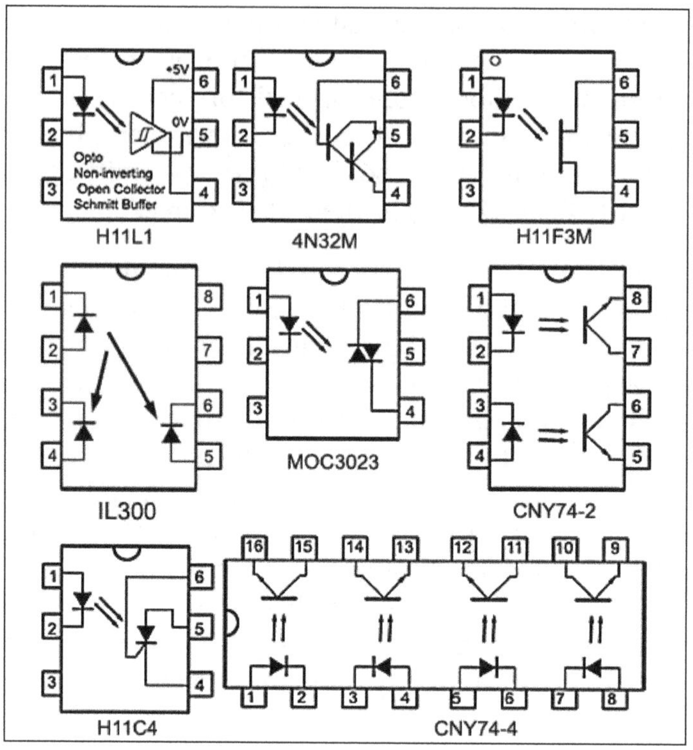

Optocoupler Examples.

Optocoupler Operation

Optocouplers/Opto Isolators

Optocouplers or opto isolators are used for passing signals between two isolated circuits using different methods, depending mainly on the types of signals being linked. A computer system and its peripheral devices may need a digital signal, such as pulse width modulation signal driving a motor. In this case the optocoupler will be used in Saturation Mode.

A switched mode power supply may need a DC sample voltage of varying value to be fed back from the output to a voltage control system in the input circuit of the power supply whilst maintaining complete electrical isolation between the input and output circuits. In this case Linear Mode will be used, as the control circuit will need to detect small changes in DC voltage. To link circuits such as audio amplifiers, where signal voltages are rapidly changing, but saturation and distortion need to be avoided, optocouplers can transfer signals using Analogue Mode so that audio can be safely transmitted, for example from an audio input device to a high powered amplifier.

Saturation Mode

In saturation mode, the optocoupler output transistor is either turned fully 'on' (saturation conditions), or fully 'off' (non-conducting). Optocouplers working in saturation mode are widely used to protect the output pins of micro controllers for example, where they may be used to drive output devices such as motors that may need more current and/or higher voltages than can be supplied directly from the micro controller port.

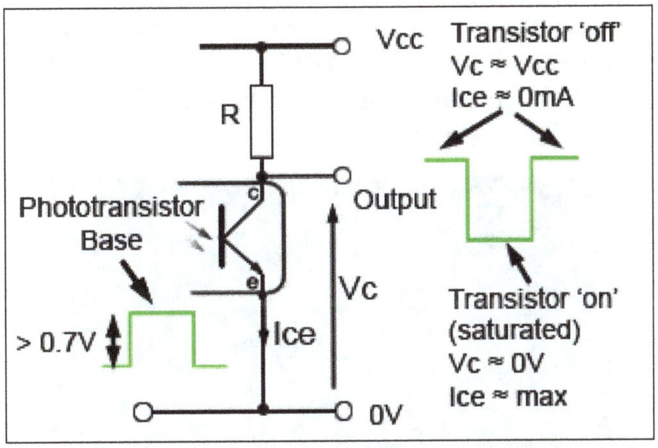

The micro controller is then effectively only driving an infrared LED, either with signals such as pulse width modulation, stepper motor data or simple on and off signals. The isolation provided by the optocoupler means that the micro controller is also protected from any externally produced high voltages, such as the back emf that may be produced when switching off an inductive load such as a motor. Optocouplers also find uses in modems providing isolation between computers and the external phone lines.

Linear Mode

Optocouplers can be used for voltage feedback in circuits such as switched mode power supplies, where the LED is illuminated by a sample of the output voltage so that any voltage variations cause a variation in the illumination of the optocoupler LED and therefore a variation in the conduction of the optocoupler's output transistor, that can be used to signify an error to the power supply control circuitry, allowing it to compensate for the output variation.

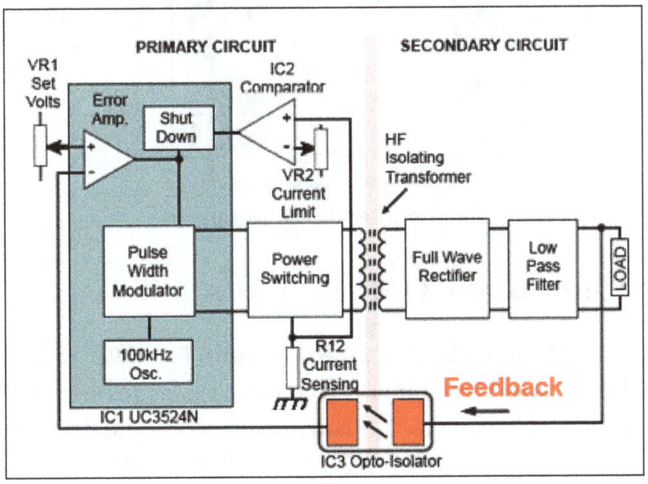

Analogue Mode

Like linear mode, the phototransistors used in analogue mode are not allowed to saturate, but a steady DC bias voltage of around half of the supply voltage is modulated by an audio, or some other rapidly varying signal. This produces a varying current in the LED, which in turn produces a varying current in the output component of the optocoupler. This may be a phototransistor or very often a photodiode. The phototransistors used in optocouplers for audio purposes may also make use of a base connection available on some optocouplers to apply a suitable bias to the phototransistor to enable an undistorted audio signal output to be obtained.

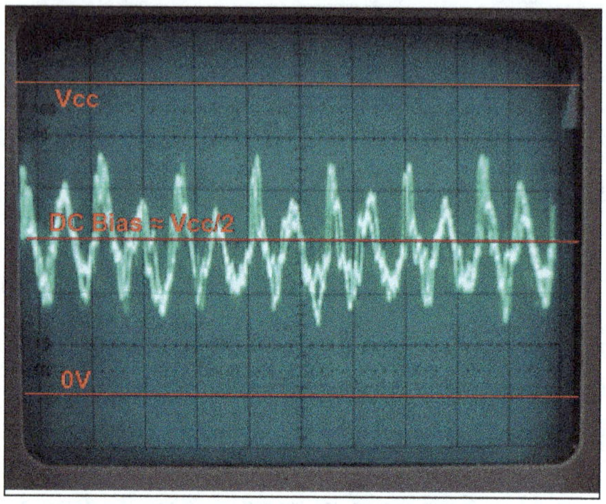

Audio Input in Analogue Mode.

Specialised audio optocouplers such as the IL300 shown in figure below may use one or more photodiodes in order to provide a more linear response than those using only phototransistors. In addition to providing a more linear (less distortion) response the second diode is used to provide (isolated) feedback to the input circuit so that the IL300 can automatically compensate for variations in CTR due to changes in temperature and/or aging of the input LED.

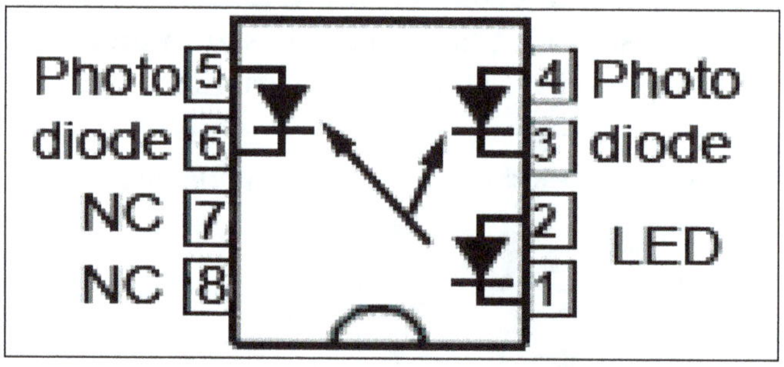

The IL300 Audio Optocoupler.

Phototransistor Vs. Photodiode Optocouplers

Optocouplers using phototransistor outputs can pass analogue audio signals up to a frequency of a few tens of kHz. Varying the infrared light beam from the LED at these frequencies then causes the amount of current generated at the base of an output phototransistor to vary, with the transistor output following and amplifying the variations at the input. However optocouplers using phototransistors do not have such as good a linear relationship between the changes in light input and output current as photodiode types therefore some distortion of the signal may occur. Photodiode output devices are preferred for use in most audio (and some digital) applications, even though their output signal amplitudes are much less than is possible with the amplification provided by a phototransistor; the reason for this is the phototransistor's distortion and poor performance at high frequencies.

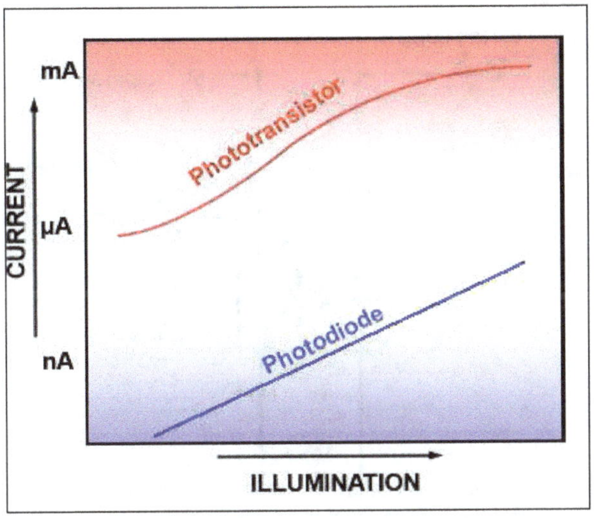

Audio Response in Analogue Mode.

This is due to the phototransistor having a much-enlarged base area, which whilst increasing the light sensitivity, also greatly increases the capacitance of the base/emitter junction. This increased capacitance is also made much worse because of the 'Miller Effect', which causes the base/emitter capacitance of a transistor to be multiplied by the current gain (h_{fe}) of the transistor. Therefore higher frequencies are progressively reduced in amplitude, because the reactance of the base/emitter capacitance reduces as frequency increases much above the audio range.

Digital signals are also affected by this effect because the square waveforms of digital signals will contain many high frequency harmonics that contribute to the fast rise and fall times of the square wave, so that the rising edges of the signal become rounded and the switching time between 0 and 1 becomes longer. High speed digital optocouplers, usable at frequencies in the hundreds of kHz and those used for audio operation usually use photodiodes as their sensing element because although some extra amplification must be provided, either externally or within the optocoupler chip itself, this is offset by having fast rise and fall times for digital operation, and a more linear response, producing less distortion when used with analogue audio.

The main function of an optocoupler, whatever type of signal is used, is to provide complete electrical isolation between the input and output circuits. An important advantage of optocouplers, compared with transformers, also often used for isolation purposes, is that optocouplers can be used with either AC or DC signals whereas transformers can only operate with AC.

Using Optocouplers

There are many different applications for optocoupler circuits, so there are many different design requirements, but a basic design for an optocoupler providing isolation for example between two circuits, simply involves the choice of appropriate resistor values for the two resistors R1 and R2. In this example a PC817 optocoupler is shown isolating a circuit using HCT logic via a 7414 Schmitt inverter gate. The Schmitt inverter at the output performs several functions; it ensures that the output conforms to HCT voltage and current specifications, it also provides very fast rise and fall times for the output, and corrects the signal inversion caused by the phototransistor being operated in common emitter mode.

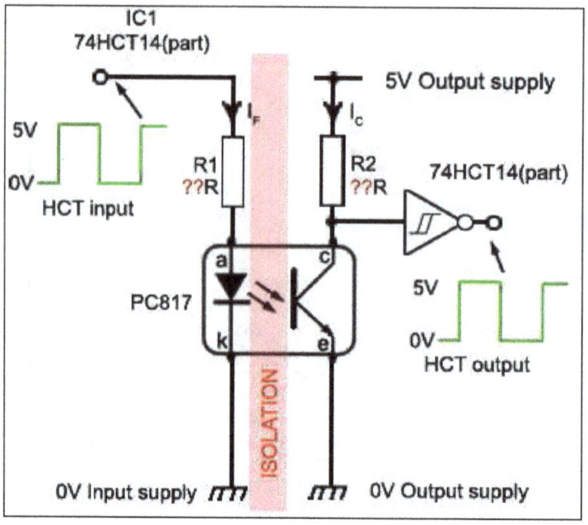

A Simple Optocoupler Interface for HCT.

Each logic family (e.g. LSTTL or CMOS types) may have different logic voltage levels and different input and output current requirements, and optocouplers can provide a convenient way of interfacing two circuits with different logic levels. What is necessary is to ensure that R1 creates an appropriate current level from the input circuit to correctly drive the LED side of the optocoupler, and that R2 creates appropriate voltage and current levels to supply the output circuit via the inverter.

Designing Optocoupler Interfaces

The main purpose of an optocoupler interface is to completely isolate the input circuit from the output circuit, which normally means there will be two completely separate power supplies, one for the input circuit and one for the output. In this simple example the input and output supplies will most likely be the same in voltage and current capabilities, so the interface is just providing isolation without any major shift in voltage or current levels. In choosing appropriate values for R1, the value for the current limiting resistor is set to produce the correct forward current (IF) through the infrared LED in the optocoupler. R2 is the load resistor for the phototransistor and the values of both resistors will depend on a number of factors.

Current Transfer Ratio

The current in each half of the circuit is linked by the Current Transfer Ratio or CTR, which is simply the ratio of output current to the input current (I_C/I_F) usually expressed as a percentage. Each optocoupler type will have a range of CTR values set out in the manufacturer's datasheet. The value of CTR also depends on a number of factors, first of all is the type of optocoupler, simple types may have a CTR value of between 20% and 100%, whilst special types, such as those that use a Darlington transistor configuration for their output phototransistor, may have CTR values of several hundred percent. Also the CTR of any particular device may vary considerably from that device's typical value by anything up to +/-30%.

Manufacturers will normally quote a range of CTR values for different output phototransistor collector voltages (V_C) and different ambient temperatures (T_A) The CTR will also vary with the age of the optocoupler, as the efficiency of LEDs decreases with age (over 1000s of operating hours). Because the CTR of an optocoupler can be expected to reduce over time, it is common practice to choose a value for IF somewhat lower than the maximum, so that the intended performance can still be achieved over the intended lifetime of the circuit. Although this example describes the design of a simple interface linking two HCT logic circuits, the difference between the results achieved here and those needed for any other optocoupler are that similar calculations can be made just using data appropriate to other voltages and currents and other optocouplers.

Calculating the Optocoupler Resistor Values

The start of the design process is to specify the input and output conditions the optocoupler is to link. Typical optocouplers can handle input and output currents from a few microamps to tens of milliamps. There are many optocouplers on the market and to find the most appropriate for a particular purpose, vendor's catalogues and manufacturer's datasheets should be studied. In this case however, a popular PC817 optocoupler from Sharp will use voltages and currents available from HCT logic. Assuming that a single HCT output is only feeding this optocoupler, a logic 1 voltage of about 4.9V can be presumed.

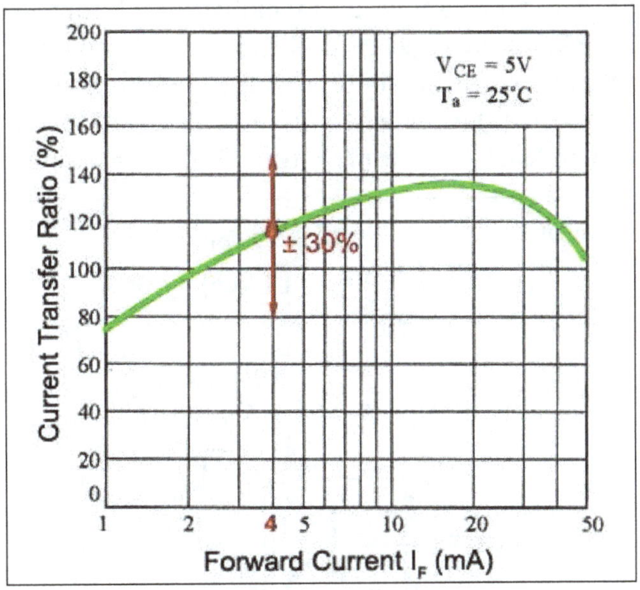

CTR vs. Forward Current for a PC817.

The output current available from a HCT gate to drive the optocoupler input is limited to 4mA, which is quite low for driving an optocoupler. The PC817 must then be capable of producing the necessary output from this low input current. The graph in Figure shows, the CTR for a PC817 with a forward (input) current I_F of 4mA will be around about 80 to 150% allowing ±30% for all the variables mentioned above). Ideally the optocoupler should in this case act as though it is invisible, that is the HCT gate connected to the optocoupler output should see an available current of up to 4mA, just as though it was connected to the output of another HCT gate. Therefore the output current of the PC817 needs also to be ideally about 4mA, with the forward current (I_F) driving the input LED at 4mA (assuming a 100% CTR). Having found an approximate figure for the CTR, which suggests that input and output conditions should be similar, at 4mA, the next task is to calculate the values of R1 and R2.

Table: Typical HCT Input and Output specifications.

Supply	Output		Input	
V_{DD} = 5V	V_{OH} = 2.7V to 4.9V	I_{OH} = 4mA	V_{IH} > 2.0V	I_{IH} = 1μA
	V_{OL} = <0.2V	I_{OL} = 4mA	V_{IL} < 0.8V	I_{IL} = 1μA

Using the data in Table and assuming minimum 4.9V to 5V input at the HCT gate output, it is possible to calculate a suitable resistance value for R1 in figure. The forward Voltage across the infrared LED with a forward current of only 4mA should be about 1.2V 5V – 1.2V = 3.8V to be developed across R1.

Therefore,

R1 = 3.8V÷4mA = 950Ω.

Using the next higher preferred resistor value,

R1 = 1KΩ

The graph of CTR vs. I_F in Figure above shows that ideally the CTR for the PC817 will be about 115% with a forward current of 4mA, which suggests that the opto output current should be about 4mA x 115% = 4.6mA. To saturate the phototransistor and produce a logic 0 (less than 0.2V) at the output, R2 must develop a voltage of 4.9 to 5V when passing a current of 4.6mA (assuming 115% CTR value). R2 must therefore be at least 5V÷4.6mA = 1087Ω or R2 = 1.2kΩ (next preferred value).

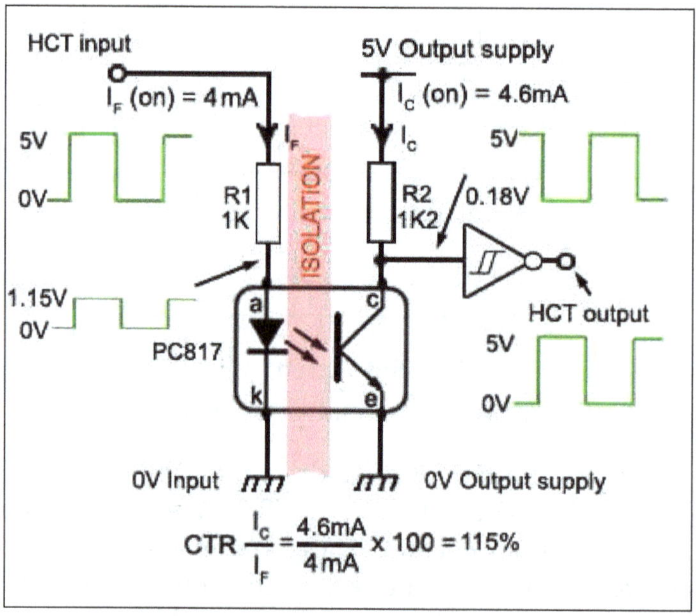

HCT to HCT Optocoupler.

If a higher value than 1.2KΩ is used, increasing this value by a few kΩ could ensure that the output has the maximum voltage swing, however increasing this value reduces the speed with which the optocoupler can respond to fast voltage changes, due to the combination of a high resistance load and a high junction capacitance of the phototransistor, which results in a rounding of the output waveform, as can be seen by comparing the waveforms in Figure a & b. Both of the waveforms shown were taken with the same input, a square wave with a frequency of 2kHz, but with two different values for R2, 1.2kΩ and 10kΩ.

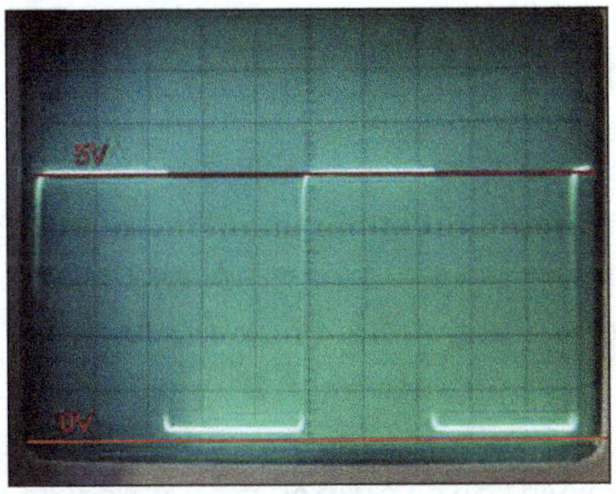

(a) Output with R2 = 1.2KΩ.

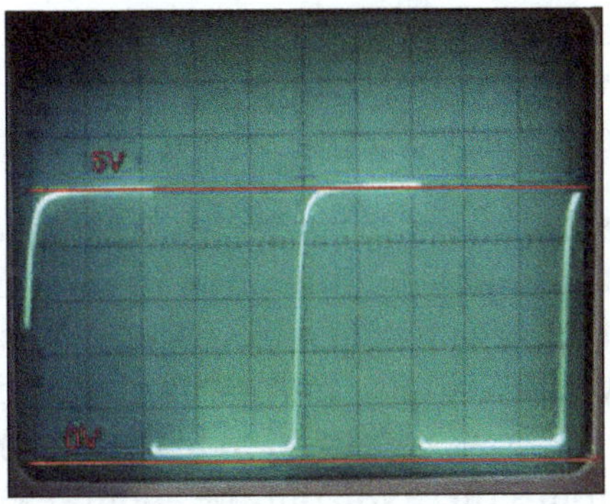

(b) Output with R2 = 10KΩ.

The rounding effect on the rise time of the pulses can be clearly seen in figure b. Also at higher frequencies amplitude of the output signal reduces markedly. Therefore for best performance the value of R2 should be kept as low as possible, but above 1kΩ. The performance of the optocoupler circuit showing the result of using the calculated values is shown in figure. Note also the effect of using a 74HCT14 Schmitt inverter at the output; any rounding of the square pulses is eliminated and although the optocoupler output only falls to 0.18V when the phototransistor saturates, the output of the Schmitt gate actually changes between +5V and 0V.

Adding a Schmitt inverter also re-inverts the output waveform, which was an inverted version of the input waveform at the collector of the phototransistor. There are of course, more useful applications for an optocoupler than simply isolating one logic IC from another. A common problem is driving a load from a computer's output port. Computers are expensive and easily damaged by mistakes made when connecting them to external circuitry. The problem is lessened by ensuring that the external circuit is fully isolated from the computer and an optocoupler such as the PC817 is a cheap and effective (assuming no major user errors) solution.

PC817 Motor Drive Circuit

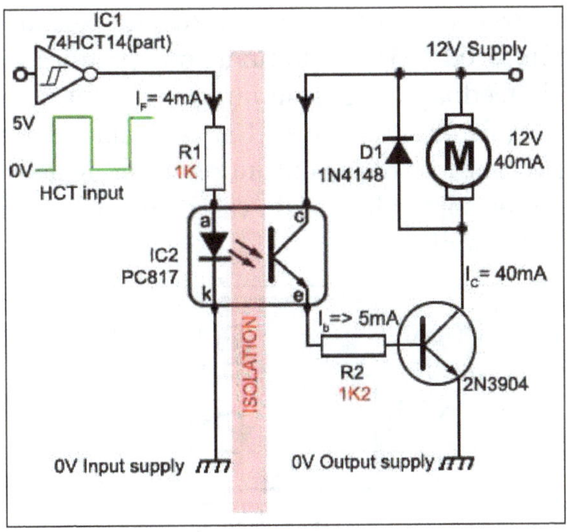

Figure illustrates a typical example where it is required to drive a 12V DC motor requiring 40mA of current from a logic circuit (or typical computer port) that can only support a few mA of current at 5V or less. As the current available from typical computer input/output ports may only be a few µA, as computer port lines are usually designed to drive some type of logic input, the input to this motor drive circuit is via a HCT Schmitt inverter gate, which only requires an input current of 1µA, with the 12V 40mA motor being driven by a 2N3904 transistor.

The optocoupler infrared LED is driven at about 4mA via a 1kΩ resistor from IC1 output. As the CTR of the PC817 is around 115% the phototransistor can supply about 9mA as the supply to the phototransistor output is now taken from the 12V motor supply. This is more than the 5mA minimum required to drive the 2N3904 into saturation. It is important that the transistor is fully saturated in order to reduce the power dissipation in the 2N3904 to a minimum, therefore although the transistor current (I_{CE}) is 40mA there will only be about 0.3V across the saturated transistor, so the power dissipation in the transistor will be 0.3V x 40mA = 12mW and the maximum dissipation for the 2N3904 is 1.5W. Although this basic interface only allows for switching the motor on or off, it could easily be adapted by changing IC1 to include a pulse width modulated speed control either from a computer, or hardware generated. This simple interface has one more safety feature; diode D1 connected across the motor will effectively prevent any nasty back EMF spikes generated by the inductive load (the motor) from causing damage to either the interface, or worse still to the computer.

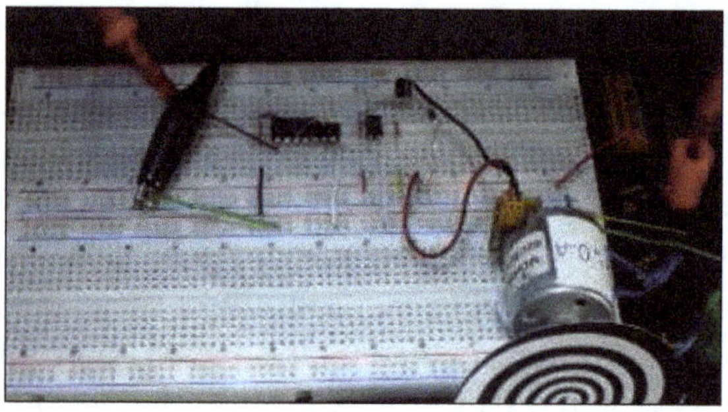

Audio Optocouplers

In audio systems, isolation between inputs and higher voltage/current equipment is usually provided by audio transformers, however it is also possible to use specialised audio optocouplers such as the IL300, which uses an infrared LED to illuminate one photodiode as an output device and a second photodiode to provide feedback, ensuring improved linearity and wider frequency range than phototransistor or photo resistor alternatives. Using feedback from a second photodiode with characteristics closely matched to those of the output photodiode also overcomes a basic problem with optocouplers. Without some sort of feedback, variations in the current transfer ratio will affect the performance of the optocoupler.

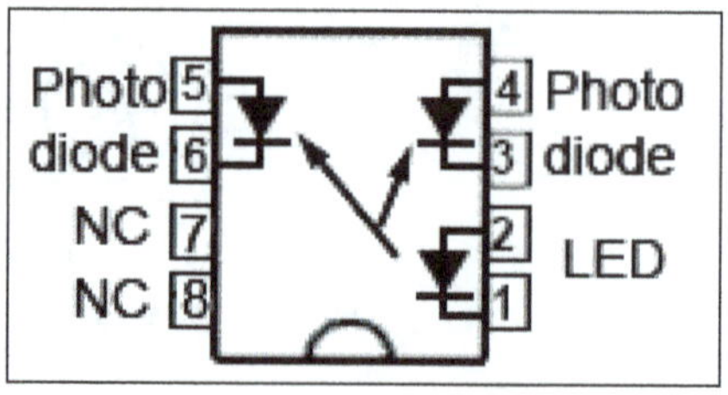

IL300 Optocoupler.

Variations can occur due to changes in ambient temperature and to ageing of the infrared LED. Optocouplers such as the IL300 from Vishay can therefore claim both better performance and stability over the lifetime of the circuit. Consequently these specialised devices are considerably more expensive than simple general-purpose optocouplers.

Using the 4N25 for Audio

However it is also possible to use some of the much cheaper, general purpose phototransistor optocouplers for audio applications, such as the popular 4N25 illustrated in Figure below also from Vishay, which is normally used for low frequency digital signals (in saturation mode) or DC applications (in linear mode) for limited audio applications.

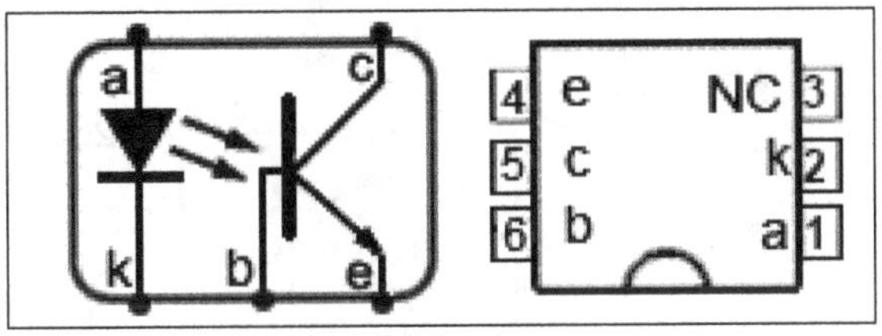

4N25 Optocoupler.

This is because the 4N25 has a connection for the phototransistor base available, as well as collector and emitter. This allows the three different methods of connection where the phototransistor can be operated in common emitter mode (a), common collector (emitter follower) mode (b), or used as a photodiode (c). When the phototransistor output is taken from the base connection (with the emitter left unconnected) the base/collector junction is being used as a photodiode. As the phototransistor now does not function as an amplifier the 'Miller Effect' (where the value of the junction capacitance is multiplied by the current gain of the transistor) does not apply, so the effectively large capacitance of the base/collector junction is greatly reduced, allowing the optocoupler to work at higher speeds, which means in the case of audio applications, a wider bandwidth is available, although this mode of operation greatly reduces the output signal amplitude.

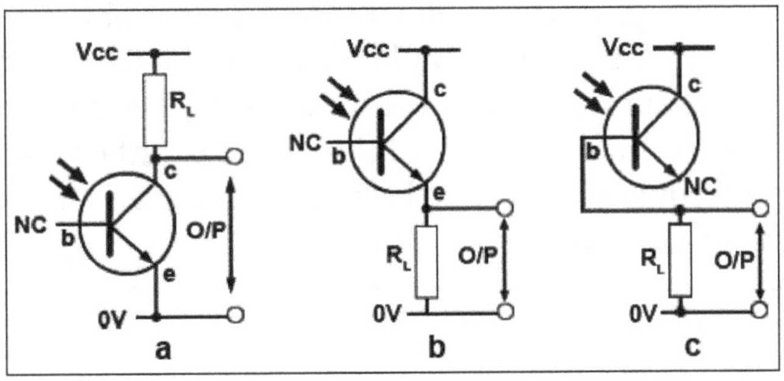

4N25 Connection Choices

The choice of load resistor has a significant effect on the output signal; the higher the value of RL the greater the amplitude of the output signals but the narrower the bandwidth, so the value of RL chosen is a compromise that depends on the purpose of the circuit. For the 4N25 to provide isolation for audio signals the input to the infrared LED must be appropriately biased with a DC voltage, so that when a modulating AC (audio) signal is applied, the current through the LED can be varied without the optocoupler output reaching either saturation or cut off. This is really an extension of the linear mode of operation, and can be applied using either one of two basic configurations, phototransistor or photodiode.

Basic Audio Optocoupler

The circuit shown adapted from a design uses the 4N25 connected as a phototransistor to pass audio signals whilst isolating the input and output circuits. An LM324 op amp is used here to drive the LED input of the 4N25. R1 and R2 form a potential divider to set pin 3 (the non-inverting input of the op amp) to half of the supply voltage and the audio input signal is superimposed on this DC bias voltage via C1 to modulate the current through the LED.

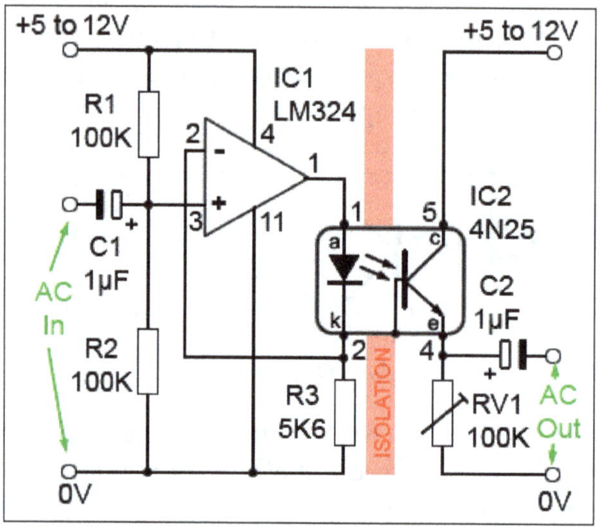

Audio Isolator using Phototransistor Mode.

Because the LM324 is connected in voltage follower mode with the LED forming part of the negative feedback loop, pin 1 exactly follows the voltage variations at pin 3, so varying the current though the LED. With no signal applied the standing current though the LED is about 0.5 to 1mA depending on the value of R3 and the supply voltage used for the input circuit.

With separate input and output supplies set at 12V and an AC input signal of 3Vpp, an output signal of around 6Vpp is obtained when RV1 is adjusted to set pin 4 of the 4N25 to about half supply (6V). Depending on supply voltages and the amplitude of signals used, fine adjustment of RV1 also provides a means of achieving minimum distortion. However, because Figure above uses a 4N25 in phototransistor mode, the useful bandwidth of the circuit is rather limited, making it useful for speech, but lacking gain at frequencies above 8kHz, a typical effect due to the large base/emitter junction capacitance of the phototransistor.

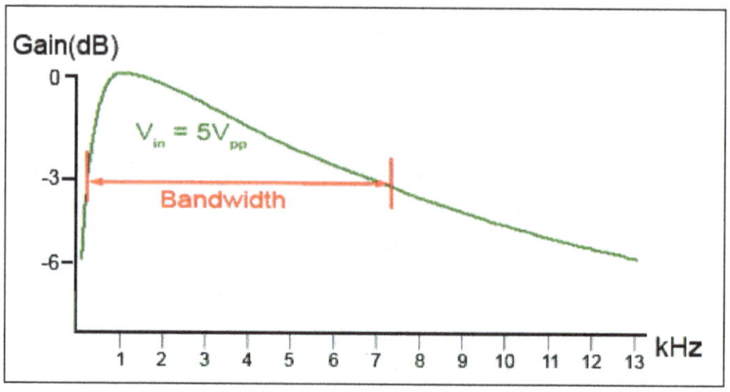

Audio Bandwidth using Phototransistor Mode.

Wide Bandwidth Audio

Figure below shows an improved circuit with a full audio bandwidth, achieved by using the 4N25 in photodiode mode where only the collector base junction is used, forming a photodiode with much less capacitance than with the full phototransistor. The downside of this method is that the 4N25 output is now reduced to milli-volts rather than volts. For this reason it is necessary to add a buffer amplifier (IC3a) having high input impedance, immediately after the optocoupler.

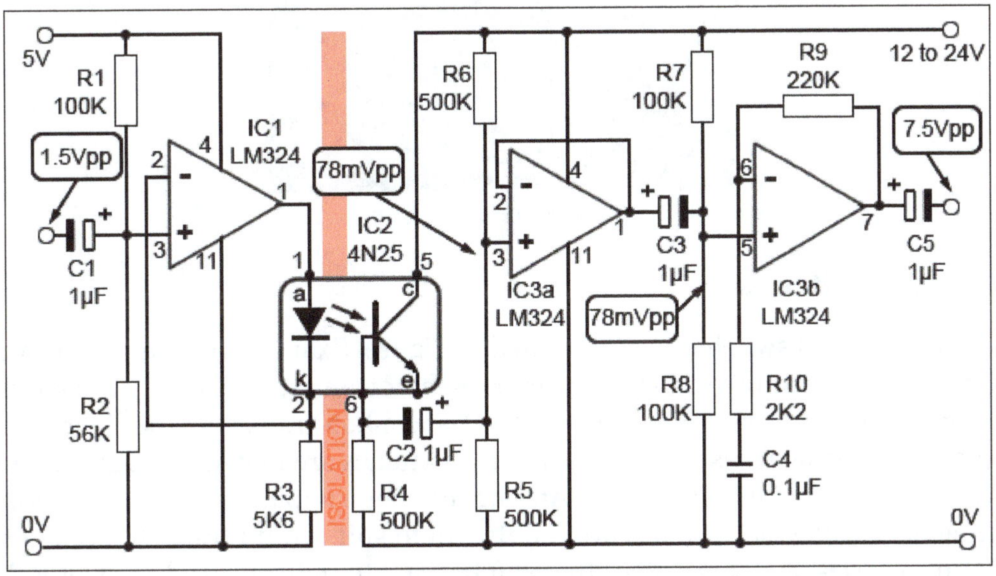

Audio Isolator using 4N25 in Photodiode Mode.

No voltage amplification takes place in the buffer stage, but this is followed by a x 100 voltage amplifier to produce an output of approximately 7.5Vpp from an original input of 1.5Vpp. The audio bandwidth of the circuit is now increased to cover a useful range from about 400Hz to 18kHz; still not hi-fi but quite useful as part of an audio amplifier for many isolation purposes.

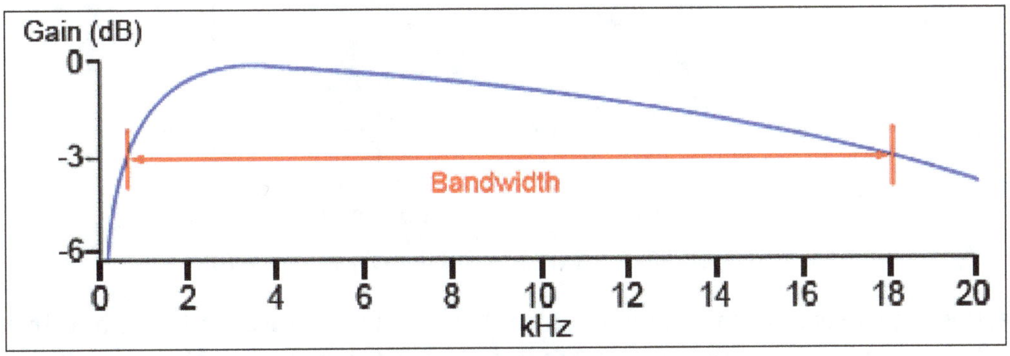

Audio Bandwidth using Photodiode Mode.

In the experimental (breadboard) version of this circuit different supply voltages between +5V and +24V were used for both input and output sections of the circuit without any problems apart from the lower supply voltages requiring signals of reduced amplitude to avoid distortion. Noise generated around the optocoupler on the breadboard could be a problem but may be reduced or eliminated by careful PCB design, and extra decoupling of the supply lines close to the ICs.

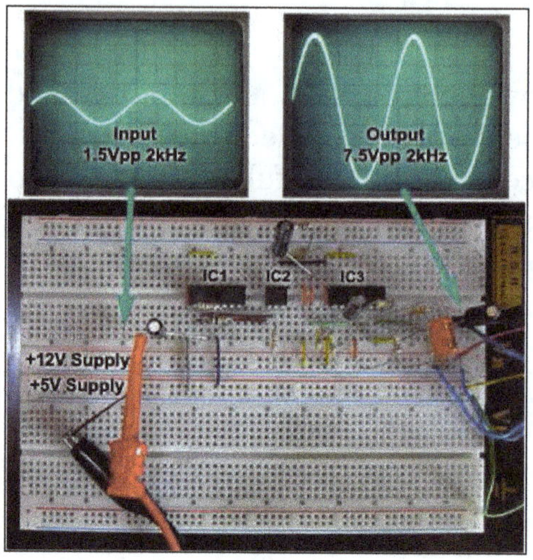

4N25 Audio Isolator Tests.

Opto Activated Switches

Two typical opto activated switches are illustrated in Figure. Example (a) is a slotted switch where a beam of infrared light from the LED illuminates a phototransistor, causing it to conduct. When an object is moved into the slot between the LED and phototransistor the light is interrupted and the phototransistor switches off. Opto activated switches are normally operated in saturation mode to provide definite on and off signals. Another common application for slotted switches is to have a rotating disc, with slots or holes around its rim to spin within the light path of the switch, thereby creating a series of on/off pulses that can be used to indicate (and electronically control) the speed of the spinning disc.

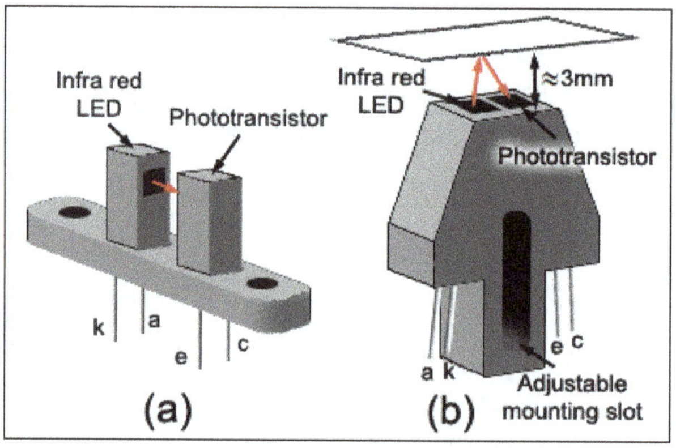

Slotted Optical Switch (a) and Reflective Object Sensor (b).

In the reflective object sensor illustrated in figure (b) the infrared LED and the phototransistor are mounted side by side in the narrow end of the switch. Here, a beam of infrared light is emitted at an angle by the LED and if some reflective material is placed about 3 to 4mm from the sensor, the light beam from the LED is reflected back onto the phototransistor causing it to conduct and produce an output current. This arrangement is often used as a proximity sensor. The phototransistor in proximity sensors also operates in 'saturation mode' where the phototransistor is caused to be either off (passing no current) or on (fully saturated, passing its maximum current) by the action of the reflected infrared light.

Slotted Optical Switch Example

When an object such as a ticket, or a small tab that is part of some mechanical system is placed in the sensor slot, the beam of infrared light from the LED to the phototransistor is blocked, turning off the phototransistor. Its emitter terminal, which has been at a high voltage of about 4.8V because the phototransistor has been in saturation and therefore conducting heavily, turns off and the emitter voltage falls to a low value of around 0.8V.

This turns off the 2N3904, which causes its emitter to fall to around 0.3V. This is seen as a logic 0 at pin 1 of IC2 (the input of one of the 6 Schmitt inverters in the IC) and its output at pin 2 changes to logic 1, lighting the green 5V LED. Removing the object from the slotted switch allows the infrared light to reach the base of the phototransistor causing it to conduct and saturate again, which now has a voltage of nearly 5V at its emitter. This switches on the 2N3904 causing a logic 1 (a voltage greater than 2V) to appear at its emitter and the Schmitt inverter input. This turns on the red LED and turns the green LED off.

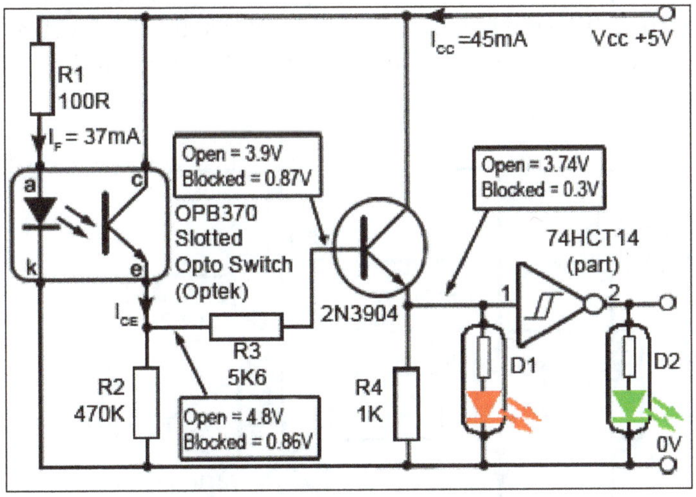

Slotted Optical Switch Circuit.

The Opto Slotted Switch Interface

The input LED of the slotted OPB370 switch chosen for this demonstration has a maximum current rating of 50mA and the data sheet for the OPB370 shows that at 20mA the forward voltage of the LED will be about 1.3V. To provide a reasonably bright illumination from the LED, it is driven via R1 with IF at about 37mA. The aim of this design is to drive a single input of a HCT Schmitt inverter gate, which will be responsible for ultimately providing an output that will have very fast rise and fall times and standard HCT voltage and current parameters.

Table: Typical HCT Input and Output specifications.

Supply	Output		Input	
V_{DD} = 5V	V_{OH} = 2.7V to 4.9V	I_{OH} = 4mA	V_{IH} > 2.0V	I_{IH} = 1μA
	V_{OL} = <0.2V	I_{OL} = 4mA	V_{IL} < 0.8V	I_{IL} = 1μA

The HCT Schmitt gate will recognise a voltage above 2.0V (V_{IH}) as logic 1 and a low voltage (V_{IL}) below 0.8Vas logic 0. The output current from the 2N3904 must also be sufficient to drive the HCT inverter input, and anything between 1μA and 4mA will be enough. R1 in the circuit is the current limiting resistor for the input LED and is chosen here to provide a current of about 37mA through the infrared input LED. Note also that the optional 5V LEDs (D1 and D2) used here have an internal current limiting resistor, but a 'normal' LED with appropriate external current limiting resistor could also be used.

Although phototransistor optocouplers produce many times the current produced by photodiode types the output from the phototransistor is still very small and is therefore further amplified by the 2N3904. In addition the 7414 Schmitt inverter provides further conditioning to make the output rise and fall times very fast, and the voltage and current levels ideal for driving HCT digital circuitry.

Reflective Object (Proximity) Sensor

These optical sensors work in a similar way to the slotted opto sensor but rely on infrared light reflected from an object (e.g. a sheet of paper in a printer) placed about 2mm to 8mm from the sensor to produce an output. The current produced by the sensor is even smaller than the output currents of the Slotted Switch and therefore also needs to be amplified by a transistor buffer stage to be useful.

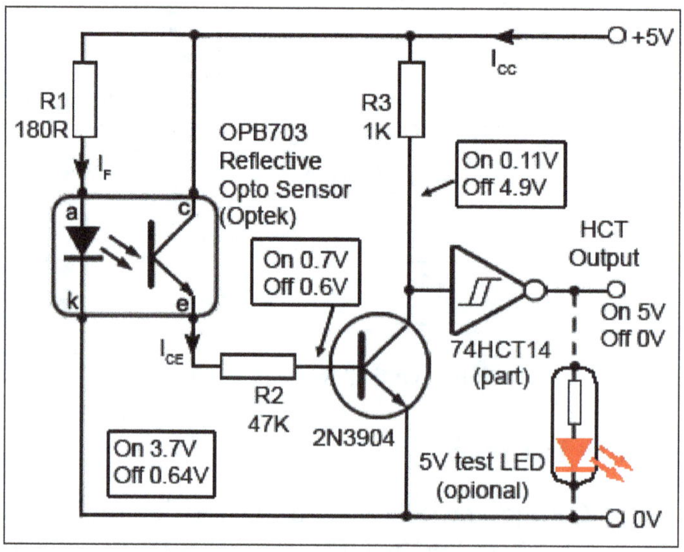

Opto Proximity Sensor Circuit.

A Schmitt Inverter is also added to the output to ensure rapid changes in logic level as a reflective object is detected within the detection range. To avoid sensing errors these sensors work best in low ambient light levels, where the only light sensed by the phototransistor is the reflected light from the infrared LED.

Notice also in both of these switching circuits that only one power source is being used, as electrical isolation between input and output is normally less important. The input to these sensors is light rather than some electrical property. The output is HCT logic, making it suitably conditioned for input to many computer or logic circuit applications.

Table: Circuit Currents.

I_{CC}	42mA	With LED
	23mA	No LED
I_F	21.2mA	
I_{CE} (opto)	64µA	R
	2µA	NR
I_{CE} (2N3904)	4.9Ma	R
	0.5mA	NR
R = Reflecting	NR = Not Reflecting	

Proximity Detector Circuit Operation

R1 is the current limiting resistor for the infrared LED and sets the LED current (IF) at approximately 20mA. The infrared light produced is reflected from an object (plain white printer paper was used in tests) and when in the detection range of about 2mm to 8mm produces a current though R2 of up to about 64µA, dropping to almost zero with no object present (and low ambient light levels). This current is sufficient to turn on the 2N3904 buffer amplifier transistor and its collector voltage falls from nearly Vcc to almost 0V.

These changes depend on the amount of light reflected into the sensor so may not be fast to change and can be of variable amplitude, however the Schmitt inverter will provide a very fast transition from low to high or high to low at any time the collector voltage of the 2N3904 transistor passes the inverter's threshold values. Where only a single sensor is required, the use of only 1 out of 6 inverters in the 7414 IC may seem wasteful. The circuit will work without the 7414 inverter, but the switching is much less definite and the output is high without a reflective object being present and low when reflection occurs.

For test purposes, a 5V red LED (with internal current limiting resistor) was used to give a clear indication of the circuit operation. However this can be left out when the circuit is used as an input to another circuit as this will considerably reduce the supply current required. Notice that the normally invisible infrared LED shows up as visible (purple) light on a digital camera. Also note that as only the inverter connected between pins 5 and 6 of the IC are being used, the five other unused inputs on the 7414 IC are all connected to 0V to prevent excessive noise being injected into the circuit.

References

- Directional-coupler-18885: techopedia.com, Retrieved 02, July 2020
- Star-coupler, engineering: sciencedirect.com, Retrieved 13, February 2020
- Distributed-Optical-Passive-Star-Couplers-3285458: researchgate.net, Retrieved 16, April 2020
- Optocoupler-devices-and-application, electronic-engineering: idc-online.com, Retrieved 29, June 2020
- Semiconductors: learnabout-electronics.org, Retrieved 05, March 2020

5

Laser Fundamentals

Laser refers to a device that generates an intense beam of light through the process of optical amplification by emitting photons from excited atoms. Some of the different types of lasers are solid laser, gas laser and liquid laser. All the diverse principles related to lasers have been carefully analyzed in this chapter.

The word "laser" is an acronym for Light Amplification by Stimulated Emission of Radiation. Lasers are finding ever increasing military applications -- principally for target acquisition, fire control, and training. These lasers are termed rangefinders, target designators, and direct-fire simulators. Lasers are also being used in communications, laser radars (LIDAR), landing systems, laser pointers, guidance systems, scanners, metal working, photography, holography, and medicine.

In this document the word laser will be limited to electromagnetic radiation emitting devices using light amplification by stimulated emission of radiation at wavelengths from 180 nanometers to 1 millimeter. The electromagnetic spectrum includes energy ranging from gamma rays to electricity. Figure below illustrates the total electromagnetic spectrum and wavelengths of the various regions.

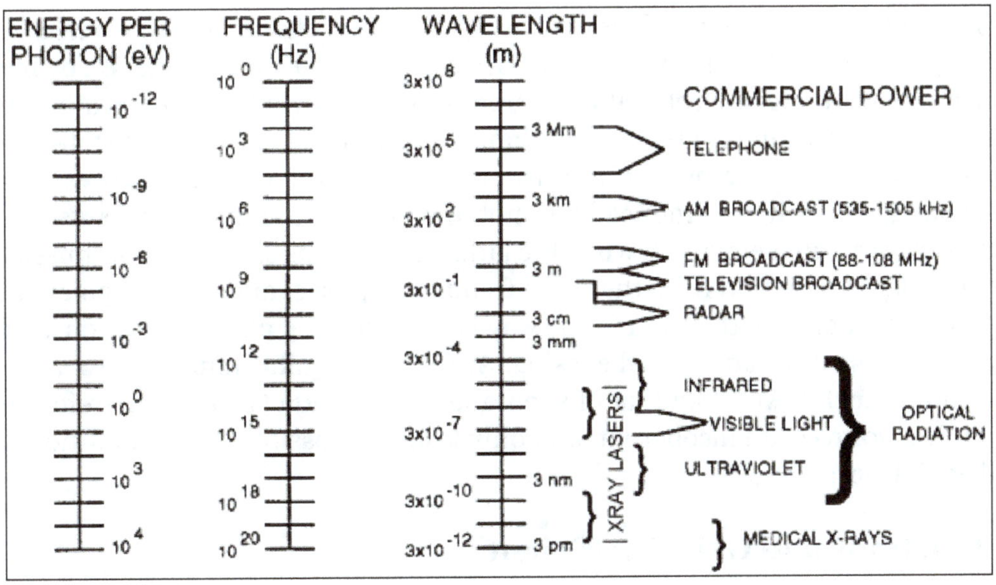

Electromagnetic Spectrum.

The primary wavelengths of laser radiation for current military and commercial applications include the ultraviolet, visible, and infrared regions of the spectrum. Ultraviolet radiation for lasers consists of wavelengths between 180 and 400 nm. The visible region consists of radiation with wavelengths between 400 and 700 nm. This is the portion we call visible light. The infrared region of the spectrum consists of radiation with wavelengths between 700 nm and 1 mm.

Laser radiation absorbed by the skin penetrates only a few layers. In the eye, visible and near infrared radiation passes through the cornea, and is focused on and absorbed by the retina. It is the wavelength of the light that determines the visible sensation of color: violet at 400 nm, red at 700 nm, and the other colors of the visible spectrum in between. When radiation is absorbed, the effect on the absorbing biological tissue is either photochemical, thermal, or mechanical: in the ultraviolet region, the action is primarily photochemical; in the infrared region, the action is primarily thermal; and in the visible region, both effects are present. When the intensity of the radiation is sufficiently high, damage to the absorbing tissue will result.

Construction and Principle of Lasing

Basic Construction and Principle of Lasing

Basically, every laser system essentially has an active/gain medium, placed between a pair of optically parallel and highly reflecting mirrors with one of them partially transmitting, and an energy source to pump active medium. The gain media may be solid, liquid, or gas and have the property to amplify the amplitude of the light wave passing through it by stimulated emission, while pumping may be electrical or optical. The gain medium used to place between pair of mirrors in such a way that light oscillating between mirrors passes every time through the gain medium and after attaining considerable amplification emits through the transmitting mirror.

Let us consider an active medium of atoms having only two energy levels: excited level E_2 and ground level E_1. If atoms in the ground state, E_1, are excited to the upper state, E_2, by means of any pumping mechanism (optical, electrical discharge, passing current, or electron bombardment), then just after few nanoseconds of their excitation, atoms return to the ground state emitting photons of energy $h\nu = E_2 - E_1$. According to Einstein's 1917 theory, emission process may occur in two different ways, either it may induced by photon or it may occur spontaneously. The former case is termed as stimulated emission, while the latter is known as spontaneous emission. Photons emitted by stimulated emission have the same frequency, phase, and state of polarization as the stimulating photon; therefore they add to the wave of stimulating photon on a constructive basis, thereby increasing its amplitude to make lasing. At thermal equilibrium, the probability of stimulated emission is much lower than that of spontaneous emission ($1 : 10^{33}$), therefore most of the conventional light sources are incoherent, and only lasing is possible in the conditions other than the thermal equilibrium.

Einstein Relations and Gain Coefficient

Consider an assembly of N_1 and N_2 atoms per unit volume with energies E_1 and E_2 ($E_2 > E_1$) is irradiated with photons of density $\rho_\nu = N h\upsilon$, where [N] is the number of photons of frequency ν per unit volume. Then the stimulated absorption and stimulated emission rates may be written as $N_1\rho_\nu B_{12}$ and $N_2\rho_\nu B_{21}$ respectively, where B_{12} and B_{21} are constants for up and downward transitions, respectively, between a given pair of energy levels. Rate of spontaneous transition depends on the average lifetime, τ_{21}, of atoms in the excited state and is given by N_2A_{21}, where

A_{21} is a constant. Constants B_{12}, B_{21}, and A_{21} are known as Einstein coefficients. Employing the condition of thermal equilibrium in the ensemble, Boltzmann statistics of atomic distribution, and Planck's law of blackbody radiation, it is easy to find out $B_{12} = B_{21}$, $A_{21} = B_{21}\ (8\pi h\nu^3/c^3)$, known as Einstein relations, and ratio, $R = \exp(h\nu/kT) - 1$, of spontaneous and stimulated emissions rates.

For example, if we have to generate light of 632.8 nm ($\nu = 4.74 \times 10^{14}$ Hz) wavelength at room temperature from the system of He–Ne, the ratio of spontaneous and stimulated emission will be almost 5×10^{26}, which shows that for getting strong lasing one has to think apart from the thermal equilibrium. For shorter wavelength, laser, ratio of spontaneous to stimulated emission is larger, ensuring that it is more difficult to produce UV light using the principle of stimulated emission compared to the IR. Producing intense laser beam or amplification of light through stimulated emission requires higher rate of stimulated emission than spontaneous emission and self-absorption, which is only possible for $N_2 > N_1$ (as $B_{12} = B_{21}$) even though $E_2 > E_1$ (opposite to the Boltzmann statistics). It means that one will have to create the condition of population inversion by going beyond the thermal equilibrium to increase the process of stimulated emission for getting intense laser light.

If a collimated beam of monochromatic light having initial intensity I_0 passes through the mentioned active medium, after traveling length x, intensity of the beam is given by $I(x) = I_0 e^{-\alpha x}$, where α is the absorption coefficient of the medium, which is proportional to the difference of N_1 and N_2. In the case of thermal equilibrium $N_1 \gg N_2$ the irradiance of the beam will decrease with the length of propagation through the medium. However, in the case of population inversion, $(N_2 > N_1) - \alpha$, will be positive and the irradiance of the beam will increase exponentially as $I(x) = I_0 e^{kx}$, where k is the gain coefficient of the medium and may be given by $k = (nN_d h\nu_{21} B_{21})/c$, where N_d is $N_2 - N_1$, c is speed of light, and n is refractive index of the medium.

Multilevel Systems for Attaining Condition of Population Inversion

Considering the case of two energy level system under optical pumping, we have already discussed that $B_{12} = B_{21}$, which means that even with very strong pumping, population distribution in upper and lower levels can only be made equal. Therefore, optical as well as any other pumping method needs either three or four level systems to attain population inversion. A three level system irradiated by intense light of frequency ν_{02} causes pumping of large number of atoms from lowest energy level E_0 to the upper energy level E_2. Non-radiative decay of atoms from E_2 to E_1 establishes population inversion between E_1 and E_0 (i.e., $N_1 > N_0$), which is practically possible if and only if atoms stay for longer time in the state E1 (metastable state, i.e., have a long lifetime) and the transition from E_2 to E_1 is rapid. If these conditions are satisfied, population inversion will be achieved between E_0 and E_1, which makes amplification of photons of energy $E_1 - E_0$ by stimulated emission. Larger width of the E_2 energy level could make possible absorption of a wider range of wavelengths to make pumping more effective, which causes increase in the rate of stimulated emission. The three level system needs very high pumping power because lower level involved in the lasing is the ground state of atom; therefore more than half of the total number of atoms have to be pumped to the state E_1 before achieving population inversion and in each of the cycle, energy used to do this is wasted. The pumping power can be greatly reduced if the lower level involved in the lasing is not ground state, which requires at least a four level system. Pumping transfers atoms from ground state to E_3, from where they decay rapidly into the metastable state E_2 to make N_2 larger than N_1 to achieve the condition of population inversion between E_2 and E_1 at moderate pumping.

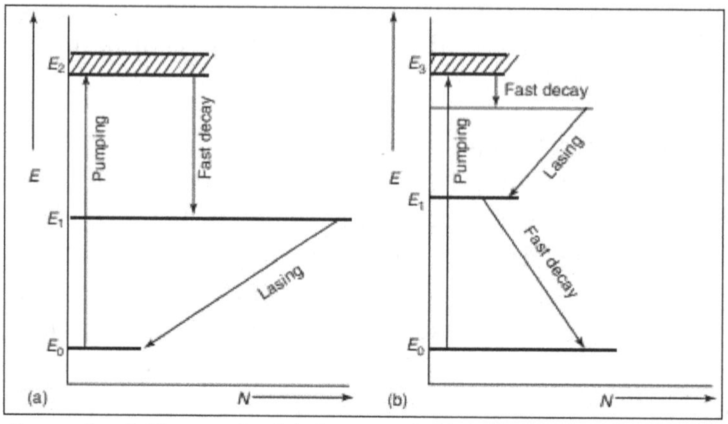

Energy level diagram for (a) three- and (b) four level laser systems.

Threshold Gain Coefficient for Lasing

Laser beam undergoes multiple oscillations (through active medium) between pair of mirrors to achieve considerable gain before it leaves the cavity through partially reflecting mirror. Laser oscillation can only sustain in the active medium if it attains at least unit gain after a round-trip between mirrors and maintains it overcoming various losses inside the cavity. If we incorporate these losses, the effective gain coefficient reduces to $k - \Upsilon$, where Υ is the loss coefficient of the medium. If round-trip gain G were less than unity, the laser oscillation would die out, while it would grow if the G value were larger than unity. Let us consider that the laser beam of intensity I_0 passes through the active medium, homogeneously filled in the length L between the space of two mirrors M_1 and M_2 with reflectivities R_1 and R_2, respectively.

The beam of intensity I_0 initiates from the surface of M_1 and attains intensity I_1($I_1 = I_0\exp(k - \Upsilon)$ L) after traveling a length L to reach at the surface of M_2. After reflection from M_2 and traveling back to M_1, the light intensity becomes I_2 ($I_1 = I_1R_2$ due to reflection and $I_2 = I_1' \exp(k - \Upsilon)L$), which finally becomes I_2 after reflection from M_1 to complete a round-trip ($I_2 = I_2R_1 = I_0R_1R_2\exp_2(k - \Upsilon)L$). Waves starting from the surface of mirror M1 and those that have completed one or more round trips are in the same phase. Now, the gain G(I_2' / I_0) attained in a round-trip should be at least unity to sustain the laser oscillation inside the cavity, therefore $R_1R_2 \exp 2(k - \Upsilon)L = 1$ is the threshold condition, which gives a value of $\Upsilon + (2L)^{-1} \ln(R_1R_2)^{-1}$ for threshold gain (kth) coefficient.

Optical Resonator

An optical resonator is an arrangement of optical components, which allows a beam of light to circulate in a closed path so that it retraces its own path multiple times, in order to increase the effective length of the media with the aim of large light amplification analogous to the positive feedback in electronic amplifiers. Combination of optical resonator with active medium is known as optical oscillator. A set of two parallel and optically flat mirrors, with one highly reflecting M_1($R \approx 100\%$) and another partially transmitting M_2($R > 95\%$), makes a simple optical oscillator as shown in Figure. Some of the pumped atoms in the excited states undergo spontaneous emission generating seed photons, which pass through the active medium and get amplified through stimulated emission. Most of the energy gets reflected from both the mirrors, passes through the active medium, and continues to get amplified until steady state level of oscillation is reached. After attaining this stage, amplification of wave amplitude within the cavity dies away and extra energy produced by stimulated emission exits as laser output from the window M_2. The gain coefficient inside the cavity should be greater than the threshold gain coefficient (kth) in order to start and maintain laser oscillation inside the cavity.

Owing to the diffraction effects, it is practically difficult to maintain a perfectly collimated beam with the combination of two parallel plane mirrors, which causes significant amount of diffraction losses. Such losses could be reduced by using a combination of concave mirrors and other optics in different optical arrangements. The optical configurations, which are able to retain the light wave inside the cavity after several transversals, are known as stable resonators. Some of the stable resonators are shown in the Figure. Laser oscillators with different geometries have their own benefits and losses. For example, in an oscillator having assembly of two parallel mirrors, it is difficult to align them in a strictly parallel manner. A slight deviation from the parallel geometry of the laser beam causes its walk away from the cavity axis after few reflections. However, it is beneficial in the sense that a large fraction of the active medium (mode volume) is pumped in this geometry. Confocal resonators are very simple to align, although lesser fraction of the active medium is being pumped.

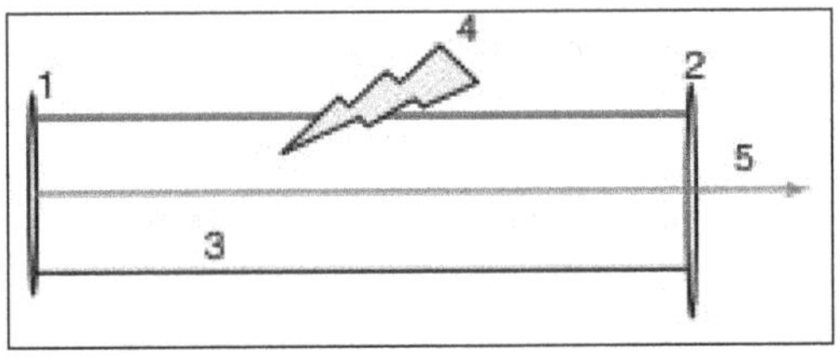

Basic geometry of laser cavity: (1) 100% and (2) 95–98% reflecting mirrors, (3) active medium, (4) pumping source, and (5) laser output.

Every laser resonator is characterized by a quantity Q termed as quality factor, which is defined by Q = (2π × energy stored)/(energy dissipated per cycle). The Q value of laser cavities lies in the range of ~10^5–10^6. Significance of higher Q value lies in the sense of capacity to store larger energy. In terms of line width ν, and frequency ν, the quality factor can be defined as Q = ν/ ν. A higher Q value associates with lower relative line width and vice versa.

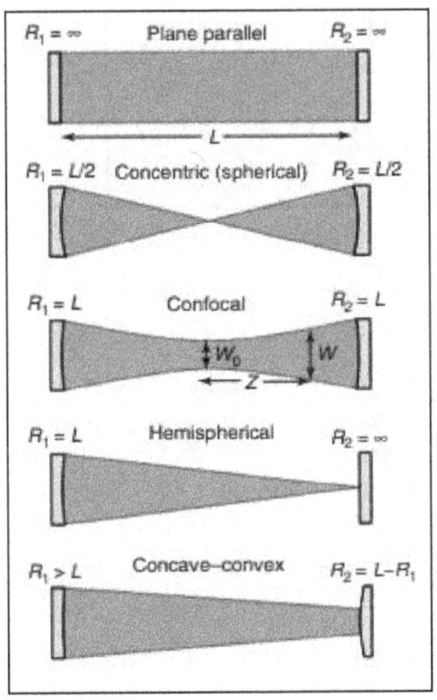

Different geometries of stable optical resonators.

A resonator that cannot maintain laser beam parallel to its axis is termed as unstable resonator. Such resonators suffer from high losses, but can make efficient use of the mode volume and have easy way of adjustment for the output coupling of the laser. Figure illustrates an unstable resonator having active medium between the mirrors. Output power of the laser and inner diameter of the annular-shaped beam can be easily adjusted by varying the distance between the two reflecting mirrors. Resonators having low irradiation volume or unstable cavities require active media with a large gain coefficients, such as CO_2 gas.

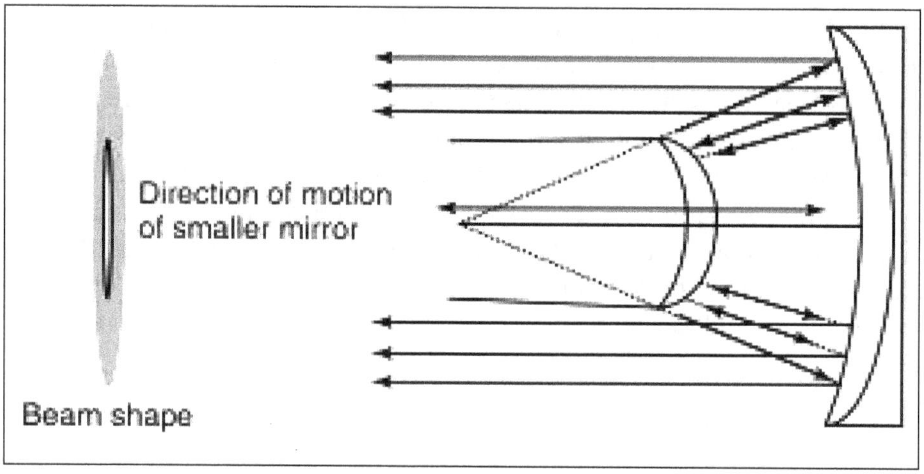

A sketch of unstable optical resonator with annular beam shape.

Laser Modes

The output of laser beam actually consists of a number of closely spaced spectral lines of different frequencies in a broad frequency range. The discrete spectral components are termed as laser modes, and coverage range is the line width of the atomic transition responsible for the laser output. Laser modes are categorized into axial and transverse modes.

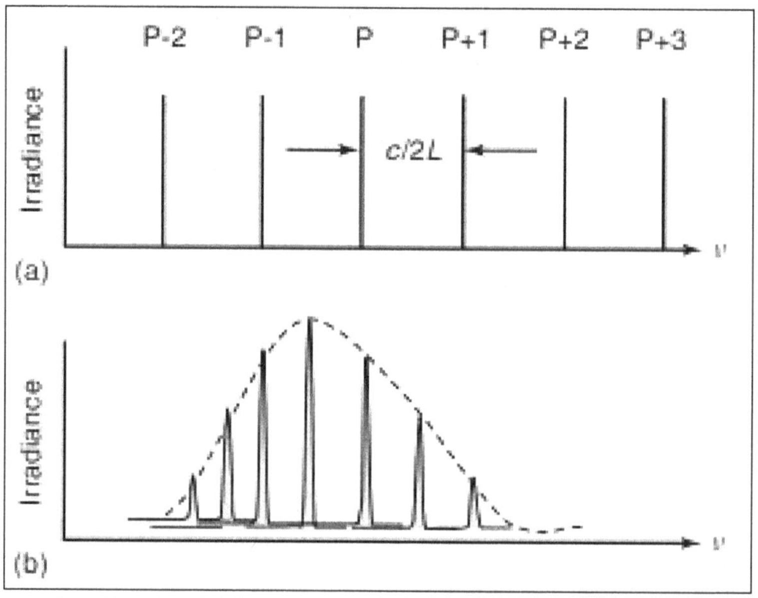

Axial laser modes (a) a simple illustration and (b) inside the laser line width, which shows that the mode at the center of the line has maximum intensity.

1) Axial modes: Let $d\phi = 2\pi/\lambda^*(2L)$ be the phase change in the laser wave after a round-trip in the cavity. In order to sustain laser oscillation inside the cavity, the phase change should be an integral multiple of 2π, that is, $2\pi/\lambda^*(2L) = 2p\pi$. In terms of frequency, this expression transforms to $v = pc/2L$; therefore separation between two adjacent p and p + 1 modes is given by $v = c/2L$. In the particular case of Nd: YAG (neodymium-ion-doped yttrium aluminum garnet) laser, $\lambda = 1064$ nm and L = 25 cm, $p = 2L/\lambda \approx 47 \times 10^4$ axial mode exists inside the laser cavity. If line width of the laser at 1064 nm is about $\Delta w = 1$ GHz, then only $w/\Delta v \approx 1$ axial mode oscillates in the cavity, while others die out. The axial modes are constructed by the light waves moving exactly parallel to the cavity axis. Light incident on a mirror and that reflected from that mirror construct a standing wave similar to a string bounded at both the ends. All the axial modes are due to the propagation of plane waves along the line joining centers of two reflecting mirrors.

2) Transverse modes: Unlike the plane waves propagating along the axis of the cavity in axial modes, there are some other waves traveling out of the axis that are not able to repeat their own path termed as transverse electromagnetic (TEM) modes. These modes can be practically seen in the form of pattern when the laser beam falls on any surface. These modes are assigned by two integers p and q in the form of TEM , where p and q are the number of minima in the horizontal and vertical directions, respectively, in the pattern of the laser beam. TEM_{00} means that there is no minimum in the beam spot, and this is known as uniphase mode. On the contrary, TEM01 shows that there is no minima in the horizontal scanning and one minima in vertical. Laser beam spots on the screen with several TEM modes are displayed in figure below.

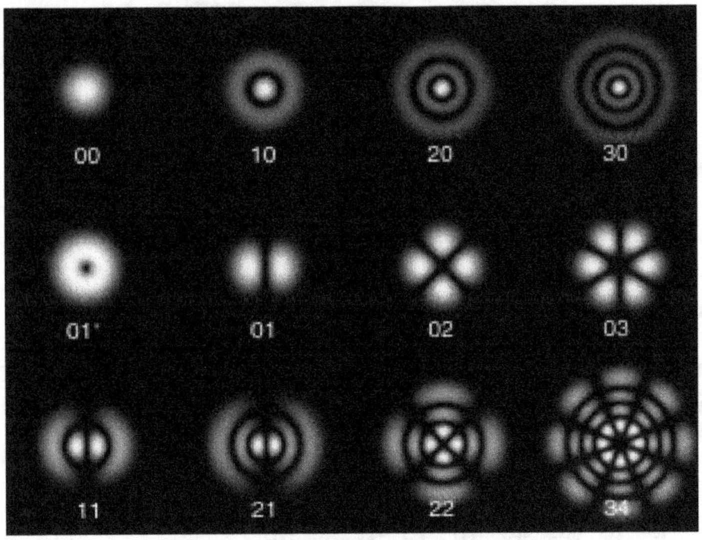

Various TEM modes of the laser.

Types of Laser

Depending on the nature of the active media, lasers are classified into three main categories, namely, solid, liquid, and gas. Scientists and researchers have investigated a wide variety of laser materials as active media in each category since 1958, when lasing action was observed in ruby crystal.

Solid Laser

Doped Insulator Laser

Solid state lasers have active media obtained by embedding transition metals (Ti^{+3} , Cr^{+3}, V^{+2}, Co^{+2}, Ni^{+2}, Fe^{+2}, etc.), rare earth ions (Ce^{+3}, Pr^{+3}, Nd^{+3} , Pm^{+3}, Sm^{+2}, $Eu^{+2,+3}$, Tb^{+3}, Dy^{+3}, Ho^{+3}, Er^{+3}, Yb^{+3}, etc.), and actinides such as U^{+3} into insulating host lattices. Energy levels of active ions are only responsible for lasing actions, while physical properties such as thermal conductivity and thermal expansivity of the host material are important in determining the efficiency of the laser operation. Arrangement of host atoms around the doped ion modifies its energy levels. Different lasing wavelength in the active media is obtained by doping of different host materials with same active ion. $Y_3Al_5O_{12}$, $YAlO_3$, $Y_3Ga_5O_{12}$, $Y_3Fe_5O_{12}$, $YLiF_4$, Y_2SiO_5, $Y_3Sc_2Al_3O_{12}$, $Y_3Sc_2Ga_3O_{12}$, $Ti:Al_2O_3$, $MgAl_2O_4$ (spinel), $CaY_4[SiO_4]_3O$, $CaWO_4$ (Scheelite), $Cr:Al_2O_3$, NdP_5O_4, $NdAl_3[BO_3]_4$, $LiNdP_4O_{12}$, $Nd:LaMgAl_{11}O_{19}$, $LaMgAl_{11}O_{19}$, $LiCaAlF_6$, $La_3Ga_5SiO_4$, $Gd_3Sc_2Al_3O_{12}$, $Gd_3Ga_5O_{12}$, $Na_3Ga_2Li_3F_{12}$, Mg_2SiO_4 (Forsterite), CaF_2, Al_2BeO_4 (Alexandrite), and so on, are some of the important hosts. Active atom replaces an atom in the host crystal lattice. Nd:YAG is one of the best lasing material and is representative of solid state lasing materials.

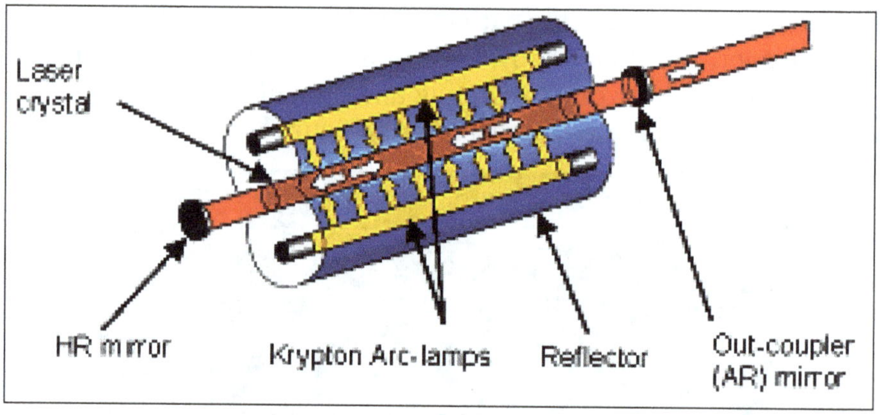

Dopant Energy Levels in the Host Matrices

Transition metal and rare earth ions have partially filled and unfilled 3d and 4f subshells, respectively. For example, the electronic configurations of trivalent Cr and Nd ions are as follows:

Cr^{+3}: $1s^2\ 2s^2\ 2p^6\ 3s^2\ 3p^6\ 3d^3$

Nd^{+3}: $1s^2\ 2s^2\ 2p^6\ 3s^2\ 3p^6\ 3d^{10}4s^2\ 4p^6\ 4d^{10}4f^3\ 5s^2\ 5p^6$

There are unshielded partially filled d electrons in the transition metal ions, while partially filled 4f electrons of the rare earth ions are shielded by 5p and 5s sub shells. Owing to the electronic shielding of inner subshells in rare earth ions, crystal field effect on the energy levels of transition metal ions are pronounced as compared to that on energy levels of rare earth ions. When one of these ions is doped into a host lattice, three main types of interactions occur: (i) columbic interaction between electrons in the unfilled shell, (ii) the crystal field, and (iii) spin–orbit interactions.

The columbic interaction between electrons causes splitting of energy levels of a single electron configuration into several levels denoted by pair of values of L and S (L and S are vector sum of angular, l, and spin, s, momenta, respectively, of electrons). Crystal field splitting dominates for transition metal, while spin–orbit interaction is the major contributor for rare earth ions in the modification of energy level of isolated host atom. The energy level diagram for Cr^{+3}:Al_2O_3 (ruby) and Nd^{+3}:YAG are displayed in figure below.

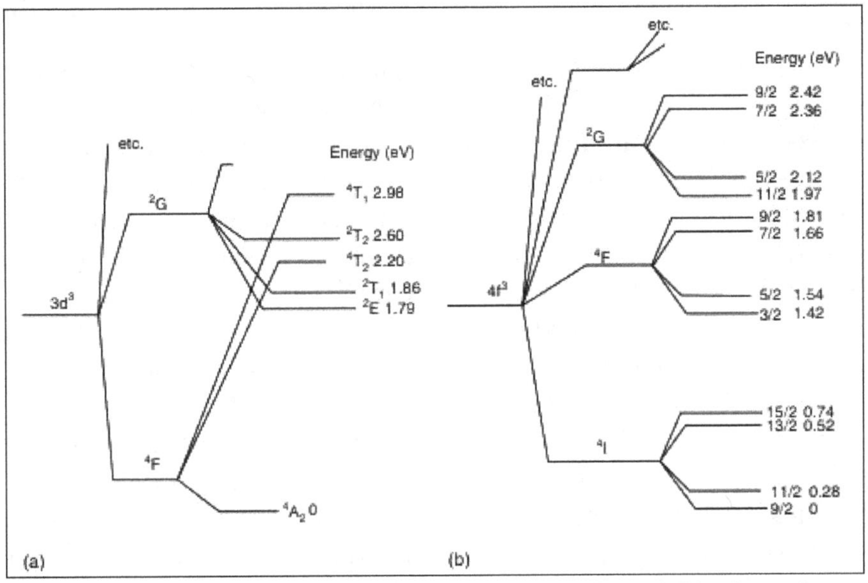

Energy level diagrams for doped insulator lasers: (a) Ruby and (b) Nd:YAG lasers.

Pumping Techniques in Solid State Lasers

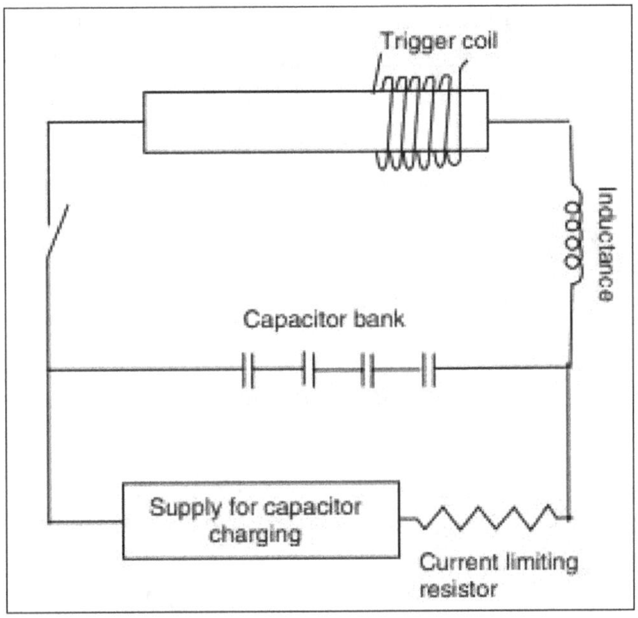

Pumping of electrons from the ground state to the excited state to achieve population inversion condition is an essential requirement for lasing. Optical pumping is the best and most efficient pumping method for solid state active media due to their broad optical absorption bands. A significant fraction of incident optical energy can be easily used for the pumping of ground state electrons using pulsed as well as continuous light sources. Excess light energy raises temperature of the laser materials; therefore pulsed light sources are more suitable for dissipation of heat through circulating water jackets. Low-pressure quartz/glass-sealed krypton/xenon lamps are mostly used for pulsed pumping light sources, while tungsten halogen lamps and high-pressure mercury discharge lamps are utilized for continuous optical pumping. An inductive, capacitive, and resistive (LCR) circuit and trigger unit as shown in Figure below is basically used for operating the flashtube.

High-voltage pulse of the trigger coil ionizes some gas in the tube and makes it conductive. This causes rapid discharge of the capacitor through the tube and generation of intense optical radiation for almost few milliseconds. A small inductor in the series protects damage of the tube due to high capacitor discharge current. Light source and active medium should be arranged in such a way that maximum pumping radiation falls on the active medium. Active media in solid state lasers are cylindrical and rod shaped with few millimeters diameter and few centimeter lengths. Several arrangements of cylindrical flash lamp and rod-shaped active media are used for optical pumping to get laser radiation. The flash lamp and active medium assembly are placed inside gold-plated reflectors of circular or elliptical cross section. In the first practical operating laser, ruby rod was pumped by helical flash lamp inside the cylindrical reflecting cavity. Such arrangement has significant uniformity of irradiation inside the rod but exhibits poor optical coupling. Side-by-side arrangement of flash lamp and laser rod inside the cylindrical gold-plated reflector or wrapping both together with a metal foil are simpler approaches having good optical coupling but poor uniformity of irradiation. An elliptical reflector having flash lamp at one focus and laser rod at the other focus is the most popular and best way of optical pumping in the solid state lasers. Light radiation leaving from the first focus gets focused close to the axis of laser rod placed at the second focus to make uniform energy distribution.

Combination of a number of elliptical reflectors having laser rod at the common foci and several flash lamps at the other foci is used for better optical pumping with more uniform energy distribution. Various geometries for the arrangement of laser rod and flash lamps are illustrated in figure below. Nd:YAG laser is widely used in the processing of materials and various characterizations.

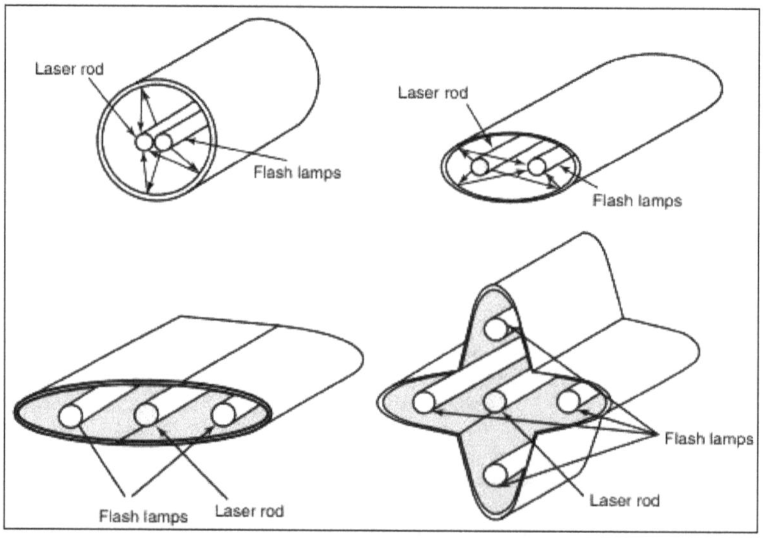

Different geometries for the arrangement of flash lamp and laser rod in solid states lasers.

Nd:YAG Laser Construction and Operation

The schematic diagram of Nd:YAG laser head as shown in Figure below, consists of oscillator section, rear mirror, quarter-wave plate, Pockel cells, polarizer, pump chambers, injection seeder, output coupler, D-Lok monitor, fold mirrors, amplifier section, harmonic generator (HG), temperature controller, dichroic mirrors, and Beam Lock pointing sensor. It may have single or multipump chambers, and each chamber consists of single or multiple flash lamps depending on the power of laser.

The laser head end panel contains coolant, output connector, coolant input connector, neutral/ground connector, control cable connector, high-voltage connector, Q-switch input connector, and nitrogen purge input connector. The HGs have potassium di-hydrogen phosphate (KDP) and beta barium borate (BBO) crystals for frequency doubling and tripling, respectively. It can be operated in long pulse and Q-switch modes. Long pulse mode has light pulses of almost 200 μs duration and separated from each other by 2–4 μs. The total energy of the pulse train is similar to that of a single Q-switched pulse. During Q-switched operation, the pulse width is less than 10 ns and the peak optical power is tens of megawatts.

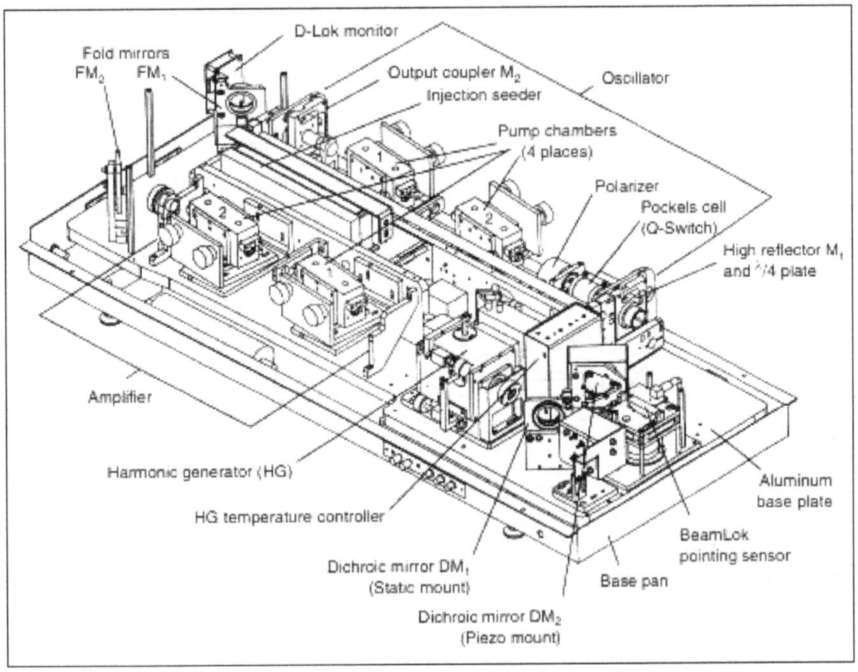

Assembly of various components in the head of an Nd:YAG laser system with four pump chambers.

The properties of Nd:YAG are the most widely studied and best understood of all solid state laser media. Its energy level diagram, optical arrangements for Q-switching and stable and unstable resonators are depicted in Figure below. The active medium is triply ionized neodymium, which is optically pumped by a flash lamp whose output matches principle absorption bands in the red and near infrared (NIR). Excited electrons quickly drop to the $F_{3/2}$ level, the upper level of the lasing transition, where they remain for a relatively longer time (~230 μs). The strongest transition is $F_{3/2} \rightarrow I_{11/2}$, emitting a photon in NIR region (1064 nm). Electrons in the I11/2 state quickly relax to the ground state, which makes its population low. Therefore, it is easy to build up a population inversion for this pair of states with high emission cross section and low lasing threshold at room temperature.

There are also some other competing transitions at 1319, 1338, and 946 nm from the same upper state, but having lower gain and a higher threshold than the 1064 nm wavelength. In normal operation, these factors and wavelength-selective optics limit oscillation to 1064 nm. A laser comprising just an active medium and resonator will emit a pulse of laser light each time the flash lamp fires. However, the pulse duration will be long, about the same as the flash lamp and its peak power will be low.

When a Q-switch is added to the resonator to shorten the pulse, output peak power is raised dramatically. Owing to the long lifetime of $F_{3/2}$, a large population of excited neodymium ions can build up in the YAG rod in a way similar to which a capacitor stores electrical energy. When oscillation is prevented for some time to build up high level of population inversion by electro-optical Q-switching and after that if the stored energy gets quickly released, the laser will emit a short pulse of high-intensity radiation.

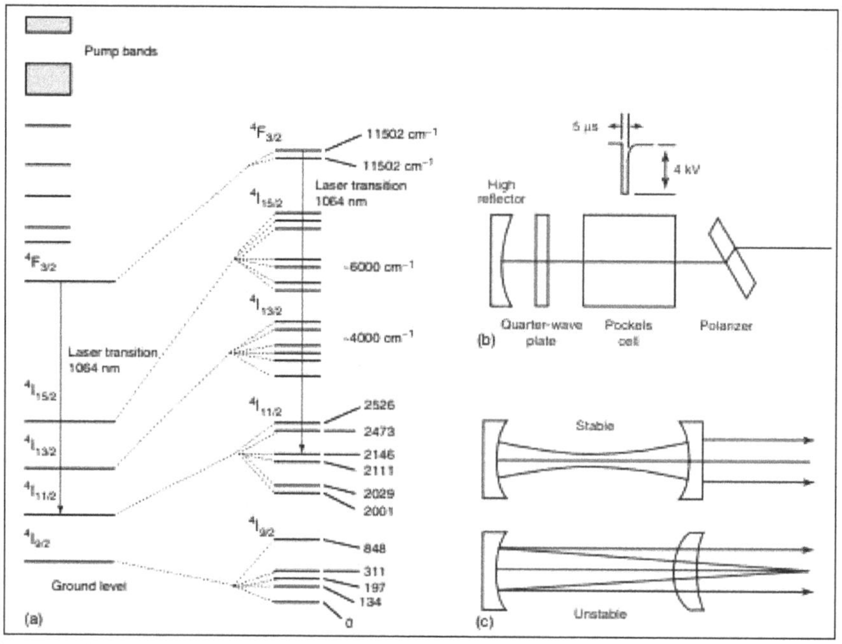

(a) Energy level diagram for the transition of Nd:YAG laser (b) The Q-switch comprises a polarizer, a quarter-wave plate, high quality reflector, and pockels cell, and (c) stable and unstable resonator configurations.

Semiconductor Laser

Semiconductor lasers also known as quantum well lasers are smallest, cheapest, can be produced in mass, and are easily scalable. They are basically p-n junction diode, which produces light of certain wavelength by recombination of charge carrier when forward biased, very similar to the light-emitting diodes (LEDs). LEDs possess spontaneous emission, while laser diodes emit radiation by stimulated emission. Operational current should be higher than the threshold value in order to attain the condition of population inversion. The active medium in a semiconductor diode laser is in the form of junction region of 2 two-dimensional layers. No external mirror is required for optical feedback in order to sustain laser oscillation. The reflectivity due to the refractive index differences between two layers or total internal reflection to the active media is sufficient for this purpose.

The diodes end faces are cleaved, and parallelism of reflecting surfaces is assured. Junction made from a single type of semiconductor material is known as homojunction, while that obtained from two different semiconductors is termed as heterojunction. Semiconductors of p and n type with high carrier density are brought together for constructing p-n junction with very thin ($\approx$1 µm) depletion layer. Figure below illustrates GaAs homojunction semiconductor diode laser. Lasing occurs in the confined narrow region, and optical feedback is done by reflections between cleaved end faces. For GaAs n = 3.6, therefore reflectivity R from the material–air interface is $R = (n - 1)^2/(n + 1)^2 = 0.32$, which is small but sufficient for lasing.

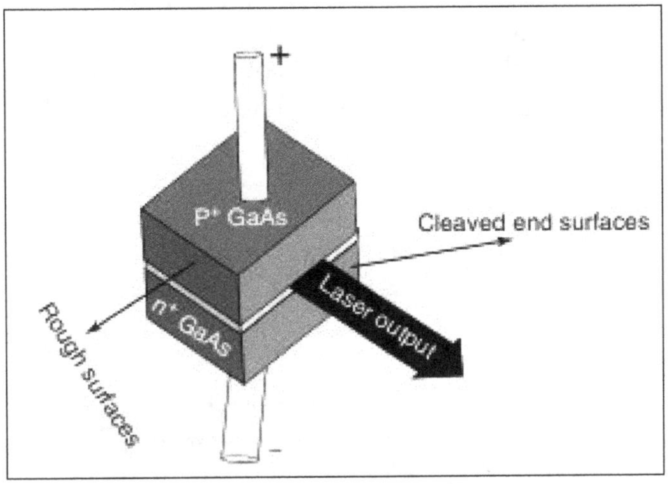

Basic geometry of semiconductor laser system.

When the operating current is small, the population inversion built compensates losses in the system and no lasing action is done. Increase of the current above a critical value named as threshold current commences lasing action, and the intensity of laser radiation increases rapidly with further increase in the operating current. Semiconductor lasers have large divergence compared to any other laser systems, which is due to their small cross section of active region. Actual dimension (d) of the active medium is of the order of light wavelength (λ), which causes diffraction and hence divergence by an angle of $\theta \approx \lambda/d$. Homojunction semiconductor lasers have some disadvantages over heterojunction lasers. Both of the laser systems should have confinement of injected electrons and emitted light in the junction region in order to initiate efficient stimulated emission process. In the homojunction laser, confinement of light is the consequence of the presence of hole and electrons close to the junction. Homojunction lasers operate under such confinement mechanism but have high threshold current density and low efficiency. Electrons have to travel different distances before they recombine with the holes. In contrast, heterojunction lasers exhibit much higher lasing efficiency and low threshold current density compared to their homojunction counterparts. Another difficulty with homojunction laser is to prevent the radiation from spreading out sideways from the gain region, which causes loss instead of gain. Therefore they can only be used in the pulsed mode.

Heterojunction lasers are constructed by sandwiching a thin layer of GaAs between two layers of ternary semiconductor compound $Ga_{1-x}Al_xAs$ with comparatively lower refractive indices and higher band gap energy. Lower refractive indices of surrounding layers causes confinement of laser radiation inside the active medium by the mechanism of total internal reflection, which makes laser oscillation to sustain in the medium. Higher band gap energy of the surrounding media creates potential barrier to prevent charge carriers to diffuse from the junction region, that is, provides a way for the confinement of charge carriers in the junction region, which enhances the condition of population inversion and hence stimulated emission. The electrical circuit for pumping the semiconductor diode lasers is similar to the doped insulator lasers.

Gas Laser

Gas lasers are widely available in almost all power (milliwatts to megawatts) and wavelengths (UV-IR) and can be operated in pulsed and continuous modes. Based on the nature of active media, there are three types of gas lasers viz atomic, ionic, and molecular. Most of the gas lasers are pumped by electrical discharge. Electrons in the discharge tube are accelerated by electric field between the electrodes. These accelerated electrons collide with atoms, ions, or molecules in the active media and induce transition to higher energy levels to achieve the condition of population inversion and stimulated emission.

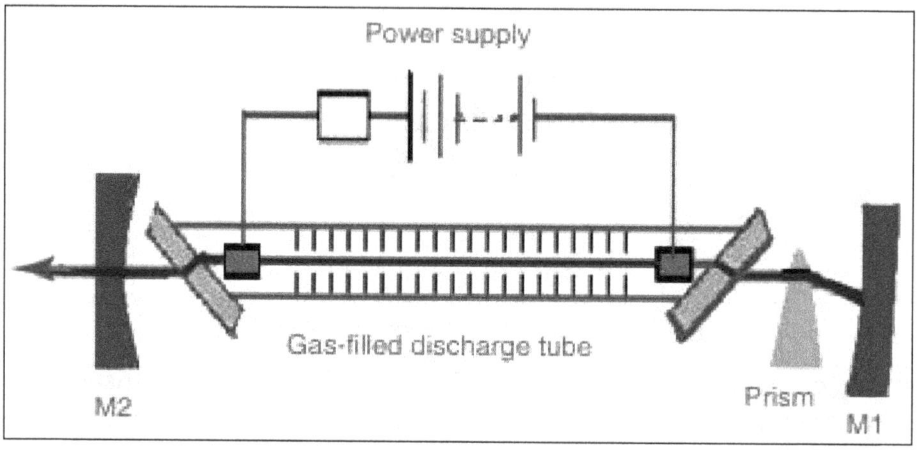

Construction of gas laser system (argon ion laser with prism-based wavelength tuning).

Atomic Gas Laser; He-Ne Laser

He–Ne laser is the simplest and representative of atomic gas lasers. The active medium is a 10:1 mixture of He and Ne gases filled in a narrow tube of few millimeter diameters and 0.1–1 m long at a pressure of about 10 Torr. Discharge tube and circuit are very similar as shown in Figure below. A resistant box is used in series with power supply in order to limit the discharge current because tube resistance falls too low once discharge is initiated. Energy levels of Ne atom are directly involved in the laser transitions, and He atoms provide an efficient excitation mechanism to the Ne atoms. Helium atoms from their ground state 1^1S, are pumped to the excited atomic states 2^1S and 2^3S by impact with accelerated electrons in the discharge tube.

Neon atoms have 3s and 2s atomic states, which are closer to the 2^1S and 2^3S states of the helium atoms, respectively. Collision between excited helium atoms in the 2_1S and 2^3S states and neon atoms in the ground states reinforce transfer of energy from helium to neon atoms. Helium atoms in the 2^1S and 2^3S states excite neon atoms from ground state to the 3s and 2s states, respectively, and return to the ground state. Excited states 3s and 2s of Ne atom have longer life times as compared to its lower (3p and 2p states), therefore they serve as metastable states and are used in achieving the condition of population inversion between s and p states. Transitions 3s → 3p, 3s → 2p, and 2s → 2p of neon atoms are consequences of lasing at 3.39 μm, 632.8 nm, and 1.15 μm wavelengths, respectively. Lifetimes of 3p and 2p atomic states are shorter; therefore Ne atoms from these states rapidly decay to the 1s state by non-radiative transitions. Neon atoms in the 1s state go to the ground state after losing energy through collision with the wall of the tube. The energy level diagram of the He–Ne laser is displayed in figure below.

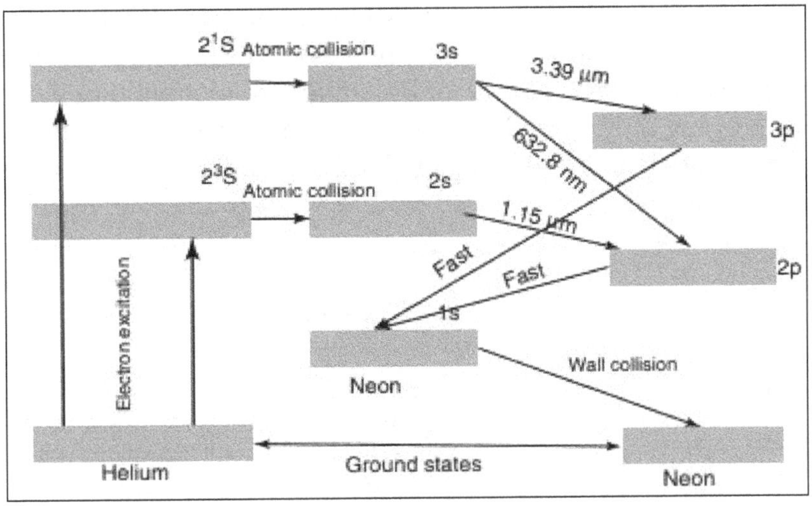

Energy level diagram for He–Ne laser system.

Ion Laser: Argon Ion Laser

Physical Construction Argon ion laser is one of the widely used ion gas lasers, which typically generates several watts power of a green or blue output beam with high beam quality. The core component of an argon ion laser is an argon-filled tube made of ceramics, for example, beryllium oxide, in which an intense electrical discharge between two hollow electrodes generates a plasma with a high density of argon (Ar^+) ions. A solenoid around the tube is used for generating a magnetic field, which increases output power of the beam by magnetic confinement of the plasma near the tube axis.

A typical device, containing a tube with a length of the order of 1 m, can generate 2.5–5 W of output power of laser beam in the green spectral region at 514.5 nm, using several tens of kilowatts of electric power. The dissipated heat is removed with a chilled water flow around the tube. The laser can be switched to other wavelengths such as 457.9 nm (blue), 488.0 nm (blue–green), or 351 nm (ultraviolet) by rotating the intracavity prism. The highest output power is achieved on the standard 514.5 nm line. Without an intracavity prism, argon ion lasers have a tendency for multiline operation with simultaneous output at various wavelengths.

Working of Ar Ion Laser

The argon ion laser is a four level laser, which facilitates to achieve population inversion and low threshold for lasing. The neutral argon atoms filled between two hollow electrodes inside the plasma tube are pumped to the 4p energy level by two steps of collisions with electrons in the plasma. The first step ionizes atoms to make ions in the 3p (E_1) state, and the second one excites these ions from the ground state E_1 either directly to the 4p4 levels (E_3) or to the $4p^2$ levels (E_4), from which it cascades almost immediately to the $4p^2$ (E_3). The 4p ions eventually decay to 4s levels (E_2), either spontaneously or when stimulated to do so by a photon of appropriate energy. The wavelength of the photon depends on the specific energy levels involved and lies in between 400 and 600 nm. The ion decays spontaneously from 4s to the ground state, emitting a deep ultraviolet photon of about 72 nm. There are many competing emission bands as shown in Figure. These can be preferentially selected using a prism in front of one of the end mirrors. This prism selects a specific wavelength to send it back through the cavity to stimulate identical emissions, which stimulates more and more emissions and make regenerative process. This facilitates laser to operate at a single wavelength. Removal of the prism allows for broadband operation, that is, several wavelengths are kept rather than keeping only a particular wavelength. The mirrors reflect a number of lines within a maximum range of about 70 nm. Energy level diagram showing various transitions of Ar ion laser is illustrated in Figure.

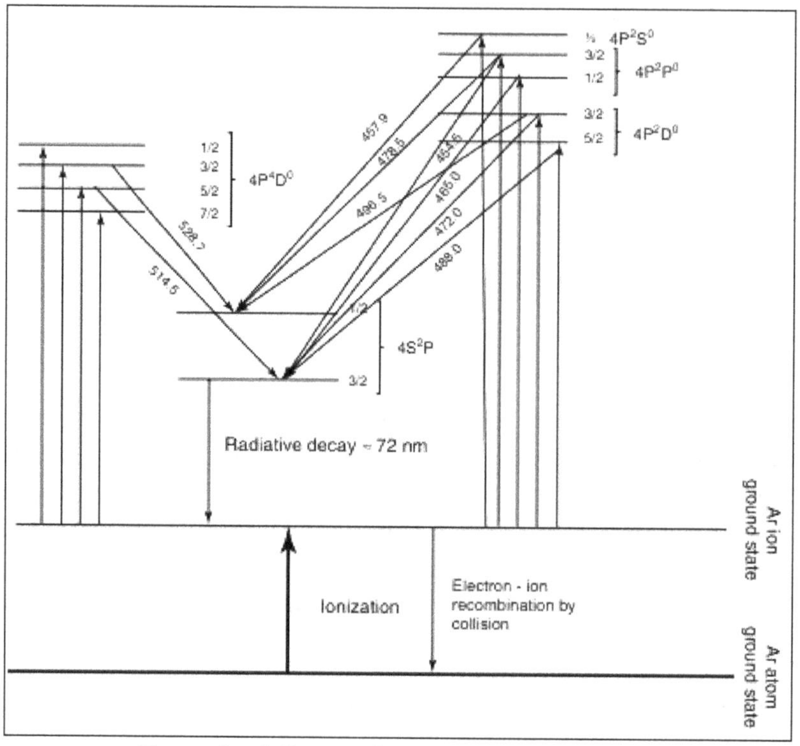

Energy level diagram for argon ion laser system.

Molecular Laser

Unlike isolated atoms and ions in atomic and ionic lasers, molecules have wide energy bands instead of discrete energy levels. They have electronic, vibrational, and rotational energy levels. Each electronic energy level has a large number of vibrational levels assigned as V, and each vibrational level has a number of rotational levels assigned as J. Energy separation between electronic energy levels lies in the UV and visible spectral ranges, while those of vibrational–rotational (separations between two rotational levels of the same vibrational level or a rotational level of one vibrational level to a rotational level from other lower vibrational level) levels, in the NIR and far-IR regions. Therefore, most of the molecular lasers operate in the NIR or far-IR regions.

Carbon Dioxide (CO_2) Laser

Carbon dioxide is the most efficient molecular gas laser material that exhibits for a high power and high efficiency gas laser at infrared wavelength. It offers maximum industrial applications including cutting, drilling, welding, and so on. It is widely used in the laser pyrolysis method of nano-materials processing. Carbon dioxide is a symmetric molecule (O=C=O) having three (i) symmetric stretching [ioo], (ii) bending [ojo], and (iii) anti-symmetric stretching [ook] modes of vibrations, where i, j, and k are integers. For example, energy level of molecules represents that it is in the pure asymmetric stretching mode with 2 units of energy. Very similar to the role of He in He–Ne laser, N_2 is used as intermediately in CO_2 lasers. The first, V = 1, vibrational level of N_2 molecule lies close to the (001) vibrational level of CO_2 molecules. The energy difference between vibrational levels of N_2 and CO_2 in CO_2 laser is much smaller (0.3 eV) as compared to the difference between the energy levels of He and Ne (20 eV) in He–Ne laser; therefore comparatively larger number of electrons in the discharge tube of CO_2 laser having energies higher than 0.3 eV are present. In addition to this, V = 1 state of N_2 is metastable, which provides longer time for the collision between excited N2 molecules and the ground state CO_2 molecules to excite them to (001) state. These two favorable conditions make it easy to attain high level of population inversion between 001 and 100, and 020 vibrational states of CO_2.

Transitions between 001 initial level to 100 and 020 final vibrational states make stimulated emissions of several IR radiations between 9.2 and 10.6 µm wavelengths. Helium gas is also mixed in the gas mixture in order to increase efficiency of lasing. Helium helps in transporting waste heats to the tube wall and de-exciting (100) and (020) energy levels by collision process. The amounts of N_2, CO_2, and He in CO_2 laser depends on the type and application of system, but usually, the amount of nitrogen and CO2 molecules are comparable, while helium concentration is higher than either. Low pressure (~10 Torr) is generally used for CW lasers, while quite higher pressure is used for high-energy and pulsed laser applications. Depending on power level and beam quality of CO_2 lasers, various internal structures are being used, and they are called sealed tube laser, gas flow laser, transversally excited atmospheric (TEA) laser, and gas dynamic laser. In the far-IR region of 10 µm wavelength, the usual optical material has large absorbance and therefore cannot be used as windows and reflecting mirrors in the cavity. Materials such as Ge, GaAs, ZnS, ZnSe, and some alkali halides having transparency in the IR region are used.

Nitrogen Laser

Lasing transition in N_2 laser takes place between two electronic energy levels, therefore this laser operates in the ultraviolet region at 337 nm wavelength. Here, upper electronic level has a shorter lifetime compared to the lower one, hence CW operation cannot be achieved, but pulsed operation with narrow pulse width is possible. The pulse width is narrow because as soon as lasing starts, population of the lower state increases, while that at upper state decreases and rapidly a state at which no lasing is possible is rapidly achieved. Such a laser system is known as self-terminating.

Excimer Lasers

Excimers are molecules such as ArF, KrF, XeCl, and so on, that have repulsive or dissociating ground states and are stable in their first excited state. Usually, there are less number of molecules in the ground state; therefore direct pumping from ground state is not possible. Molecules directly form in the first excited electronic state by the combination of energetic halide and rare gas ions. Condition of population inversion can be easily achieved because the number of molecules in the ground state is too low as compared to that in the excited state. Lasing action is done by transition from bound excited electronic state to the dissociative ground state. Population in the ground state always remains low because molecules dissociate into atoms at this point. Usually a mixture of halide such as F_2 and rare gas such as Ar is filled into the discharge tube. Electrons in the discharge tube dissociate and ionize halide molecules and create negative halide ions. Positive Ar^+ and negative F^- ions react to produce ArF^* molecules in the first excited bound state, followed by their transition to the repulsive ground state to commence lasing action. Various excimer lasers are developed in the wavelength range of 120–500 nm with 20–15% efficiency and up to 1 J peak and 200 W average powers. These lasers are widely used in materials processing and characterizations as well as for the pumping of dye lasers.

Liquid Laser

Liquids are more homogeneous as compared to solids and have larger density of active atoms as compared to the gasses. In addition to these, they do not offer any fabrication difficulties, offer simple circulation ways for transportation of heat from cavity, and can be easily replaced. Organic dyes such DCM (4-dicyanomethylene2-methyl-6-p-dimethylaminostyryl-4H-pyran), rhodamine, styryl, LDS, coumarin, stilbene, and so on, dissolved in appropriate solvents act as gain media. When the solution of dye molecules is optically excited by a wavelength of radiation with good absorption coefficient, it emits radiation of longer wavelength, known as fluorescence. The energy difference between absorbed and emitted photons is mostly used by nonradiative transitions and creates heat in the system. The broader fluorescence band in dye/liquid lasers facinates them with the unique feature of wavelength tuning. Organic dye lasers, as tunable and coherent light sources, are becoming increasingly important in spectroscopy, holography, and in biomedical applications.

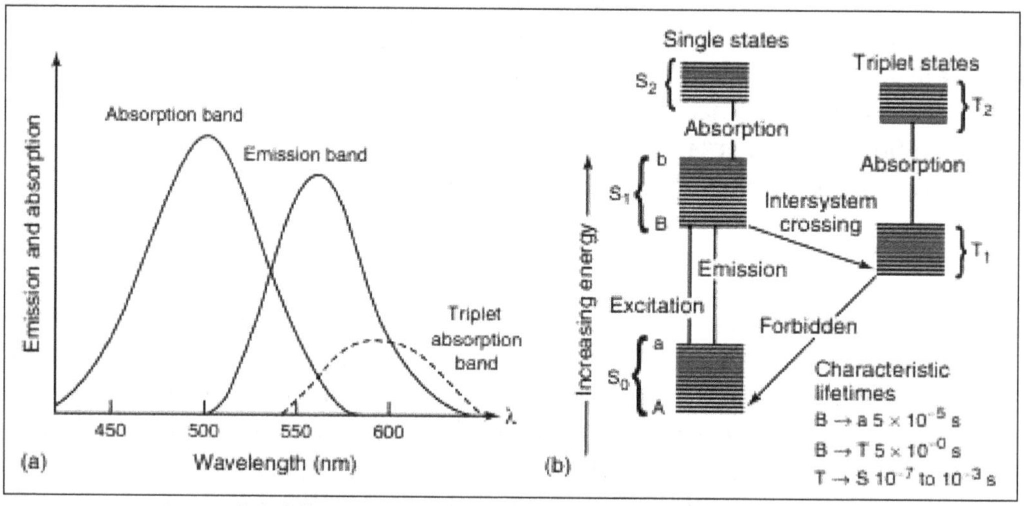

Energy level diagram for carbon dioxide (molecular) laser system.
Inset shows three mode of vibration of CO_2 molecule.

A recent important application of dye lasers involves isotope separation. Here, the laser is used to selectively excite one of several isotopes, thereby inducing the desired isotope to undergo a chemical reaction more readily. The dye molecules have singlet (S_0, S_1, and S_2) and triplet (T_1 and T_2) group of states with fine energy levels in each of them. Single and triplet states correspond to the zero and unit values of total spin momentum of electrons, respectively. According to selection rules for transitions in quantum mechanics, singlet–triplet and triplet–singlet transitions are quite less probable as compared to the transitions between two singlet or two triplet states. Optical pumping of dye molecules initially at the bottom of S_0 state transfers them to the top of S_1 state.

Collisional relaxation of these molecules takes them to the bottom of S_1 state, from where they transit to the top of S_0 state with stimulated emission of radiation. Most of the states in the complex systems are usually neither pure singlet nor pure triplet. Singlet states have small contribution of triplet and vice versa. In the case of most of the dye molecules, unfortunately, T_1 state lies just below the S_1, therefore few molecules transit from S_1 to T_1 by losing some energy through non-radiative transitions. Difference between T_1 and T_2 states is almost same as the wavelength of lasing transition, therefore emitted lasing radiation gets absorbed, which reduces laser gain and may cease the laser action. Therefore, some of the dye lasers operate in the pulsed mode with the pulse duration shorter than the time required to attain a significant population in the state T_1. Some of the dyes also absorb laser radiation corresponding to the transition from S_1 to upper singlet transitions. Therefore, one should select the dye molecule so that energy differences between these states do not lie between the ranges of laser radiation.

Methods of Producing EUV/VUV and X-ray Laser Beams

EUV and VUV coherent light sources, that is, EUV/VUV lasers, are in high demand in order to continue the validity of Moore's law in future, in the high-density data writing on disks, materials synthesis and characterizations, and spectroscopic investigations. For example, in photolithography for making the micro/nano-patterns on microelectronic chips, the width of the pattern is proportional to the wavelength of laser light used. Therefore, we would reach a limit for further miniaturization of electronic devices, which causes failure of the well-known Moore's law if shorter wavelength lasers would not be developed. Similarly, for data writing on the optical discs, if we have shorter wavelength lasers, larger amount of data can be written on the same size of disc. In addition to these, such sources are the future of 3D high-density data writing, microscopy at the atomic scale, crystallography, and medical sciences. Following are the methods for developing short wavelength lasers.

Free Electron Lasers (FEL)

In contrast to the other laser sources, free electron lasers (FELs) have an active medium that consists of a beam of free electrons, propagating at relativistic velocities in a spatially periodic magnet (undulator). Here, electrons experience the Lorentz force, execute transverse oscillations, and emit synchrotron radiation in the forward direction.

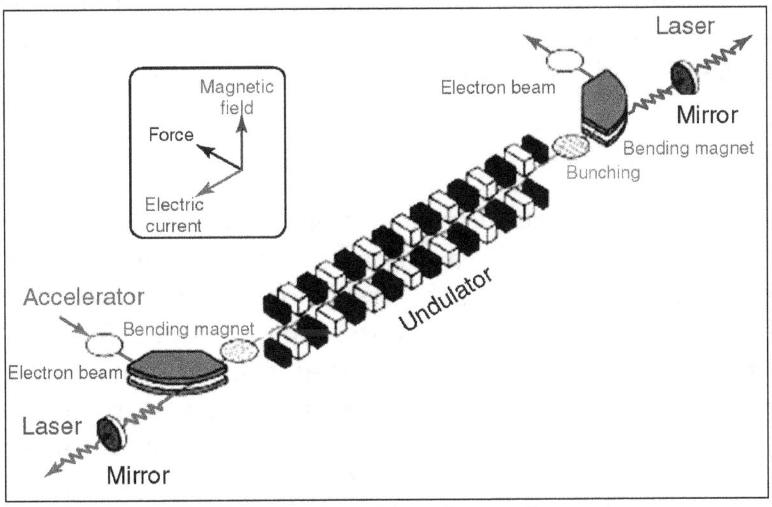

Construction geometry of free electron laser system.

We know that an accelerated charged particle moving with relativistic velocity emits radiation, which may be considered as spontaneous. The emitted spontaneous radiation interacts with the electron beam in order to stimulate laser radiation. Since electric field of light is perpendicular to the direction of its propagation, electrons cannot interact with photons unless they also have a velocity component parallel to the light electric field. Hence, electron beam is allowed to travel through wiggler magnet, which generates a spatially periodic magnetic field along the direction of propagation of electron beam. After passing through such field, electrons undergo transverse oscillations as shown in Figure.

In the present case, wavelength of emitted radiation is given by the expression $\lambda = \lambda_u/2\gamma^2$ where λ_u is the undulator length and γ is the ratio of total electron energy to its rest energy. Similar to other lasers, emitted radiation oscillates between two mirrors of the laser cavity and interacts with the electron beam to enhance power of the laser radiation before leaving the cavity as laser output. Since mirrors have very low reflectivity in the X-ray region, therefore there is severe problem in achieving intense X-ray laser beam. In another mechanism known as self-amplified stimulated emission (SASE), radiation of the spontaneous emission from relativistic electron interacts with the same electron to stimulate it for stimulated emission. Following are the salient features of FELs:

- Tunability: FELs generate coherent, high-power radiation that is widely tunable, currently ranging in wavelength from microwaves, through terahertz radiation and infrared, to the visible spectrum, to soft X-rays. They have the widest frequency range of any laser type. Wavelength of FELs can be easily tuned by varying electron energy and period and amplitude of magnetic field. More shorter wavelength can be produced by harmonic generation.
- Pulse duration: Based on the electron beam time structure pulse duration ranges from CW to ultra-short pulsed regime (fraction of picoseconds).
- Coherence: It is transverse and longitudinal for oscillators and coherent seed amplifiers, while only transverse for SASE.
- Brilliance: Depending on the status of the art of the electron beam technology, the FEL brilliance can be larger, in some spectral regions (in particular in VUV-X), by many orders of magnitude than the brilliance of the existing sources (lasers and synchrotron radiation).

X-ray Lasers

X-ray lasers provide coherent beam of electromagnetic radiation with the wavelength range from ~30 nm down to ~0.01 nm. The first X-ray laser beam was initiated in 1980 by underground nuclear explosion at Nevada test site, while the first laboratory demonstration of X-ray laser was made in 1984 in the form of Nova laser. Most of the experimental demonstration of light amplification in this spectral range has come from the high-density plasma produced by interaction of high-energy laser with the solid target. Like the above case of FEL, unavailability of good-quality of cavity mirrors has utilized the mechanism of SASE in X-ray lasers. In such case intensity of the laser beam depends on the amplifier length. Gain of the active medium depends on the mass and temperature of the ions and multiplicity and population of the upper/lower levels.

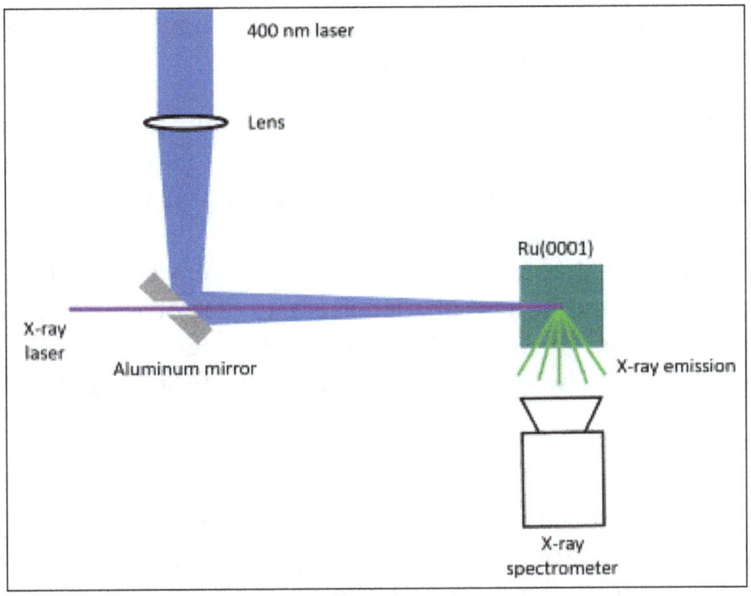

Most of the X-ray lasers utilize transitions among L, M, N, and so on, shells of highly ionized atoms. The short life times of the excited states (picoseconds) require very large pumping rates in order to achieve and maintain the condition of population inversion. Laser produced plasmas (LPPs) have high electron temperatures (~0.1–1.0 keV) and densities ($\sim 10^{18-21}$ cm^{-3}), which is required for higher degree of ionization and excitation in X-ray lasers. Therefore LPPs are used as primary medium for X-ray lasers.

In addition to these, LPPs have uniform density and temperature and thus can provide good media for amplification and propagation of X-rays. Of the various processes suggested, two main (i) collisional excitation and (ii) electron recombination are responsible for attaining the population inversion condition. In case of collisional excitation upper state of the lasing has lower probability of decay with dipole radiation, compared to the lower state, which creates the condition of population inversion between them. In contrast, the rate of population at any state by three-body recombination process is proportional to the forth power of the principle quantum number. The combination of preferential population of upper levels and fast radiative decay of lower states leads to an inversion amongt the N = 2, 3, 4, ... states.

EUV/VUV Lasers through Higher Harmonic Generation

Higher harmonic generation (HHG) is a nonlinear optical process used for the generation of shorter wavelength laser light from the interaction of high-intensity longer wavelength lasers source with nonlinear optical medium. The obtained new frequencies are integral multiples (nω) of the fundamental (ω) frequency of original laser light. This phenomenon was first observed in 1961 by Franken et al. with ruby laser and quartz as nonlinear medium.

The first HHG result was found in 1988, which shows that intensity of the spectra decreases with the increase of order, reaches a plateau, where the intensity remains constant for several orders, and finally ends abruptly at a position called high harmonic cutoff. They are portable sources of EUV/soft X-rays, synchronized with the fundamental laser and operated at the same frequencies with much shorter pulse width. These are more spatially coherent compared to X-ray lasers and cheaper than FELs. The harmonic cutoff increases linearly with increasing laser intensity up to the saturation intensity I_{sat} where harmonic generation stops.

The saturation intensity can be increased by changing the atomic intensities of lighter noble gases, but these have lower conversion efficiency. HHG strongly depends on the driving laser field; therefore the produced harmonics have similar spatial and temporal coherence. Owing to the phase matching and ionization conditions required for HHG, the produced new pulses are with shorter pulse duration compared to the driving laser. Mostly, harmonics are produced in very short time frame, when phase matching condition is satisfied. Instead of shorter temporal window, they emit co-linearly with the driving laser pulse and have very tight angular confinement. Gaseous media and LPP on the solid surfaces are two sources used as nonlinear active media for the generation of harmonics.

The shortest wavelength producible with the harmonic generation is given by cutoff of the plateau, which is given by the maximum energy gained by ionized electron from the light electric field. The energy of cutoff is given by $E_{max} = I_p + 3.17U_p$, where U_p is the ponder-motive potential from the laser field and I_p is the ionization potential. It is assumed that electron is initially produced by ionization into the continuum with zero initial velocity and is accelerated by the laser electric field. After half period of the laser electric field, direction of motion of electron is reversed and it is accelerated back to the parent nuclei. After attaining high kinetic energy of the order of hundreds of electron volts (depending on the intensity of laser field), when electron enters into the parent nuclei, it emits Bremsstrahlung-like radiation during a recombination process with the atoms as it returns back to the ground state. The three-step model of ionization, acceleration of electron, and its recombination with parent nuclei by the emission of EUV photons is shown in Figure below.

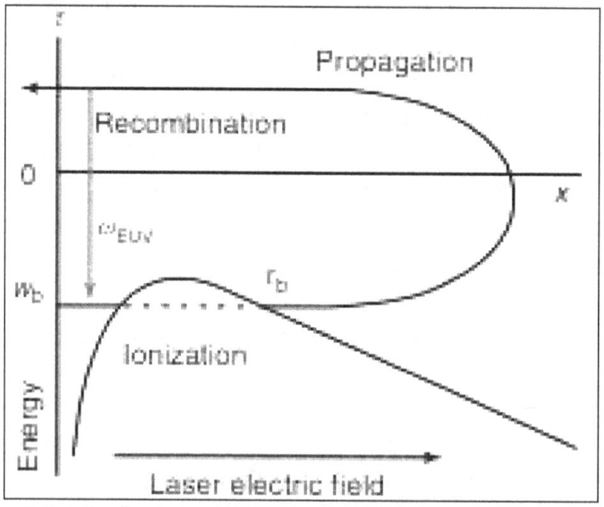

Three-step model of ionization and acceleration of electron and its recombination with parent nuclei by the emission of EUV photons.

Wide wavelength ranges of laser systems starting from hundreds of micrometers (molecular liquid lasers) to X-ray regions (FELs, X-ray lasers, and HHG lasers) with CW and pulsed lasers having various pulse width from milliseconds to atto-seconds and repletion rates from single pulse to the megahertz are available nowadays. Figure below represents the spectrum of available laser system with wavelength range and active media.

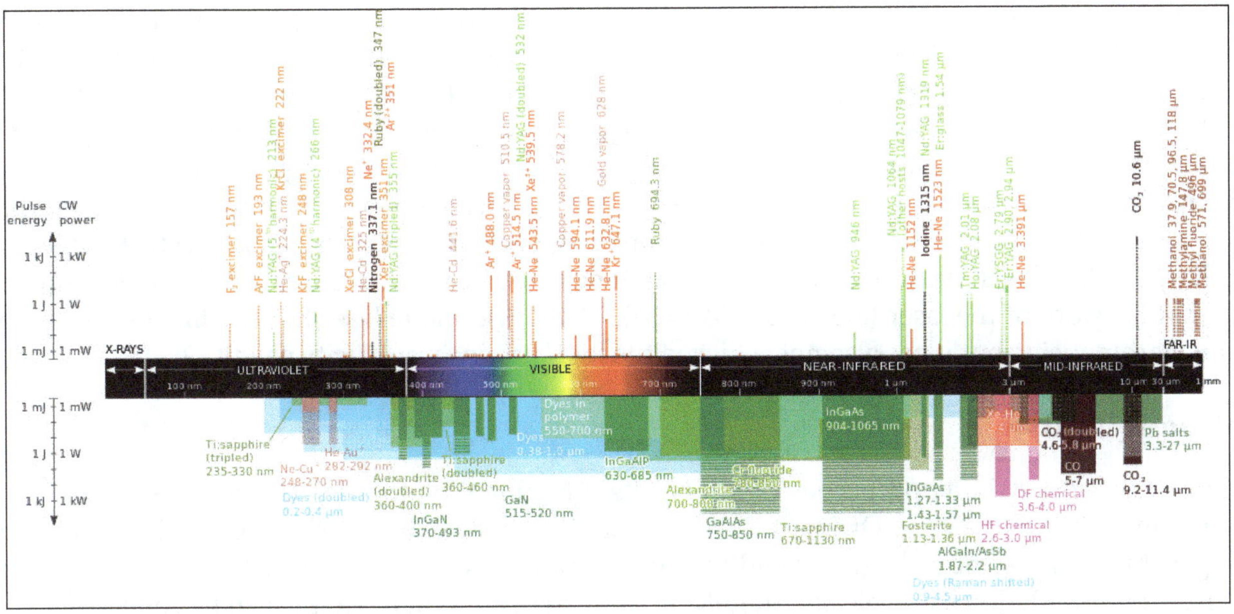

Spectrum of available laser systems, their wavelength range and active media.

Properties of Laser Radiation

Light produced from the lasers have several valuable characteristics not shown by light obtained from other conventional light sources, which make them suitable for a variety of scientific and technological applications. Their monochromaticity, directionality, laser line width, brightness, and coherence of laser light make them highly important for various materials processing and characterization applications.

Monochromaticity

Theoretically, waves of light with single frequency ν of vibration or single wavelength λ is termed as single color or monochromatic light source. Practically, no source of light including laser is ideally monochromatic. Monochromaticity is a relative term. One source of light may be more monochromatic than others. Quantitatively, degree of monochromaticity is characterized by the spread in frequency of a line by ν, line width of the light source, or corresponding spread in wavelength λ. For small value of $\Delta\lambda$, frequency spreading, ν, is given as $\nu = -(c/\lambda^2)\ \lambda$ and $\lambda = (c/\nu^2)$. The most important property of laser is its spectacular monochromaticity. Based on the type of laser media, solid, liquid, or gas and molecular, atomic, or ions, and the type of excitation, produced laser line consists of color bands that range from broad (as dye laser $\Delta\lambda \sim 200$) to narrow (for gas discharge lines, $\Delta\lambda \sim 0.01$ nm). Utilizing suitable filters one can get the monochromaticity as good as a single line of lasing transition. But such a single line also contains a set of closely spaced lines of discrete frequencies, known as laser modes. Suppressing all other modes excluding the central intense mode using mode locking one can increase monochromaticity of laser line. Suppressing of modes is possible by increasing the separation ($\nu_{sep} = c/2L$) between two modes, which can be done by reducing the cavity length.

When the axial mode separation approaches the line width, Δv, of the lasing transition it is possible that only single mode oscillates. Now, the line width of the laser is equal to that of the single longitudinal mode, which is too narrow. The width of the laser line is directly related to the quality factor Q of the cavity and is given by $Q = v/\Delta v$. The quality factor Q actually defines ($Q = 2\pi \times$ energy stored in the resonator at resonance/energy dissipated per cycle) the ratio of energy stored in the cavity at resonance condition and energy dissipated per cycle. High degree of monochromaticity of laser line is required in the diagnosis of closely spaced rotational levels of molecules using selective excitation of that level. The laser line would be absolutely monochromatic if it is oscillating with single frequency, that is, width of the laser line is zero ($\Delta v = 0$). The single-mode laser has the highest degree of monochromaticity, but it has also not achieved the ideal monochromaticity condition.

Directionality

One of the most striking properties of laser is its directionality, that is, its output is in the form of an almost parallel beam. Owing to its directional nature it can carry energy and data to very long distances for remote diagnosis and communication purposes. In contrast, conventional light sources emit radiation isotropically; therefore, very small amount of energy can be collected using lens. Beam of an ideal laser is perfectly parallel, and its diameter at the exit window should be same to that after traveling very long distances, although in reality, it is impossible to achieve. Deviation in the parallelism of practical laser beam from the ideal is not due to any fault in the laser design, but due to diffraction from the edges of mirrors and windows. From the theory of diffraction, we know that circular aperture has angle of diffraction given by $\theta = \sin^{-1}(1.22\lambda/D)$. The spreading of the beam does depend on the physical nature of aperture and on the type of transverse mode oscillating inside the cavity. In the particular case of TEM_{00} mode oscillating inside the nearly confocal cavity, the value of minimum diameter at the center of cavity is given by $wo = (\lambda r/2\pi)^{1/2}$, where r is the radius of curvature of cavity mirrors. The beam diameter, w, varies with distance, z, from the point of minimum diameter and is given by $w = w_0(2z/r)$. The beam radius at any point inside and outside the cavity is determined by the distance from the cavity axis where intensity reduces 1/e times of its maximum value.

Coherence

Coherence is one of the striking properties of the lasers, over other conventional sources, which makes them useful for several scientific and technological applications. The basic meaning of coherence is that all the waves in the laser beam remain spatially and temporarily in the same phase. Photons generated through stimulated emission are in phase with the stimulating photons. For an ideal laser system, electric field of light waves at every point in the cross section of beam follows the same trend with time. Such a beam is called spatially coherent. The length of the beam up to which this statement is true is called coherence length (L_C) of the beam.

Another type of coherence of the laser beam is temporal coherence, which defines uniformity in the rate of change in the phase of laser light wave at any point on the beam. The length of time frame up to which rate of phase change at any point on the laser beam remains constant is known as coherence time (t_C). Let P_1 and P_2 be two points on the same laser beam at time t and $t + \tau$. The correlation function between phases at these two points is termed as mutual coherence function, which is a complex number with magnitude between 0 and 1 (0 for completely incoherent beam and 1 for ideally coherent). Coherence time (t_C) is also defined as the time taken by the atoms/molecules in active medium to emit a light wave train of length L_C.

These two coherences are thus related by $t_C = L_C/c$. The coherence time of the laser beam is almost inverse t $1/\Delta\nu$ of the width, ν, of the laser transition. The lasers operating in the single mode (well-stabilized lasers) have narrow line width, therefore they exhibit higher coherence time and coherence length compared to those operating in multimode. Spatial and temporal coherences of continuous laser beams are much higher compared to those of pulsed laser systems because temporal coherence in the pulse lasers are limited by the presence of spikes within the pulse or fluctuation in the frequency of emission.

Brightness

Lasers are more intense and brighter sources compared to other conventional sources such as the sun. A 1 mW He–Ne laser, which is a highly directional low divergence laser source, is brighter than the sun, which is emitting radiation isotropically. Brightness is defined as power emitted per unit area per unit solid angle. In the particular case of 1 mW He–Ne laser with 3.2×10^{-5} rad beam divergence and 0.2 mm spot diameter at exit window the solid angle ($\pi(3.2 \times 10^{-5})^2$) is 3.2×10^{-9} sr and spot area ($\pi(2 \times 10^{-4})^2$) at the exit window is $1.3 \times 10^{-7} m^2$. Thus the brightness of the beam is given by $((1 \times 10^{-3})/(1.3 \times 10^{-7}\ m^2 \times 3.2 \times 10^{-9}\ sr)) = 2.4 \times 10^{12}\ W/m^{-2}\ sr^{-1}$, which is almost 106 times brighter than the sun ($1.3 \times 10^{6}\ W/m^{-2}\ sr^{-1}$).

Focusing of Laser Beam

In practice, every laser system has some angle of divergence, which increases the spot size of laser beam and reduces its brightness. If a convergent lens of suitable focal length is inserted in the path of the laser beam, it focuses laser energy into small spot area at focal point. If w_L is the radius of the beam and f is the focal length of convergent lens, then radius of the spot at focal point is given as $r_s = \lambda f/\pi w_L$, where λ wavelength of laser radiation. If D is the lens diameter and the whole aperture is illuminated by laser radiation (i.e., $w_L = D/2$) then $r_s = 2\lambda f/\pi D$ or rs = $2\lambda F/\pi$, where $F = f/D$ is f number of the lens.

In the case of a particular Nd:YAG laser operating at 1064 nm wavelength and 35 mJ/pulse energy, and 10 ns pulse width. It is focused by a convex lens of F = 5, and whole of the lens area is illuminated by laser. The spot diameter at the focal point is given by $r_s = 2 \times 1.064 \times 10^{-6} \times 5/\pi = 3.4\ \mu m$. The irradiance of the laser beam is given by $I = E(J)/\text{pulse width (s)} \times (\pi r_s^2) = 35 \times 10^{-3}/(10 \times 10^{-9}\ s \times \pi(3.4 \times 10^{-6})^2) \approx 10^{16}\ W\ m^{-2}$.

Modification Techniques in Laser Structure

Addition of some electronic, optical, or electro-optical systems between the active media and mirror to modify the pulse width, pulse shape, and energy/pulse and generation of integral multiple of laser frequency is important for advanced technological applications. Mode locking or phase locking, Q-switching, pulse shaping, pulse compression and expansion, frequency multiplication, and so on, are some commonly used methods in advanced laser technology.

Mode Locking

Basic Principle of Mode Locking

Mode locking is a technique in optics by which a laser can be made to produce light pulses of extremely short duration of the order of picoseconds (10^{-12} s) or femtoseconds (10^{-15} s). The basis of this technique is to induce constant phase relationship between the modes of laser cavity. Simply, same phase of δ can be chosen for all laser modes. Such a laser is called mode-locked or phase-locked laser, which produces a train of extremely narrow laser pulses separated by equal time intervals. Let N modes are oscillating simultaneously in the laser cavity with $(A_o)_n$, ω_n, and δ_n being the amplitude, angular frequency, and phase of the nth mode. All these parameters vary with time, therefore modes are incoherent. The output of such laser is a linear combination of n different modes and is given by $A(t)=\sum_{n=0}^{N}(A_o)_n\, e^{i(\overline{\omega}_n t+\delta_n)}$ expression. For simplicity, frequency of the nth mode can be written as $\omega_n = \omega - n\Delta\omega$, where ω_n is the mode with highest frequency and $\Delta\omega = c\pi/L$ is the angular frequency separation between two modes. If all the modes have same amplitude and we force the various modes to maintain same relative phase δ to one another, that is, we mode lock the laser such that $\delta_n = \delta$, then the expression for resultant amplitude will be $A(t)=A_o e^{i(\overline{\omega}_n t+\delta)}\sum_{n=0}^{N} e^{-i\pi nct/L} = A_o e^{i(\overline{\omega}_n t+\delta)} \sin(N\phi/2)/\sin(\phi/2)$, where $\phi = \pi ct/L$.

The irradiance of the laser output is given by $I(t) = A(t)A(t)^* = A_o^2 \sin^2(N\phi/2)/\sin^2(\phi/2)$, which is the periodic function of the period $\phi = 2\pi$ in the time interval $t = 2L/c$ (time of round-trip inside the cavity). The maximum value of irradiance is $N^2 A_o^2$ at $\phi = 0$ or $2p\pi$ (p is integer). Irradiance has zero value for $N\phi/2 = p\pi$, where p is an integer with values neither zero nor a multiple of N. This makes $\phi = 2p\pi/N = \pi ct/L$ or $t = (1/N)(2L/c)p$. Therefore, separation between two consecutive minima, that is, pulse width of a single laser pulse is $\Delta t = (1/N)(2L/c)$. Hence, the output of a mode-locked laser has sequence of short pulses of pulse duration $(1/N)(2L/c)$ separated in time by $2L/c$. The ratio of pulse separation to the pulse width is equal to the number of modes N, which shows that there should be a large number of modes in the cavity in order to get high-power short duration (picoseconds and femtoseconds) laser pulses.

Mode Locking Techniques

Active Mode Locking

We have discussed that mode locking is achieved by inducing the longitudinal mode to attain the fixed phase relationship, which may be exploited by varying the loss of the laser cavity at a frequency equal to the intermode separation $c/2L$. Let us consider a shutter between the active medium and output mirror, which is closed for most of the time and is opened after every $2L/c$ seconds (corresponding to the time of round-trip) and remains open for short duration of $(1/N)(2L/c)$ seconds. If the laser pulse train is as long as the shutter remains opened and arrives at the shutter exactly at the time of its opening, the pulse train is unaffected by the presence of shutter. The segment of the pulse, which arrives before opening and after closing of the shutter, will be clipped. Thus phase relationship of the modes is maintained by periodic oscillation of the shutter. An electro-optical or acousto-optical crystal operating on the principle of Pockels or Kerr effect respectively, may be used as a periodic shutter.

In the former case of electro-optical switching, a polarizer and an electro-optical crystal are arranged in between the active medium and laser exit mirror so that laser beam passes through the polarizer before entering into the crystal. Laser light from the active medium passes through the polarizer and gets plane polarized. When this polarized beam passes twice through the crystal (with appropriate electric field along the direction of light propagation) before returning to the polarizer, the plane of polarization is rotated by an angle of 90° , which does not allow the light beam to enter into the active media through the polarizer. In other words, the shutter is effectively closed.

If there is no field along the crystal in the direction of propagation of light, there will be no rotation of the plane of polarization of light and it can pass though the polarizer and enter into the active medium (shutter is open). Similarly in acousto-optical switching, a 3D grating pattern is established in a medium (water or glass, not active medium of lasing) by the incident and reflected sound waves created by piezoelectric transducer attached at one end of the medium. This grating diffracts a part of the laser beam and creates a high loss.

Passive Mode Locking

Passive mode locking method consists of placing a saturable absorber inside the cavity. Saturable absorbers are molecules having a nonlinear decrease in absorption coefficients with the increase in the irradiance of light. The saturable absorber is placed between active laser medium and mirror. If an intense pulse of light passes through the saturable absorber placed inside the laser cavity, the low-power tails (weaker modes) of the pulse are absorbed because of the absorption of dye molecules. The high-power peak of the pulse is, however, transmitted because the dye is bleached. Owing to this nonlinear absorption, the shortest and most intense fluctuation grows, while the weaker dies out.

Q-switching

Q-switching, sometimes known as giant pulse formation, is a technique by which a laser can be made to produce a pulsed output beam. The technique allows the production of light pulses with extremely high (of the order of approximately gigawatts) peak power, much higher than would be produced by the same laser if it were operating in a CW mode. Compared to mode locking, another technique for pulse generation with lasers, Q-switching leads to much lower pulse repetition rates, much higher pulse energies, and much longer pulse durations. Both techniques are sometimes applied at once. In contrast to mode locking, where we achieve a train of pulses, Q-switching provides a single strong and short pulse of laser radiation.

Q-switching is achieved by putting a variable attenuator inside the laser's optical resonator. When the attenuator is functioning, light that leaves the gain medium does not return and lasing cannot begin. This attenuation inside the cavity corresponds to a decrease in the Q factor or quality factor of the optical resonator. A high Q factor corresponds to low resonator losses per round-trip, and vice versa. The variable attenuator is commonly called a "Q-switch," when used for this purpose. Initially the laser medium is pumped while the Q-switch is set to prevent feedback of light into the gain medium (producing an optical resonator with low Q). This produces a population inversion, but laser operation cannot yet occur since there is no feedback from the resonator. Since the rate of stimulated emission is dependent on the amount of light entering the medium, the amount of energy stored in the gain medium increases as the medium is pumped.

Owing to losses from spontaneous emission and other processes, after a certain time, the stored energy will reach some maximum level; the medium is said to be gain saturated. At this point, the Q-switch device is quickly changed from low to high Q, allowing feedback and the process of optical amplification by stimulated emission to begin. Because of the large amount of energy already stored in the gain medium, the intensity of light in the laser resonator builds up very quickly; this also causes the energy stored in the medium to be depleted almost as quickly. The net result is a short pulse of light output from the laser, known as a giant pulse, which may have very high peak intensity. Similar to mode locking, active and passive techniques are used for Q-switching. Electro-optical and opto-acoustic (active Q-switching) switches, and saturable absorbers (passive Q-switching) are used to prevent feedback signal into the active media.

Pulse Shaping

Pulse shaping is a technique of optics, which modifies temporal profile of a pulse from laser. It may lengthen or shorten pulse duration, can generate complex pulses, or generate multiple pulses with femtosecond/picosecond separation from a single laser pulse. A pulse shaper may act as a modulator. Modulating function is applied on the input pulse to get the desired pulse. The modulating function in pulse shapers may be in time or a frequency domain (obtained by Fourier transform of time profile of pulse). In the pulse shaping methods optical signal is converted into electronic signal, where the presence and absence of pulses are designated by 1 and 0, respectively. There are two well-known pulse shaping techniques (i) direct space to time pulse shaper (DST-PS) and (ii) Fourier transform pulse shaper (FT-PS). In DST-PS, electrical analog of the desired output signal is used as a modulator with the electrical analog of input pulses. In contrast to DST-PS, the FT-PS uses a modulating function, which is a Fourier transform of the required sequence. In other words, a specific temporal sequence of pulses, the modulating function, is its Fourier transform and acts in frequency domain.

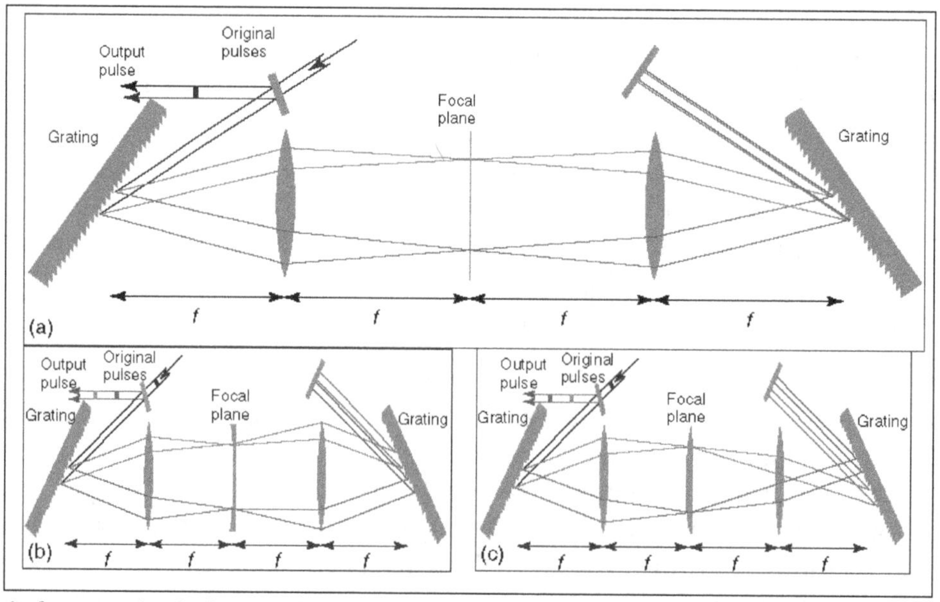

(a) Basic optical geometries for laser pulse shaping, (b) positive chirped, and (c) negative chirped pulse shaping.

Optical design is a common experimental arrangement consisting of a combination of two sets of grating, a pair of mirrors, a pair of convex lenses, and a pulse shaper. By placing different types of optical components we can design various types of laser pulses. For example, employing suitable filter at focal plane can remove optical radiation of undesired frequency. A slab of transparent material with varying thickness can offer different path lengths for different frequency components and can decide which component of light will come out first. A concave lens at the focal plane provides longer path length for higher frequency component, and vice versa. This causes the lower frequency component to come first and the higher frequency one comparatively later, resulting in positively chirped) output pulse (frequency increases with time). In contrast to this, a planoconvex or biconvex lens makes longer path length for lower frequency, and vice versa, which makes higher frequency to come out first and shorter frequency after that, that is, negative chirped output. Both positive and negative chirping lengthens the pulse duration, that is, stretching of the laser pulse. Removing or reducing the chirp causes compression (reduction of pulse duration) of the pulse.

Lasers in Industry

Lasers are used in industry in a huge variety of applications. These applications can be divided between those that involve the processing of materials and all other applications. Materials processing includes cutting, drilling, welding, etc., and generally involves the use of high-powered lasers.

Materials Processing

The advantages of using lasers in materials processing include:

- There is no contact tool required, unlike normal machining in which the tool bit must be sharpened and often replaced.
- Brittle or very pliable materials that are very difficult or impossible to machine with tools can be processed using lasers. In fact, one of the first materials processing applications of a laser in the 1960's was to make the holes in rubber baby bottle nipples, a difficult task to do otherwise with drilling.
- Using fiber optics, access to previously inaccessible locations is available.
- Laser processing is easily automated to allow computer and robot control.
- New types of processing has and continues to be developed, including producing new surface alloys and hardening of materials.

Some of the problems with using lasers for processing include:

- Certain tasks, such as drilling large diameter holes, are difficult to accomplish with lasers.
- Lasers and optics work best in clean and vibration-free environments, often difficult to achieve on the manufacturing floor.
- Initial costs can be large and so laser systems have been limited to large industry.

Let's now discuss how a laser interacts with materials differently as the laser power is increased. At low power, the surface of the material gets heated, depending on the material properties for absorption and conduction of heat. A mirror clearly reflects most of the light, while a dull black surface absorbs most of the light. Better conductive properties means that the heat will flow more quickly and the local temperature will not get as hot.

When the laser power increases sufficiently to increase the temperature enough, the surface begins to melt, turning from solid to liquid. If the power is considerably higher, the material may directly vaporize, the surface atoms turning to a gas. At even higher powers these gas molecules that form become ionized to form what is known as a plasma, a cloud of charged ions. At this point, the efficiency dramatically drops because the plasma begins to block the incident laser beam.

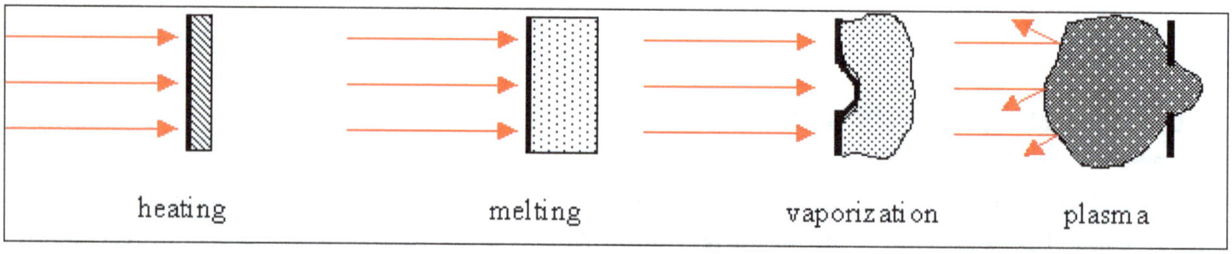

Depending on the kind of material processing needed, the intensity range of the laser, as well as its pulse duration, since most applications use pulsed lasers, will be determined.

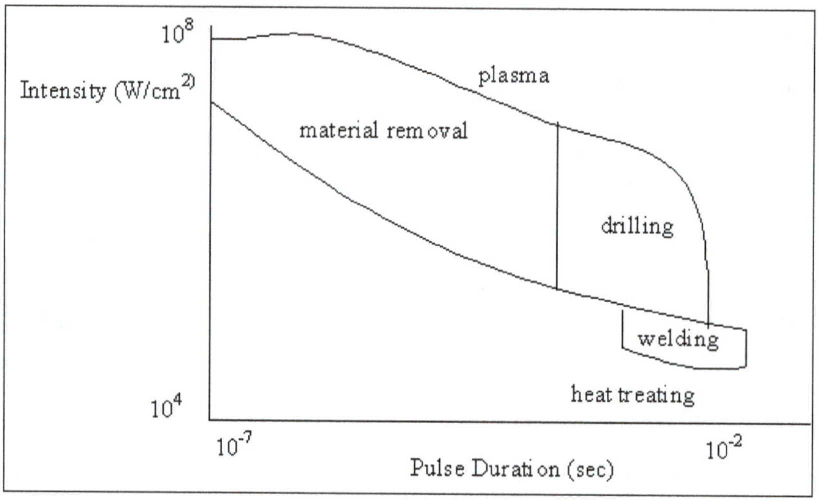

A major low intensity application is heat treating surfaces. Local heating changes the crystal properties of the material and can actually harden it. This is known as transformational hardening and it is just like the blacksmith's used to do in heating horseshoes and pounding on them to make them harder. Another example is the production of artificial diamonds by taking graphite, a form of carbon used in pencil lead, and putting in under extremely high pressure. Today, carbon dioxide lasers are used to harden steel for the automotive industry for cylinders in the engine and other parts.

At somewhat higher intensities melting will occur and this opens up a large area of welding applications. Welding is the fusion of two different metals together into a solid joint. Most of you have seen this when you get a car muffler replaced and two pieces of metal tubing need to be joined rigidly in an air-tight manner. A solder material is used and the metal surfaces are heated until the metals actually melt and fuse together. There are a wide range of welding applications, ranging from making ship hulls to micro-joints inside glass vacuum tubes or on printed electronic circuit boards.

At higher intensities still there is direct vaporization of material and both drilling and cutting applications abound. In the case of drilling, examples of applications that would be difficult to drill using mechanical drills include making holes in ceramics, which are brittle and easily shattered (these are often used in the microelectronics industry as electrical insulators), in rubber, which is also difficult to drill because of its elastic properties, and in diamond, the hardest material known, which previously had to be drilled with diamond tool bits that wore out rapidly. Two other interesting examples include making the holes in aerosol can valves and making the microscopic holes in time-release capsules, where the number and size of the holes regulates the rate of time-release.

Drilling with a laser does have some problems. Because such high intensities are needed to vaporize metal, the laser beams must be focused and therefore the holes produced, especially if deep, tend to be somewhat conical since the beam tapers at the focal point. Holes can only be drilled in metals to about 10 times the focused beam diameter before the intensity becomes too weak, although much larger holes can be drilled by moving the beam and actually cutting the border of the hole rather than vaporizing the entire plug of metal from the hole. Another problem is that often the vaporized metal tends to re-deposit on the walls of the hole and so usually a jet of gas is blown over the hole to remove the metal vapors.

Cutting is similar to drilling - the beam is simply moves along the material, or vice versa. For non-reflective materials (wood, plastic, cloth) this works directly but if the surfaces are shiny, often oxygen is blown over the surface near the focused beam and the oxygen is used to burn away the metal in a process known as laser-assisted cutting.

Non-materials Processing Applications

These applications can be divided into two classes: those that simply use the laser beam as a straight line tool and those involving inspection and scanning. Straight line applications include those in the construction industry for surveying and building. The simplest application of a laser is its use in "snapping a straight line" or in defining a horizontal plane in construction, whether it be new homes, tunnels, bridges, etc.

Laser beams are used in surveying to measure distances extremely accurately. There are two methods used, namely triangulation or time-of-flight. In triangulation, a distant object is sighted from one location and then re-sighted from a nearby location a known distance D away from the first location. Then, using trigonometry and knowing the distance D and the angle between the two sightings, the distance to the object can be calculated. In the time-of-flight measurement, a pulsed laser is reflected from the object and the round trip time determined. Knowing the speed of light, the distance to the object can be found from distance = speed x (round-trip time/2). Using this method, aerial mapping of topography can be done covering several hundred square miles per day. If done on land with many people, such a mapping might take a year or more.

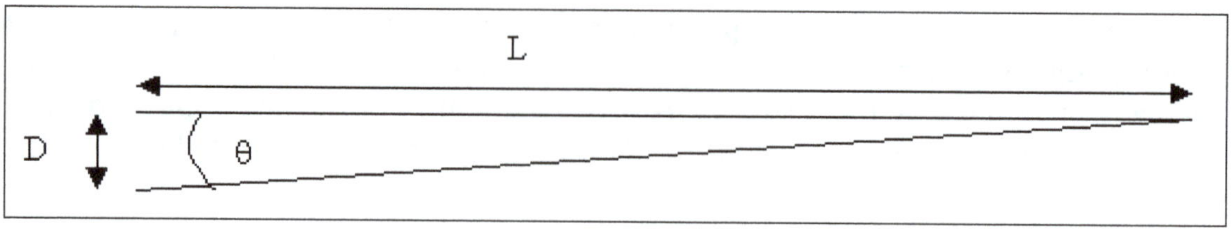

Another interesting application is used in monitoring geological motions on the earth using the LAGEOS satellite, in a stationary orbit over the earth. This satellite is named the LAser GEOdynamic Satellite and consists of 426 retroflectors (optical devices that reflect any incident beam directly back on itself no matter what the incident angle). Two detection stations on the earth can beam pulses of laser light up the LAGEOS and by monitoring the time-of-flight measurements, the relative positions of the two stations can be very precisely determined. When used across a fault line, for example, with two stations that might be 1 - 10 km distant, the separation distance can be monitored to better than 1 part in a million and so distance changes of 0.1 mm can be detected.

Inspections of manufactured goods can be done by examining laser light reflected from the surfaces of goods on an assembly line. Examples include cloth and lumber inspections at rates up to several meters per second. Another example is the inspection of "solder" joints - welded micro-joints - made in the microelectronics industry. It turns out that solder joints have a thermal signature in the infrared that indicates the difference between a good and bad (non-conductive) connection. This is a very important reliability check used by the microelectronics industry to insure high quality electronic components. A final application of this type is the supermarket scanner, using a holographic method to read the UPC bar code labels on goods sold. This system uses 10 lines with the first 5 coded for the manufacturer and the other 5 for the type of item. The scanned information is used to generate pricing and inventory records.

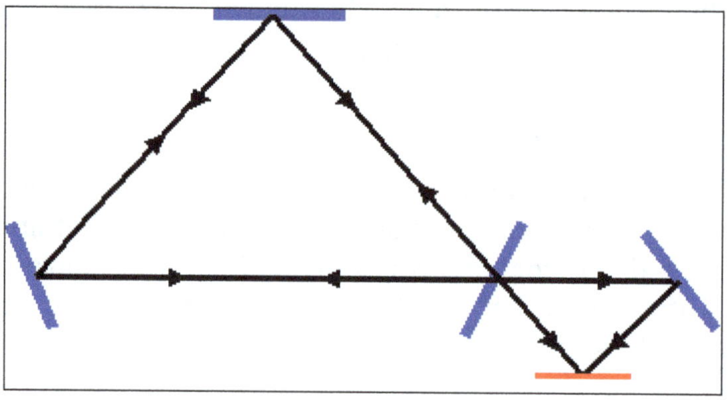

A ring laser with its two counter-propagating beams interfering at the red detector.

As a final example let's consider the laser gyroscope, used in the aerospace industry to detect stability and rotational motion of planes, rockets, and satellites. The gyroscope consists of a ring laser (shown schematically below) with counter-propagating beams (clockwise and counter-clockwise) that leak out of the laser and are made to interfere. If the laser is motionless, then the two beams will produce an interference pattern when they overlap. If the ring laser is made to rotate either clockwise or counter-clockwise, then the two beams have their frequencies shifted slightly up and down due to the opposite Doppler shifts of the two beams, and the fringe pattern will change.

Monitoring the fringe patter gives an extremely high precision monitor of rotational motion - capable of detecting better than 0.0010 rotation per hour. Three such laser, mounted to probe rotations in all possible directions, are used in airplanes, satellites and rockets in place of the older mechanical gyroscopes, super versions of the spinning top you may have used as a child.

Lasers in Medicine

The utilities of lasers in medical treatments consist mainly of photothermal therapy and PDT. Today, the laser is applied in both cases to almost all disciplines of medicine including dermatology, ophthalmology, dentistry, otolaryngology, gastroenterology, urology, gynaecology, cardiovascular system, neurosurgery and orthopaedics. Although photothermal therapy still plays a major role in medical laser treatment, PDT has been established as a treatment modality for a number of non-oncological and oncological diseases. At least seven photosensitizers and porphyrin precursors have been approved by regulatory health agencies in many countries for PDT and photodetection of diseases.

Table: Approved photosensitizers and porphyrin precursors in PD and PDT.

Compound	Trade mark	λ (nm)	Indication
Porfimer sodium	Photofrin	632	Superficial bladder, endobronchial, esophageal, gastric, cervical cancers.
Benzoporphyrin derivative monoacid ring A (BPD-MA)	Visudyne	690	Choroidal neovascularization (CNV) secondary to wet type of age-related macular degeneration (AMD), pathological myopia or presumed ocular histoplasmosis.
Tetra(meta-hydroxyphenyl) chlorin (mTHPC)	Foscan	652	Head and neck cancer.
5-Aminolevulinic acid (ALA)	Levulan	blue	Actinic keratosis of face and scalp.
5-Aminolevulinic acid (ALA)	Giolan	375–400	Intraoperative PD of residual malignant glioma.
Methyl aminolevulinate	Metvix	632	Actinic keratosis, basal cell carcinoma.
Hexyl aminolevulinate	Hexvix	370–400	Fluorescence detection of bladder cancer.

In addition, several second-generation photosensitizers are being investigated in clinical trials. The major advantage of PDT over photothermal therapy is the selective effect without hyperthermia, so that the surrounding healthy tissues are undamaged.

Table: Photosensitizers and porphyrin precursors used in PD and PDT clinical trials.

Compound	Trade mark	λ (nm)	Indication
5-Aminolevulinic acid (ALA)		630	Urethral condylomata acuminata
5-Aminolevulinic acid (ALA)		630	Acne
Pd-bacteriopheophorbide	Tookad	762	Prostate cancer
Tin ethyletiopurpurin (SnET2)	Purlytin	660	CNV due to AMD
			AIDS-related Kaposis
			Prostate cancer

2-(1-Hexyloxyethyl)-2-devinyl pyropheophorbide-a (HPPH)	Photochlor	660	Barrett's esophagus Esophageal and lung cancer Skin basal cell carcinoma
Lutetium-texaphyrin (Lu-tex)	LUTRIN ANTRIN	732	Locally recurrent breast cancer Photoangioplasty of peripheral arterial diseases
Boronated porphyrin	BOPP	628	Brain cancer
Mono-N-Aspartyl chlorine e6 (NPe6)	Talaporfin	664	Head and neck, lung, rectal, prostate cancer
Chlorin e6 conjugated with polyvinyl pyrrolidone	Fotolon	665	Bowen's disease
Hypericin	VIMRxyn	600–1000	Glioblastoma. HIV-infected diseases. Cutaneous T-cell lymphoma, Kaposi's sarcoma, warts and psoriasis
Hematoporphyrin derivative	Hiporfin	630	Skin, lung, bladder, liver, gastrointestinal and other cancer
Sulphonated aluminium phthalocyanine (AlPcS)	Photosense	675	Skin, breast, lung, gastrointestinal and other cancer
Porphycene (ATMPn)		630	Psoriasis
Dibromorhodamine derivative	TH9402	515	Bone marrow purging

Table: Comparison between photothermal therapy and PDT.

Laser (most) Laser wavelength Laser power	Nd : YAG 805–1064 nm 3–5 W	Diode 510–530 or 630–730 nm 0.1–0.3 W or higher for hollow organs
Light fluence rate	>300 mW cm^{-2}	<200 mW cm^{-2}
Tissue penetration of light	>4 mm	<4 mm
Nature of effect	Thermal	Photochemical
Destruction of connective tissue	Yes	No
Selective effect	None	To a high extent
Healing	Scarring	Often no scarring
Toxicity	None	Skin photosensitization (weeks)

Factors Influencing Laser Choice

Choosing a suitable laser for a given application depends on (a) the absorptive characteristics of the tissue to be destroyed, (b) the wavelength of the emitted radiation, (c) the temporal parameters of the delivered energy including the level of applied power (power density), the total energy delivered over a given surface area (energy density) and the rate and duration of the exposure (pulse repetition) and (d) the mode of beam energy delivery to a target tissue (i.e. continuous versus pulsed energy and direct contact or no contact with the target tissue).

Laser Delivery Systems

The laser delivery system transmits the laser beam to a targeted tissue. Several laser delivery systems are available and their choice depends on wavelength, operating power desired spot size and accessibility of the treatment site, and transmission/absorption characteristics of optical elements. Generally, the systems can be classified into two types: noncontact and contact. In the non-contact type only the laser beam interacts with targeted tissue with a predominant optic effect.

There are mainly three various systems of this type. (a) Articulated arm and lens system: an articulating arm is a series of hollow tubes and mirrors. The laser beam is transmitted through a tube and then reflected into the next tube by an appropriately angled mirror. This system can be adapted to an operating microscope or handpiece providing excellent precision. (b) Fibre optics: fibre optics are thin, flexible optical fibres coated with opaque nylon or metal casings, which can transmit visible and near infrared radiation by reflection off the opaque casing. The optical fibres can be used in conjunction with rigid or flexible scopes to provide minimally invasive surgery. Recently flexible, biocompatible silica optical fibres with side-firing and diffusing tips for high power delivery of laser irradiation have been shown to be promising for a number of medical indications. (c) Waveguide: waveguides are semiflexible steel tubes lined by ceramic tile. Laser energy is reflected down the tube by bouncing the beam off the lateral walls. In the type of contact, the laser energy is mainly concentrated at a tip to induce a cautery effect.

It should be pointed out that high energy lasers can damage optical fibres during their delivery. Often, the laser-induced damage occurs on either of the end faces, or at the point of the first focus, a point at which the light has its highest intensity inside the fibre. Such damage is associated with gradual heating of the fibres partly as a result of fibre-beam misalignment, and largely limits the applications of endoscopic cauterization systems.

Laser Safety

As lasers are potentially hazardous tools, devastating thermal injuries can occur with improper use. Regulatory agencies worldwide have thus proposed safety guidelines for the operation of lasers by setting limits for laser exposure and maximum permissible exposure (MPE). In US and UK regulations have been established since the mid-1980s and they are continuously updated. As various tissues tolerate different levels of laser power at different wavelengths, various MPE standards have been made for different tissues such as skin and eye.

Based on the MPE standard for a specific laser with a known beam diameter and divergence, the safety working distance for the specific laser beam in clinics and laboratories can be calculated as a laser safety document. All personnel involved must be fully conversant with the safety document and also obtain laser safety practical training. Although laser safety regulations do not apply to the exposure of the patient at the target site, accidental exposure to the patient from misdirection of the laser beam that can injure the skin and eye should be avoided.

Generally, the skin is an easy target for a laser beam because of its high chance of laser exposure. Once damaged, the laser thermal effect-induced collagenation in the dermis can last for several months, that further delays healing and promotes fibrosis. Before the reepitheliation process starts heat damaged epithelial tissue must be remodelled, either by sloughing of coagulative necrosis or by biochemical repair of sub-lethally injured cells. As long as the dermis is exposed to air, more and more fibroblasts are activated, resulting in an increase in collagen deposition. Fortunately, the skin is relatively resistant to light due to the protection of the pigmented epidermal layer.

The most vulnerable organ of the human body is the eye. Unlike the skin, the cornea of the eye does not have the externally protective layer of pigmented cells. In particular, the ocular lens focuses the incident laser beam on the retina with a very small spot to produce a high power density. This can burn a hole on the retina. Since the laser beam irradiance at the retina can be multiplied by a factor of 100 000 compared with the irradiance on the cornea, even a low power laser can result in retinal damage without harming the cornea or other eye tissues. It should be pointed out that only the visible and near infrared lasers can damage the retina, as the light of such wavelengths can penetrate the watery layers in front of the retina and finally reach the pigmented epithelium of the retina. The UV and far infrared lasers, on the other hand, can be absorbed by the cornea giving rise to corneal damage when the power density of a laser is over the threshold. Thus, the MPE standard for the retina is different from those for the cornea and other eye tissues. Today, laser safety goggles for a specific laser are available and the use of eye goggles is mandatory, because the blink reflex of the eyes is not fast enough to prevent the laser light reaching the delicate retina. In addition, windows of the laser room should be covered and doors be locked during laser use. Appropriate hazard warning signs should be posted in the vicinity of the laser room.

Lasers can produce potentially hazardous airborne contaminants from the photovaporization of tissue. The vaporized tissue has been referred to as laser plume. The laser-induced pyrolytic products are similar to those resulting from the barbecue of meats. They contain toxic by-products and known carcinogens such as nitrosamines. A local exhaust ventilation system should be required in a laser operating room to reduce the risk of chronic breathing of the laser plume.

Therapeutic Applications of Lasers

Lasers in Dermatology

Lasers were introduced in the specialty of dermatology in the mid-1960s. Because of the accessibility of the skin to examination and study by lasers, dermatologists play an extremely important role in defining the clinical usefulness and limitations of laser systems. They have also helped to define the specificity of laser–tissue interaction that further improved the usefulness of these devices. When the absorptive characteristics of a targeted tissue are precisely matched with an ideal laser wavelength, the maximal specificity of the laser–tissue interaction can occur. The optical property of the skin is an important determinant for the selectivity of laser effects. There are two main chromophores in the skin: oxygenated hemoglobin with three absorption peaks at 418, 542 and 577 nm; and melanin which has a very broad range of absorption. In addition, water is the key component of the skin tissue that can affect the quality of the thermal effects ranging from structural changes of protein at temperatures of 42–45 °C, to coagulation at 50–60 °C and vaporization at above 100 °C.

There are a large number of lasers used for various types of dermatological indications including visible wavelengths of pulsed dye, ruby, KTP, diode, alexandrite and infrared wavelengths of CO_2, Nd:YAG. The major effect of lasers on skin tissue is photothermolysis and the common skin indications treated with lasers are vascular lesions, benign and malignant tumours, infectious lesions, pigmented lesions and tattoos and a number of cosmetic conditions. Generally, lasers used correctly cause limited side-effects with little discomfort, no risk of infection and no scarring.

One of the most common cutaneous vascular lesions treated with lasers is the port-wine stain (PWS). PWSs with an incidence of about 0.4% of newborns are benign vascular birthmarks consisting of superficial and deep dilated capillaries in the skin resulting in a reddish to purplish discoloration. PWS can cause significant psychological trauma and a reduction in the quality of life. Treatment of PWS includes ionizing radiation, cryosurgery, skin grafting, but lasers provide the treatment of choice for most patients today. Being absorbed by intravascular oxyhaemoglobin within the visible-light range a laser can induce photothermolysis to destroy the diseased blood vessels selectively without affecting the surrounding tissues and causing scarring. Various types of lasers have been used for PWS, but the flashlamp-pumped pulsed dye laser (PDL) is the treatment of choice with the wavelengths from 585 to 600 nm. In addition, the KTP laser at the shorter wavelength of 532 nm is still used for the PWS treatment with the advantage of causing less purpura. PDT is now also applied for PWS. Unlike PDL-induced coagulation, PDT uses continuous low light irradiation with no photothermal effect to activate a photosensitizer that has accumulated in diseased vessels and thus selectively destroys the vessels.

Although thermal laser therapy is used for the treatment of skin tumours, PDT with ALA or its methylester has recently been applied for the treatment of cutaneous premalignant and malignant lesions . The principle of ALA-PDT is that in the initial step of the heme biosynthetic pathway in cells ALA is formed from glycine and succinyl CoA. The last step is that the rate-limiting enzyme, ferrochelatase, catalyzes the insertion of ferrous iron into PpIX (a potent photosensitizer) in the mitochondria. By adding exogenous ALA, the naturally occurring PpIX accumulates with a high degree of selectivity in tumours possibly because of limited capacity and/or low activity of ferrochelatase. Such a selectivity has therefore been exploited for its application in photodetection and PDT of tumour as an alternative to administration of exogenous photosensitizers. The benefits of ALA-based PDT include reduced skin photosensitivity (1 or 2 days compared with 1 or 2 months with some other photosensitizers), easy administration (topical or oral) and repeat use if necessary. PDT with topically applied ALA or its methyl ester has already been approved by regulatory agencies of many governments for the treatment of actinic keratosis and basal cell carcinoma.

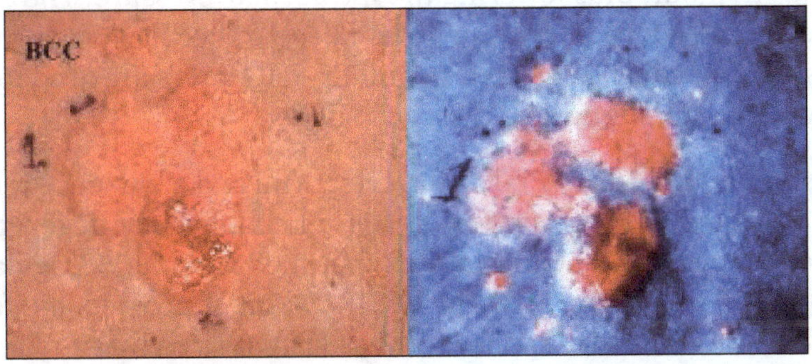

Highly selective ALA methylester (Metvix)-induced PpIX fluorescence (right) in human skin basal cell carcinoma.

Q-switched lasers with short-pulsed and high intensity are used to bleach the dyes in unwanted tattoos and fade the pigments in age spots and moles. The 308 nm excimer laser is effective to clear psoriasis. In addition, lasers are applied for resurfacing, hair removal and wound healing.

Lasers in Ophthalmology

No field has seen greater accomplishments with lasers than ophthalmology. The advantage in this field is the ability of a laser beam to enter the eye without causing injury. In the 1940s, Gerd Meyer-Schwickerath, a German ophthalmologist, pioneered to use the energy of the sun to 'weld' the retina to the underlying epithelium. This was the first eye surgery with light coagulation of the retina. Ophthalmology was perhaps the first subspecialty in medicine to use laser light to treat patients. Immediately after the invention of the first laser in 1960, Campbell used a confocal laser transmission system for the first retinal laser coagulation in 1961. Today, lasers are indispensable for effective and minimally invasive microsurgery of the eye. The focusing system of the cornea and lens brings laser beams to a sharp focus inside the eye. This actually carries a risk of injury, but also has considerable therapeutic possibilities. Generally, laser energy has four different effects of light–tissue interaction on the eye: photo-disruption, photo-ablation, photocoagulation and photochemical reactions.

Photo-disruption uses the mechanical energy of a laser to create micro-explosions, expanded plasma formation, acoustic waves and cavitations in order to cut intraocular structures with minimal thermal damage. The Nd:YAG laser, that can induce a photo-disruptive effect to make tiny holes, is applied for iridotomy in the pupillary-block glaucoma caused by increased pressure in the eye, and also for lens capsulotomies. Photo-ablation involves minimal thermal damage. Hyperopia (long-sightedness), myopia (shortsightedness) and astigmatism can be corrected by photorefractive keratectomy (PRK) with non-thermal excimer laser ablation in the far ultraviolet. The laser-assisted in situ keratomileusis (LASIK) is a common ype of the PRK and is done in the stromal bed by creating a plano-hinged flap with a keratome. Myopia up to 12 dioptres and hyperopia up to 5 dioptres, each with a certain degree of astigmatism, can be treated effectively in this way. Patients choose LASIK as an alternative to wearing corrective eyeglasses or contact lens. In addition, the laser ablation can be used as phototherapeutic keratectomy to remove superficial corneal opacifications in scars and dystrophies and to close the epithelium in non-infectious corneal ulcers.

Photocoagulation uses the thermal energy of a laser to seal leaking blood vessels. The argon laser is a standard modality for retinal coagulation of diseases such as diabetic retinopathy and vascular diseases. In the treatment of glaucoma, focal coagulation can be induced within the trabecular meshwork of the anterior-chamber angle with the argon laser using a high energy to cause the outflow of the aqueous humour, thus lowering the intraocular pressure. The excimer and Er-YAG lasers can also be used to produce a few tiny holes in the trabecular meshwork. In vascular and pigmented eyelid lesions including tumours CO_2 and Er-YAG lasers are used for tissue coagulation to reduce bleeding, making the surgical field easy to survey.

The photochemical effect generated by PDT with verteporfin and a low-energy laser is used to treat choroidal neovascularization (CNV) secondary to age-related macular degeneration (AMD). Until recently, the only treatment proven beneficial was thermal laser photocoagulation. However, the thermal laser damages the retina overlying the CNV, making it problematic for lesions involving the foveal centre. The photochemical application has shown to reduce the risk of moderate and severe vision loss in patients with predominantly classic subfoveal CNV secondary to AMD.

A scanning laser ophthalmoscope is a diagnostic tool for eye disorders. The principle is that a focused low-level laser beam is used to scan the fundus of the eye to provide a digital and fluorescent image of the retina, in which the depth of the optic cup and the thickness of the retinal nerve fibre layer can be measured.

Lasers in Dentistry

The common lasers used today in different fields of clinical dentistry are argon, KTP, HeNe, diode, Nd:YAG, ErCr:YSGG, Er:YAG and CO_2. Among them He–Ne laser and diode laser (632 nm) with a low power are applied for photo-detection and PDT. These fields mainly include periodontology, endodontics, hard tissue applications, soft tissue surgery and esthetic dentistry.

Periodontal diseases are among the most prevalent infectious diseases, with 75% of people aged between 35– 44 and 95% of the population over 65-year olds being affected. Most pathogenic bacteria in the oral cavity are present as complex aggregates (biofilms) on the surfaces of the teeth (known as dental plaques). The accumulation of these bacterial biofilms on the tooth surface above the level of the gingival margin causes the progressive dissolution of enamel and the underlying dentine to induce caries.

Treatments of carious lesions include the elimination of infected dentine by drilling and restoration of the tooth by filling with a variety of materials. However, instead of the removal of the infecting microorganisms by drilling, a more attractive alternative approach is to kill the bacteria in situ. Recent studies have shown that the bacteria present in dental plaques and caries are susceptible to lethal photosensitisation of the photosensitizer Toluidine Blue O or aluminium disulfonated phthalocyanine to a laser. This would help not only to prevent dental caries but also to eliminate infected dentine. Periodontitis is a chronic inflammation due to the accumulation of bacterial biofilms on the root surface of the tooth below the level of the gum margin (subgingival plaque). The bacterial biofilm induces destruction of the periodontal ligament and a gap between the tooth and the gum (periodontal pocket), resulting in the loosening and eventually loss of the tooth. The physical removal of the subgingival biofilms combined with topical application of antimicrobial agents to the periodontal pocket is the conventional therapy. However, it is difficult to maintain therapeutic concentrations of the antimicrobial agents due to the dilution by the saliva and gingival crevicular fluid. PDT is feasible, as a photosensitizer can be directly applied to the periodontal pocket followed by light irradiation either through the thin gingival tissues or via an optical fibre placed into the pocket. A number of studies have shown that PDT can effectively kill periodontopathogenic bacteria.

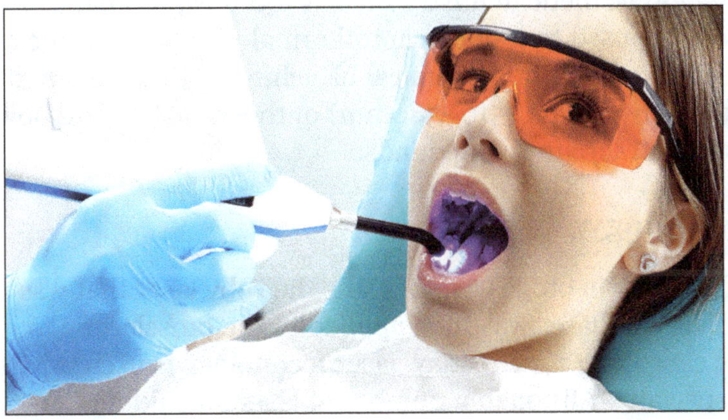

In the case of infected root canal system the routine endodontic treatment procedures are to mechanically enlarge the canals followed by irrigation with an antibiotic agent and filling of the affected space. However, the complex anatomy of the root canals makes it difficult or virtually impossible for the routine procedures to completely remove the bacteria. PDT may offer an alternative to disinfect and sterilize the canals due to the easy access of a photosensitizer and light.

Overall, the advantages of PDT over conventional antimicrobial agents are (a) a faster process of killing bacteria (seconds to minutes), (b) can be repeated without inducing drug-resistance and (c) sparing surrounding tissues. Argon and diode lasers with a low power are used to excite hydroxyapatite and bacterial by-products for fluorescence detection and quantification of incipient occlusal and dental carious lesions in pits and fissures. This fluorescence technique has greater sensitivity than conventional visual and tactile methods. Since laser exposure lacks the risks of ionizing radiation, this allows its frequent use for monitoring dental lesions. Apparently, the best is the combination of a fluorescence detection system with a therapeutic laser to allow diagnosis and treatment of dental diseases in a single device.

ErCR:YSGG and Er:YAG lasers, operating at the wavelengths of 2780 nm and 2940 nm, respectively, are used for dental hard tissues including caries removal and cavity preparation without significant thermal effects, collateral damage to tooth structure and patient discomfort. This is because normal dental enamel contains sufficient water and these Er-based laser wavelengths correspond to the peak absorption of water, so that they can achieve effective ablation at temperatures well below the melting and vaporization temperatures of the enamel. For example, the most uncomfortable component in dental treatment is probably the drill (handpiece). Er:YAG and CrEr:YSGG lasers can drill into the enamel and dentin, and may replace the dental drill with the advantages of minimal pain and no noise or vibration. The CO_2 laser is also highly absorbed by dental hard tissues, but it is not suitable for drilling and cutting enamel and dentine because its deeper thermal absorption can damage the dental pulp.

Numerous surgical procedures for soft tissues can be performed with lasers including gingivectomy, gingivoplasty, removal of gingival pigmentation, frenectomy, vestibuloplasty, aphthous removal, tumour excision, etc. Argon, KTP, Nd:YAG, ErCr:YSGG and Er:YAG lasers are usually used for minor soft tissue surgery, whereas CO_2 lasers are suitable for major soft tissue procedures. This is due to the efficient absorption of many commonly used dental lasers by water and haemoglobin in oral tissues. The main advantages of laser soft tissue surgery are reduced bleeding and less pain. Lasers can also be used for aesthetic dentistry such as tooth whitening. Diode (810–980 nm) and CO_2 (10 600 nm) lasers are used to induce photothermal bleaching; while argon (514.5 nm) and KTP (532 nm) lasers are used for photochemical bleaching due to the fact that their wavelengths fit well with the maximal absorption (510–540 nm) of those chelate compounds formed between apatites, porphyrins and tetracycline compounds.

Lasers in Otolaryngology

Experimentally, the first laser in otolaryngology was the ruby laser that was used on the inner ear of pigeons in 1965. A Nd:YAG laser was then applied to the otosclerotic footplate in 1967 and an argon laser to the otic capsule in 1972. Today, laser technology is used in virtually all areas of the ear, nose and throat specialty. Although a number of lasers including CO_2, argon, KTP, Ho:YAG, Nd:YAG, diode lasers are available today, only the CO_2 laser is in routine use as a workhorse laser in otolaryngology because of its excellent cutting properties with little lateral tissue damage. The major advantage of the laser over a scalpel is, for example, that any damage to the voice box by larynx surgery can severely impair speech; while the laser enables lesions to be vaporized without additional damage to the larynx. Lasers have also been used to treat snoring by trimming away part of the uvula.

The CO_2 laser, with the help of a microlaryngoscope as well as a delivery system, offers a distinct advantage in the management of polyps, nodules, cysts, leukoplakia, subglottic stenosis, webs, capillary hemangiomas, etc in laryngology. Treatment of these conditions needs careful attention to preserve the phonatory mechanism, as the vocal ligament may be involved. Micro-point manipulators on the newer generation of CO_2 lasers provide an excellent precision in the management of certain conditions such as haemorrhagic polyp and Reinke's oedema. With an appropriate power setting in a pulsed mode, the capillaries are sealed off during surgery, facilitating the incision and dissection. The bilateral abductor paralysis, a neuromuscular disorder causing dyspnoea, can be treated with a laser to obtain satisfactory airways on phonation. Treatment of laryngeal tumours with a laser has been the subject of much debate and controversy, although lasers are the method of choice for treating recurrent respiratory papillomatosis.

In rhinology, almost any lesion in the nasal cavity can be treated with a laser. The nasal mucosa is extremely vascular and conventional surgical procedures (punch and grasping forceps) in the nose cause profuse bleeding, which obscures the view and makes the procedures become somewhat 'blind'. The relatively bloodless field offered by a laser and an endoscopic application device provides the surgical procedure with a better visual control of the surgical field and surrounding structures. Generally, effective preoperative decongestion of the nasal mucosa with a laser is almost universally useful. The benefits of nasal laser management are dependent on the nature of the lesion, laser energy as well as wavelength, surgical technique and the operator's experience.

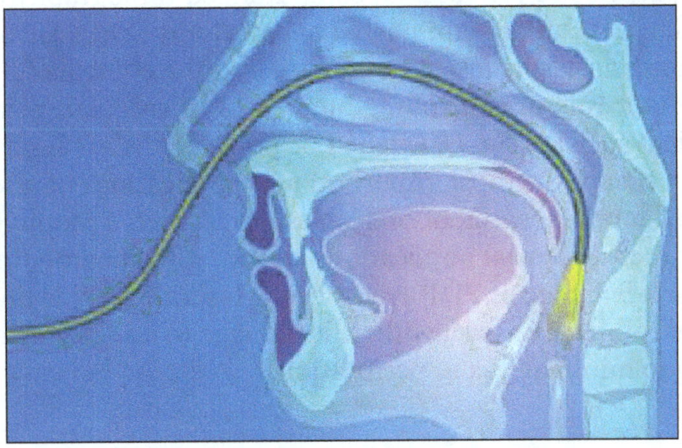

In otology, laser applications for ear surgery are still very limited. The operative procedure on the ear largely involves gross bone removal. The energy provided by a laser may be utilized to undertake certain steps of the procedure where an extreme degree of finesse is required or where conventional procedures can produce unwanted effects. At a low intensity, argon laser radiation penetrates through the bone without altering it, and may thus be used to devitalise extensions of cholesteatoma within the cell spaces of the mastoid bone. At a high energy, it vaporizes the bone, and has been used to undertake fenestration of the footplate in otosclerosis. PDT has been tried to treat head and neck cancer for more than two decades, and Foscan as a photosensitizer has recently been approved for such PDT treatment.

Lasers in Gastroenterology

Gastroenterology was one of the earliest specialties to examine the use of lasers in the early 1970s for the arrest of gastrointestinal haemorrhage. Because of the developments in gastrointestinal endoscopic techniques and the optic fibres inserted through the instrument channels of endoscopes, laser light can be easily delivered to the upper and lower gastrointestinal tracts in a safe and relatively non-invasive fashion .

The most important laser used today in gastroenterology is Nd:YAG. Short shots from this laser obtain good haemostasis due to thermal contraction of soft tissue. Longer shots at high powers can vaporize tissue and coagulate the underlying layers for effective debulking of advanced tumours; while those at much lower powers can coagulate a larger volume of tissue without vaporization. Thermal lasers in current practice are used for palliation of advanced, inoperable cancers of the upper and lower gastrointestinal tract. Under direct vision with the laser fibre held away from the surface of the target, nodules of exophytic tumour can be vaporized and the underlying tumour coagulated either to relieve obstruction or to reduce blood loss. Laser therapy can improve dysphagia in patients with cancers of the esophagus and gastric cardia, but several treatments and the introduction of expanding stents are often needed to achieve optimum recanalization.

The tip of the laser fibre can also be directly inserted into a targeted tissue with a much lower power to gently 'cook' the targeted area over a period of several minutes. This is called interstitial laser photocoagulation (ILP). ILP can be used for the percutaneous treatment of small hepatic metastases under the guidance of ultrasound, computerised tomography or magnetic resonance imaging in patients who are unsuitable for partial hepatectomy . Because of the simple and cheap sclerotherapy for haemostasis, lasers are not often used for the control of haemorrhage today, but they still play an important role in controlling blood loss in vascular lesions such as hereditary telangiectasia and angiodysplasia. The Qswitched or pulsed Nd:YAG laser is often used for fragmenting biliary stones. Perhaps the most important new applications of lasers in gastroenterology are PDT being developed for the treatment of dysplasia in Barrett's esophagus and small tumours in the gastrointestinal tract. Potential applications of PDT also include tumours of the pancreas and bile duct. The major advantage of PDT over thermal laser therapy is the selective effect on the mucosal layer of the gastrointestinal tract with little damage to underlying connective and muscular tissues, thus leading to minimal risk of perforation.

Lasers in Urology

Since the first report on the use of a ruby laser to fragment urinary calculi in 1968, this technology has been extensively applied by urologists. Laser lithotripsy in the 1980s and now laser prostatectomy have dominated laser usage in urology. Because of improvements in endourology, nephroscopy, perinephroscopy, laparoscopy, pelviscopy and retroperitoneoscopy, almost all urologic organs have become visually accessible via fibreoptic laser delivery. Laser applications in urology are based on several mechanisms for laser–tissue interactions including photomechanical, photothermal and photochemical effects.

Clinical applications of laser lithotripsy began in the mid1980s with the introduction of the pulsed dye laser. The laser operates with a short pulse of 1μs at a wavelength of 504 nm. This is based on a photomechanical mechanism. The optical fibre is placed in direct contact with the stone, and a short laser pulse is delivered, resulting in a shockwave at the stone surface to fragment most types of stone. However, due to its large size and high cost as well as maintenance, the dye laser has been replaced with the Ho:YAG laser. Unlike the dye laser, the Ho:YAG laser operates with a long pulse of approximately 500μs, generating a photo-thermal mechanism of stone fragmentation with chemical decomposition of the irradiated calculus components.

The Ho:YAG laser wavelength of 2120 nm is strongly absorbed by water in the tissue, allowing the laser to cut and coagulate tissue with a minimum zone of 0.3–0.4 mm. This property makes the laser also ideal for a range of soft tissue procedures including benign prostate hyperplasia (BPH) and bladder tumour as well as strictures. BPH, or enlargement of the prostate gland, is a common benign disease that occurs with increasing age in the male population. Recently, the frequency doubled KTP and Nd:YAG lasers with a high power have been introduced for vaporization of BPH. The KTP laser wavelength of 532 nm is strongly absorbed by haemoglobin and therefore provides excellent haemostatic properties during vaporization and coagulation of the prostate, although with a larger thermal damage zone (1–2 mm) than the Ho:YAG laser. Thus, the KTP laser represents a higher power, less expensive, more reliable solid-state alternative.

Other common applications of lasers in urology include Ho:YAG laser incision of urethral strictures caused by surgery-induced tissue trauma, Ho:YAG laser ablation of superficial bladder transitional cell carcinoma , Nd:YAG and CO_2 laser ablation of penile carcinoma and Nd:YAG and Ho:YAG laser incision of ureteroceles.

Although PDT is still under clinical development, it has shown promising results in the treatment of bladder cancer and small prostate and penile cancers with several photosensitizers including Photofrin and Tookad. Hexvix, a porphyrin precursor, has recently been approved by the EU for the photodetection of bladder cancer.

Lasers in Gynaecology

Since the CO_2 laser was first used in gynaecology more than 20 years ago, a number of other lasers have been used in this field such as Nd:YAG, KTP, dye and diode lasers. As more gynaecological procedures move to laparoscopic, colposcopic and hysteroscopic surgery, the use of lasers continues to increase. Advances in laser technique have improved precision and minimized thermal damage. These make it possible to use a laser for incision, excision, resection, ablation, vaporization, coagulation and haemostasis of soft tissues in the field of gynaecology. With colposcopy and CO_2 laser ablation, for example, condyloma, leukoplakia and high-grade cervical, vulvar and vaginal intraepithelial neoplasia (CIN, VIN and VAIN) can be treated. A number of reports have shown this laser to be effective at treating VIN with a success rate of approximately 85%. By laparoscopy/hysteroscopy and photovaporizing diseased or excess tissue, CO_2 and KTP lasers have also been used to treat ectopic pregnancies, dysmenorrhea, endometriosis, ovarian cysts, etc. In addition, lasers have been employed to perform hysterectomies and to reconstruct damaged or blocked fallopian tubes, thus allowing infertile women to conceive. The major advantage of laser treatment is the eradication of any volume of diseased epithelium, with negligible risk of scarring. Recently, fluorescent-dyed tissue samples with porphyrin precursors have been shown to reveal abnormal cells when excited by laser light, leading to early detection of cancer such as cervical cancer.

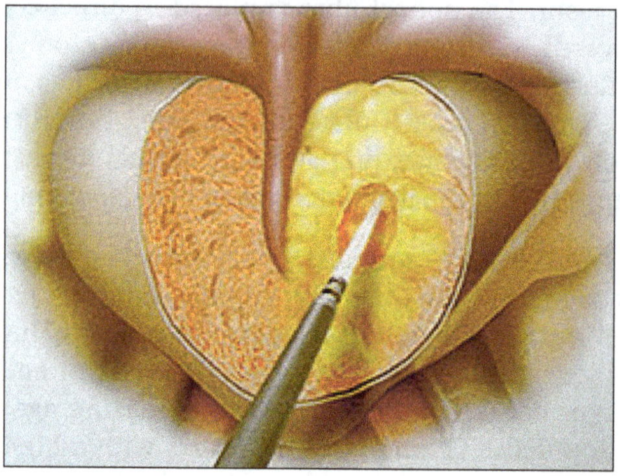

Lasers in the Cardiovascular System

By fibreoptics laser radiation can be transmitted to anywhere in the cardiovascular system. This makes laser-based modalities attractive, and today laser technology is available for the treatment of artery disease, ventricular and super-ventricular arrhythmias, hypertrophic cardiomyopathy and congenital heart disease. The first laser application to treat cardiovascular diseases was made by McGuff et al in 1963. They used a laser to vaporize atherosclerotic plaques. Choy et al used argon laser radiation to treat thrombosis in animals, and then performed the first clinical coronary argon laser angioplasty with a bare fibre in 1983.

In contrast to balloon angioplasty where plaque material is fractured, compressed, or displaced, laser angioplasty vaporizes the plaque material and thereby has a high success rate for treating chronic coronary artery occlusions. This approach is often referred to as laser thermal angioplasty, since the tips of optical fibres can convert the laser light energy into heat energy to achieve recanalization by mechanical compression and tissue vaporization of the plaque material. The results reported for excimer laser angioplasty show success rates of 82–85% with major complication rates of only 5–7%. Failures of laser angioplasty occur often due to the inability to advance the catheter to the lesion because of pre-stenotic vessel tortuosity.

A laser can also ablate thrombi and emboli by photo vaporization. This is because light absorption by haemoglobin in thrombi is larger than that by vascular tissue at 482 nm, thus providing a degree of selective laser ablation of the thrombi without damage to vascular walls. Gregory et al reported selective laser thrombolysis of coronary artery thrombi in 17 out of 18 patients with significantly improved coronary blood flow.

Transmyocardial laser revascularization is a technique for the treatment of patients with chronic angina pectoris. This procedure employs a laser to create multiple transmyocardial channels in the ischemic areas. The procedure can also be performed by percutaneous myocardial laser revascularization with a less-invasive advantage. Clinical trials have shown a significant improvement in angina class, but it is too early to draw any clear conclusion, as the techniques have not yet indicated any significant increase in survival and myocardial functions.

Laser ablation has been studied in the treatment of ventricular and supraventricular arrhythmias. Saksena et al and Isner et al used argon and CO_2 lasers to achieve superficial vaporization of endocardial tissue responsible for ventricular tachycardia. Compared with argon and CO_2 lasers for tissue vaporization, the Nd:YAG laser has been proposed for antiarrhythmic therapy by in situ photocoagulation of the inciting focus. One of the advantages of Nd:YAG laser photocoagulation is that the treated tissues are left intact, preserving structural integrity of the myocardium. The major benefit in contrast to cryoablation is that Nd:YAG laser photocoagulation can be performed on the normothermic beating heart during ventricular tachycardia.

Although laser technology has been evaluated for the treatment of a number of cardiovascular disorders for several decades, it has not performed as well as hoped. The procedure may create dangerous blood clots and perforation of vascular wall due to laser ablation. Catheters have been reported to cause mechanical trauma and are also too stiff to pass through convoluted blood vessels.

Lasers in Neurosurgery

Since initial experimental studies on the effects of ruby, argon and CO_2 lasers on the central nervous system (CNS) of rats, dogs and pigs in the period 1965–1970, dramatic progress has been made with respect to both technical and surgical applications in brain and nervous tissues. The major effect of a laser on neural tissue is thermal. Today, the CO_2 and Nd:YAG lasers, perhaps also new high power diode lasers, are effective in the treatment of CNS tumours and vascular malformations. Generally, laser-induced tissue vaporization is suited for resection of intracerebral, extra-axial, skull base and spinal tumours including acoustic neuromas, pituitary tumours, spinal cord neuromas, intracerebral gliomas and metastases; and also for dissection of intracranial, spinal cord and intra-orbital meningiomas; while laser-induced tissue coagulation is used for the resection of vascular malformations such as arteriovenous malformations and cavernomas. The benefits of such laser therapy are due to the accurate and noncontact instruments that can reduce surgical brain trauma.

It should be pointed out that normal and abnormal brain tissues may have very different optical properties. For example, most brain tumours (meningioma, neuromas, highgrade gliomas, metastases) are highly vascularised with a high haemoglobin level and thus demonstrate high absorption coefficients in the 400–800 nm wavelength band. This indicates that argon and diode lasers may be effective for these types of tumours.

Stereotactic techniques are more often used for neurosurgery because of their smaller openings, reduced brain injury, decreased morbidity and a shorter postoperative period. In fact, any fibreoptic-guided laser can be used for stereotactic neurosurgical procedures including imaging-guided (MRI, CT or angiography) resection of superficial and deep-seated tumours and vascular lesions or endoscopic resection of tumours and cysts in ventricles. Laserinduced interstitial thermotherapy (LITT) is also a minimally invasive neurosurgical approach to the stereotactic treatment of tumours in poorly accessible regions. MRI serves as the most promising technique for on-line monitoring of LITT.

PDT is a relative new technique that has been used for the treatment of brain tumours with HPD, Photofrin or Foscan.

Lasers in Orthopaedics

Patients suffering from herniated discs and unable to recover using physical therapy, can now be treated with lasers. Over 500 000 Americans undergo low-back pain treatment with a laser each year. Lasers can vaporize tissue in a disc, creating a vacuum. This causes the disc to shrink away from the pressed nerve, relieving pain. Such surgery eliminates the need for cutting, scarring, hospitalization, postoperative instability, immobility and even general anaesthesia.

Diagnostic Applications of Lasers

'Optical biopsy' or 'optical diagnostics' is a technique whereby light energy is used to obtain information on the structure and function of tissue without disrupting it. Today, this technique employs a number of spectroscopic and imaging methods including absorption, fluorescence, reflectance, elastic scattering and Raman scattering to distinguish malignant from benign tissue, monitor metabolic state and measure local blood flow as well as drug concentration.

The techniques of laser-induced fluorescence (LIF) spectroscopy and imaging are based on the findings that endogenous auto-fluorescence spectroscopic/imaging patterns differ between normal and premalignant or malignant tissues. For application of LIF to superficial lesions of internal hollow organs optical fibres are inserted through an endoscope, and tissue auto-fluorescence can be both laser-excited and captured for spectroscopic and imaging analyses. For example, to view mucosal abnormalities or tumours in the trachea and bronchi, lung imaging fluorescence endoscopy (LIFE) uses a fibreoptic endoscope with a blue laser light source to stimulate the natural fluorescence of tissue. Normal and abnormal tissues respond to this illumination differently, so LIFE images reveal minor abnormalities that would otherwise remain invisible. LIFE can detect early lung cancers or precancerous lesions as small as one millimetre across before they grow into invasive lung cancers. However, due to the complex structure of biological tissue, the low intensity of natural fluorescence signals and the artefacts from scattering and fluorescence reabsorption, photo-detection and interpretation of the tissue autofluorescence may be complicated, particularly in the case of inflammatory conditions that may cause false-positive results. These optical diagnostic techniques may be significantly improved by exogenously administered fluorescence compounds or their precursors that selectively localize in specific lesions, such as those photo-sensitisers used for PDT. ALA- or its derivative-induced endogenously fluorescent PpIX has already shown the enhanced discriminating potential of fluorescence spectroscopy and imaging.

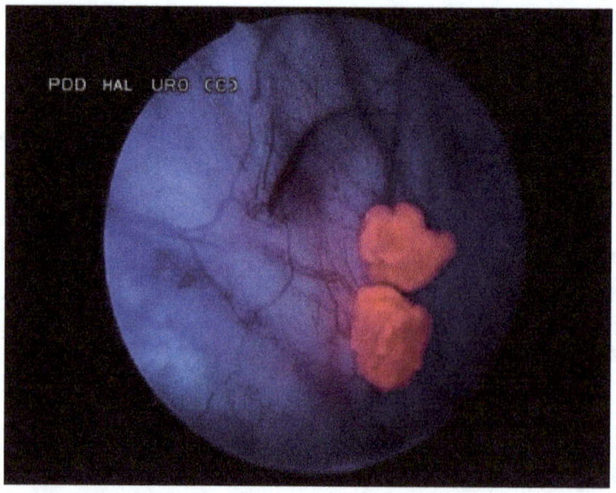

Highly selective ALA hexylester (Hexvix)-induced PpIX fluorescence in human bladder cancer.

Furthermore, the endogenous PpIX can serve as a potent photosensitizing agent for PDT. Most importantly, as a result of rapid advances in methodology and technology of tissue fluorescence imaging and PDT, endoscopists will likely in the coming years have a new armamentarium of diagnostic and therapeutic tools for detection and treatment of superficial premalignancies and malignancies of internal hollow organs during a single procedure. A standard endoscope equipped with an imaging system and a therapeutic illumination system may meet such a requirement for initial photodetection and subsequent PDT of lesions within one treatment session after employing PpIX precursors or other photosensitizers. The combination of endoscopy with the techniques of photodetection and PDT may revolutionize endoscopic technology.

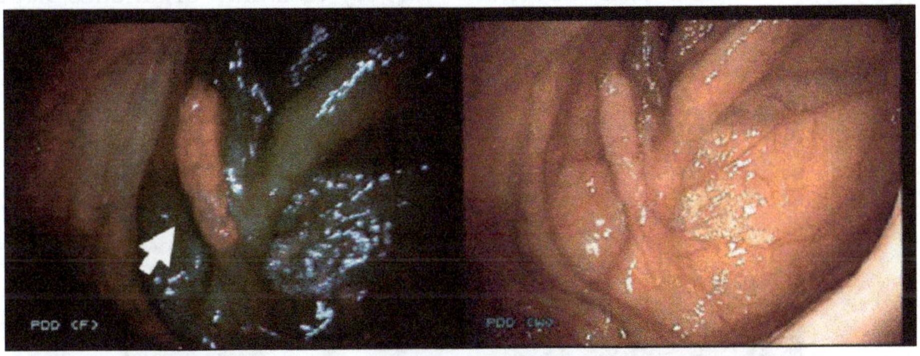

Highly selective ALA hexylester (Hexvix)-induced PpIX fluorescence (arrow in the left) in human rectal adenoma.

Confocal microscopy is a novel and non-invasive tool that allows for real-time imaging of tissue in vivo or fresh biopsies ex vivo without the fixing, sectioning and staining necessary for routine histopathology. A confocal microscope consists of a laser light source that illuminates a small spot in tissue. Reflected or fluorescent light from the illuminated spot of the tissue is then imaged through a small pinhole aperture, producing an image of the plane only in focus in the tissue. The confocal microscopy provides fast, high (axial) resolution and high contrast imaging of live tissue that has potential diagnostic applications for lesions.

Optical coherence tomography is a promising diagnostic method that uses low-coherence interferometry to produce a two-dimensional image of optical scattering from internal tissue microstructures in a way that is analogous to ultrasonic pulse-echo imaging. Cross-sectional images of tissue may be obtained in vivo with a better spatial resolution than confocal microscopy. This technique is potentially useful for noninvasive diagnosis of the retina and skin tumours with a penetration depth of 0.5–1.5 mm.

Laser Doppler velocimetry (LDV), also called laser Doppler flowmetry (LDF), is a simple and non-invasive method enabling the monitoring of microvascular blood flow, a very important marker of tissue health. In principle, a monochromatic laser beam is directed at the skin surface. Light that is reflected off stationary tissue undergoes no shift whilst light that is reflected off moving cells (like red blood cells) undergoes Doppler shift. The degree of Doppler shift increases with the velocity of the cells. This light is randomly reflected back from the tissue to a photodetector which calculates the average velocity of cells within the tissue.

As biotechnology evolves, a number of new techniques using a laser such as flow cytometry and mass spectrometry are more often applied to discover new biomarkers for detection of diseases as well as for evaluating the results of treatment.

Laser Nanosurgery in Cell Biology

During the past 10 years the quality and availability of pulsed lasers have significantly improved. Very high light intensity can be delivered in an ultrashort period ranging from nanoseconds (10–9 s) to femtoseconds (10–15 s). By tightly focusing the femtosecond laser pulse with a microscope objective within an extremely small volume, it is possible to achieve ablation on the level of a single cell or a subcellular compartment with a precision in the range of hundreds of nanometres without damaging surrounding structures. Since 1995 when Konig et al first demonstrated the dissection of isolated human chromosomes, picosecond UV and femtosecond IR pulsed lasers have increasingly been employed for cellular and subcellular surgery to ablate organelles without affecting cell viability. Figure below shows, for example, the entire ablation of a single mitochondrion by a femtosecond laser with precision targeting.

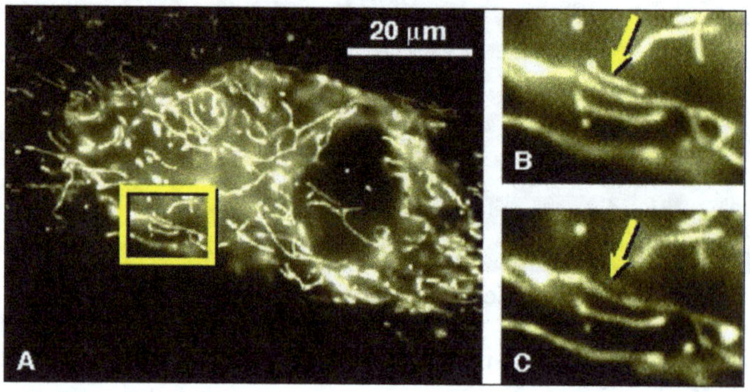

In figure, Ablation of single mitochondrion in a cell with a femtosecond pulsed laser. (a) Fluorescence image showing multiple mitochondria before the femtosecond pulsed laser application. Ablation of a single mitochondrion before ((b), arrow) and after ((c), arrow) laser pulses. Note that neighbouring mitochondria are unaffected.

Other subcellular targets include actin filaments and microtubules. Femtosecond lasers have even improved transfection efficiency of DNA in cells for gene therapy . In fact, femtosecond lasers can function as a pair of 'nanoscissors' in subcellular surgery and have potential applications in a single organelle or chromosome dissection, inactivation of specific genomic regions on individual chromosomes and highly localized gene and molecular transfer. The major advantage of pulsed laser nano-surgery is the well-controlled and non-invasive capability of severing subcellular structures with high accuracy in time and three-dimensional space.

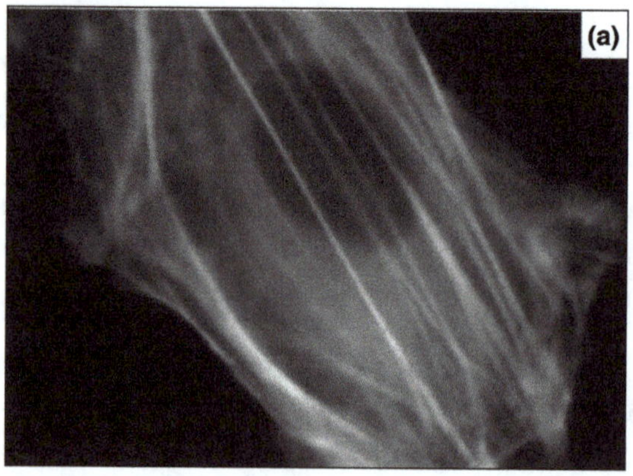

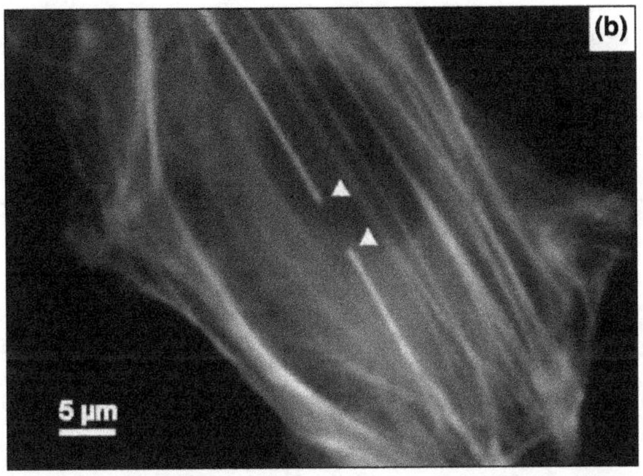

Ablation of actin filaments in a cell with a femtosecond pulsed laser. Fluorescence images of actin filaments before (a) and after (b) laser dissection (arrowheads) of an actin fibre.

Future Directions

The future development of medical lasers will largely depend upon the demands of patients and doctors for cost-effective methods. Although lasers have found their place in medical diagnosis and therapy since their creation about half a century ago, they will find many more medical applications as the progress in optical technology and better understanding of the interactions between laser irradiation and tissue are made. A recent market study by Frost & Sullivan, a research firm in London, predicts that the European market for medical lasers will nearly triple by 2013 to $862 million from $329 million in 2006. The major driving force for the growth is the ophthalmic, cardiovascular and urologic applications. On the horizon, a number of newer types of lasers are under investigation for possible medical uses.

Table: Common types of medical lasers and their major indications.

	Type of lasers	Major indications
Dermatology	Pulsed dye, ruby, KTP, diode, alexandrite, argon, CO_2, Nd:YAG, excimer	Vascular lesions Benign and malignant tumours infectious lesions Pigmented lesions and tattoos cosmetic conditions
Ophthalmology	Ruby, argon, Nd:YAG, diode, excimer	Diabetic retinopathy Age-related macular degeneration Glaucoma Corneal disorder
Dentistry	Argon, KTP, HeNe, diode, Nd:YAG, ErCr : YSGG, Er:YAG, CO_2	Caries Periodontitis Infected root canals Cavity preparation Soft tissue surgery Tooth whitening

Otolaryngology	CO_2, KTP, argon, Nd:YAG, Ho:YAG, diode	Polyps, nodules, cysts leukoplakia Subglottic stenosis Webs, capillary hemangiomas,
Gastroenterology	Nd:YAG, diode	Haemostasis Vascular lesions Dysplasia in Barrett's esophagus Cutting and debulking of tumours Fragmentation of gall stones
Urology	Pulsed dye, Ho:YAG, KTP, Nd:YAG, diode	Lithotripsy Benign prostate hyperplasia Prostate tumour Bladder tumour
Gynaecology	Nd:YAG, CO_2, KTP, dye, diode	Condyloma Leukoplakia CIN, VIN, VAIN Ectopic pregnancies Dysmenorrhea, endometriosis ovarian cysts Hysterectomies
Cardiovascular system	Argon, excimer, Ho:YAG, CO_2	Atherosclerotic lesions Thrombi and emboli Transmyocardial revascularization Percutaneous myocardial revascularization
Neurosurgery	CO_2, Nd:YAG, diode, argon	Meningiomas Acoustic neuromas Spinal tumours Tumour metastases Vascular malformations Stereotactic neurosurgery
Orthopaedics	Nd:YAG, Ho:YAG	Cutting and ablating soft/hard tissue Smoothing cartilage Knee surgery Lumbar disc decompression

Generally, new lasers are becoming smaller, more flexible and less expensive with specific applications. Femtosecond pulsed lasers with high power and low energy will allow surgeons to perform precise microsurgery without significant heating of the tissue. New fibre lasers with various wavelengths in a continuous or pulsed way will be developed without the need for a conventional resonant cavity. More powerful solid-state pumped lasers will have greater power and cover all the visible and near infrared regions. Mid-infrared semiconductor lasers with the capability of determining exact chemical specificity would likely enable optical characterization of a disease. Intense pulse light sources and LEDs will replace the laser in some applications. Perhaps the future of the laser will include multi-wave systems in a single or compact unit with multiple applications. In particular, multi-functional diagnostic laser facilities including in vivo spectroscopic and imaging systems can be combined with therapeutic laser procedures using an effective photosensitizer.

References

- Laser-Fundamentals: fas.org, Retrieved 25, July 2020
- Lasers-in-industry: union.edu, Retrieved 07, April 2020
- Lasers-in-medicine-231009441: researchgate.net, Retrieved 21, January 2020

Permissions

All chapters in this book are published with permission under the Creative Commons Attribution Share Alike License or equivalent. Every chapter published in this book has been scrutinized by our experts. Their significance has been extensively debated. The topics covered herein carry significant information for a comprehensive understanding. They may even be implemented as practical applications or may be referred to as a beginning point for further studies.

We would like to thank the editorial team for lending their expertise to make the book truly unique. They have played a crucial role in the development of this book. Without their invaluable contributions this book wouldn't have been possible. They have made vital efforts to compile up to date information on the varied aspects of this subject to make this book a valuable addition to the collection of many professionals and students.

This book was conceptualized with the vision of imparting up-to-date and integrated information in this field. To ensure the same, a matchless editorial board was set up. Every individual on the board went through rigorous rounds of assessment to prove their worth. After which they invested a large part of their time researching and compiling the most relevant data for our readers.

The editorial board has been involved in producing this book since its inception. They have spent rigorous hours researching and exploring the diverse topics which have resulted in the successful publishing of this book. They have passed on their knowledge of decades through this book. To expedite this challenging task, the publisher supported the team at every step. A small team of assistant editors was also appointed to further simplify the editing procedure and attain best results for the readers.

Apart from the editorial board, the designing team has also invested a significant amount of their time in understanding the subject and creating the most relevant covers. They scrutinized every image to scout for the most suitable representation of the subject and create an appropriate cover for the book.

The publishing team has been an ardent support to the editorial, designing and production team. Their endless efforts to recruit the best for this project, has resulted in the accomplishment of this book. They are a veteran in the field of academics and their pool of knowledge is as vast as their experience in printing. Their expertise and guidance has proved useful at every step. Their uncompromising quality standards have made this book an exceptional effort. Their encouragement from time to time has been an inspiration for everyone.

The publisher and the editorial board hope that this book will prove to be a valuable piece of knowledge for students, practitioners and scholars across the globe.

Index

www.ingramcontent.com/pod-product-compliance
Lightning Source LLC
LaVergne TN
LVHW080720210926
848485LV00003B/494